THE LOGICAL FOUNDATIONS OF STATISTICAL INFERENCE

SYNTHESE LIBRARY

MONOGRAPHS ON EPISTEMOLOGY,

LOGIC, METHODOLOGY, PHILOSOPHY OF SCIENCE,

SOCIOLOGY OF SCIENCE AND OF KNOWLEDGE,

AND OF THE MATHEMATICAL METHODS OF

SOCIAL AND BEHAVIORAL SCIENCES

VOLUME 65

HENRY E. KYBURG, JR.
University of Rochester

THE LOGICAL FOUNDATIONS OF STATISTICAL INFERENCE

D. REIDEL PUBLISHING COMPANY

DORDRECHT-HOLAND / BOSTON-U.S.A.

Library of Congress Catalog Card Number 72–92530

Cloth edition: ISBN 90 277 0330 2
Paperback edition: ISBN 90 277 0430 9

Published by D. Reidel Publishing Company,
P.O. Box 17, Dordrecht, Holland

Sold and distributed in the U.S.A., Canada and Mexico
by D. Reidel Publishing Company, Inc.
306 Dartmouth Street, Boston,
Mass. 02116, U.S.A.

Printed in The Netherlands by D. Reidel, Dordrecht

TABLE OF CONTENTS

PREFACE

Everyone knows it is easy to lie with statistics. It is important then to be able to tell a statistical lie from a valid statistical inference. It is a relatively widely accepted commonplace that our scientific knowledge is not certain and incorrigible, but merely probable, subject to refinement, modification, and even overthrow. The rankest beginner at a gambling table understands that his decisions must be based on mathematical expectations – that is, on utilities weighted by probabilities. It is widely held that the same principles apply almost all the time in the game of life.

If we turn to philosophers, or to mathematical statisticians, or to probability theorists for criteria of validity in statistical inference, for the general principles that distinguish well grounded from ill grounded generalizations and laws, or for the interpretation of that probability we must, like the gambler, take as our guide in life, we find disagreement, confusion, and frustration.

We might be prepared to find disagreements on a philosophical and theoretical level (although we do not find them in the case of deductive logic) but we do not expect, and we may be surprised to find, that these theoretical disagreements lead to differences in the conclusions that are regarded as 'acceptable' in the practice of science and public affairs, and in the conduct of business.

The present work is an attempt to try to make sense of these disputes, and to find a universal framework, based on relatively simple and innocuous principles, within which these disputes can rationally be argued, if not settled.

The first section of the book consists in relatively thorough examinations of several philosophical and statistical points of view regarding statistical inference. These chapters are didactic, as well as critical, on the grounds that many people who have been 'brought up' in one school of statistical thought will not find it insulting to have their memories refreshed concerning other schools of thought. There is a background chapter on the probability framework, designed primarily for philosophers; the appendix, which contains an embedding of the relevant theory of measure and integration in Zermelo-Frankel set theory is also primarily for philosophers, although there are some novelties of notation concerning joint distributions that may be of interest to statisticans as well as philosophers. The last chapter of the first part of the book contains a comparison of various views discussed previously, and an examination of the claims of interreducibility that have been made by proponents of various schools.

The fresh start begins with Chapter Seven; we define certain useful syntactical notions for a formal language L; we then define a rational corpus or body of knowledge to be a set of statements satisfying certain conditions. In Chapters Nine, Ten, and Eleven the basic ideas of the epistemological interpretation of probability are laid out formally. The epistemological interpretation is then compared to other interpretations of probability, and it is argued that there are circumstances under which these other interpretations are valid, but that under these special circumstances other interpretations can be construed as special cases of the epistemological conception. The epistemological interpretation has certain consequences for statistical inference, and these consequences are explored briefly for Confidence Methods, for Bayesian Inference, and for Fiducial Inference. Finally we investigate some further properties that bodies of knowledge might be expected to have, and examine the connection between probability and acceptability – i.e., worthiness for inclusion in a body of knowledge.

This work has taken so long and been supported in one way or another by so many institutions, that full and detailed acknowledgement is probably precluded by the frailty of human memory. Among the institutions to which gratitude is owed is first and foremost the National Science Foundation, which has contributed to the support of this work by a continuous series of grants running from 1962 to the present. In 1961 the American Council of Learned Societies helped me to get started on the project by awarding me a summer research grant. During the academic year preceding that summer, the work was supported by the Rockefeller Institute, and since then time and encouragement have been provided by the University of Denver, Wayne State University, the University of Western Ontario, and the University of Rochester. The

University of Rochester also contributed two semesters of sabbatical leave.

Among the individuals who have encouraged or corrected me – usually both – I must mention first and foremost Isaac Levi, whose penetrating criticisms, invariably accompanied by encouragement, have forced me to rethink a number of the points contained herein. I have also benefited from discussion and argument with Richard Jeffrey, Wesley Salmon, Patrick Suppes, Jaakko Hintikka, Ron Giere, and my colleagues Keith Lehrer and Rolf Eberle. I would never have embarked on this long and often tedious work, had it not been for the inspiring example and encouragement of my teacher, Ernest Nagel.

A number of graduate students have been invaluable in saving me from both typographical and logical errors, as well as from infelicities of expression. I am indebted on this ground to Charles Cardwell, William Harper, Gary Merrill, and John Cassidy. To Professor Harper, I am also indebted for making it possible to go through a large part of this manuscript in a graduate seminar at the University of Western Ontario. John Cassidy has worked hard at preparing an index. Those graduate students who have raised questions and difficulties as well as those who have offered help and advice in the course of a variety of seminars that have touched on one part or another of this work are also deserving of thoughtful thanks.

To Mrs Ruth Spall, Mrs Karen DeVisser, and Mrs Roslyn Yaravitz who have severely suffered through many typings of the manuscript, I should like to express my appreciation: Thanks for being enormously patient and thoughtful.

CHAPTER ONE

THE PROBABILITY FRAMEWORK

1. There are three fundamentally distinct ways of interpreting probability: personalistic, logical, and empirical. At least one of these conceptions of probability underlies any theory of statistical inference (or, to use Neyman's phrase, 'inductive behavior'). Each interpretation admits a variety of subinterpretations, but the three basic kinds of interpretation may be distinguished by the truth conditions associated with statements of probability.

Personalistic interpretations of probability take probability statements to assert somēthing about the degree of belief of a particular person at a particular time. These degrees of belief are idealized, in a sense to be discussed later, but they are usually taken to reflect the actual opinions of an actual person. A degree of belief, in turn, is interpreted as a propensity to take certain actions, make certain choices, or accept certain bets. A statement of probability, for example, 'The probability of rain tomorrow is $\frac{1}{3}$', is always construed as having two supressed parameters: person *p*, and time *t*. It is true just in case *p*'s degree of belief in the event of rain tomorrow (or in the statement, 'It will rain tomorrow') is $\frac{1}{3}$ at time *t*. 'Degree of belief' is given some behavioral interpretation, such as: at *t*, *p* will offer two to one odds against the occurrence of rain tomorrow, or, alternatively, *p* will be indifferent at *t* between receiving a prize of one dollar whatever happens, and receiving a prize of three dollars if it rains and nothing otherwise. A probability statement is thus an emprirical psychological statement which is true or false according to the actual or idealized state of the person's opinions at a certain time.

While personalistic interpretations take the probability of an event or a statement describing that event to be a function of a particular person and a particular time, logical interpretations take the probability of a statement or state of affairs to be a function of a body of evidence. Thus the statement 'The probability of rain tomorrow is $\frac{1}{3}$' is construed as having one suppressed parameter: the body of evidence *e*. On this interpretation, the truth conditions for a probability statement are logical or semantic: they are built into the language in the same way that the truth conditions for logical entailment are built into the language. Given the statement 'It will rain tomorrow' and given a body of evidence *e*, there is exactly one real number *r*, determined by logical conditions alone, such that the probability of 'It will rain tomorrow', relative to the body of evidence *e*, is *r*. A probability statement is either logically true (if it mentions the right number) or logically false (if it mentions the wrong number).

There is a close connection between these two interpretations of probability, since the body of evidence that is of interest to us is generally the body of evidence available to a particular person at a particular time. But they are fundamentally different: Given a body of evidence and a statement, the logical rules of the language determine the logical probability of that statement relative to that body of evidence, regardless of whether that body of evidence represents any actual person's body of evidence. Given a person, a time, and a statement, the personalistic probability of that statement for that person at that time is a reflection of his opinions at that time, regardless of what evidence he has for those opinions.

Empirical interpretations of probability, which underlie the classical approach to statistical inference, take probability statements to represent statistical truths about the world. A probability statement is a statement about the world, and its truth or falsity has nothing to do with any person's opinions, or any body of evidence, but only with the state of the universe. To say that the probability that a birth is a birth of a boy is 0.51 is precisely to say no more and no less than that 51% of births are – or would be, in the long run – births of boys, or alternatively, to say that human births have a certain *propensity*, measured by the number 0.51 and revealed by long run frequencies, to be the births of boys.

Needless to say, there are many circumstances under which any of the three interpretations make good sense, and which therefore give us no grounds for choosing among them. For example, 'The probability of getting a head on a toss of this coin is $\frac{1}{2}$', may be interpreted in any of the following three ways:

Personalistic Interpretation: The person (to whom reference is, on this interpretation, implicit) would just as soon (now) stake a gain on the occurrence of heads on a toss of this coin as on the occurrence of tails. That is, given the two alternatives: win a prize P if heads occur, win nothing otherwise; and win a prize P if tails occur, win nothing otherwise, the person would be quite indifferent as to which alternative he chose.

Logical Interpretation: The body of evidence (to which reference is, on this interpretation, implicit) is such that it supports the hypothesis that future toss t will yield heads precisely to the degree 0.5000.... Given the body of evidence and the rules of the language, we may compute that this hypothesis is confirmed to just this degree.

Empirical Interpretation: There is a certain class of events in the world comprising tosses of the coin, or a certain physical system, consisting of the coin and a mechanism for tossing it (reference to which is implicit) which has a certain physical property expressed by the statement in question. This property may reveal itself in the long run behavior of the coin as it is tossed, but the class of events or the physical system has this property whether or not we ever know it, or believe it, or have evidence either for or against it. To say that the class or the system is such that the probability of heads is $\frac{1}{2}$ is to make an empirical statement no different in character from the statement that the coin weighs half a gram.

Each of these general types of interpretation of probability yields a number of sub-types. There has been a particularly large variety of empirical interpretations. A thimbleful of history may be of interest. In 1866 John Venn[1] proposed an explicitly empirical interpretation of probability, defining it as a ratio in an infinitely large class. The probability of heads is a half, because in the infinitely long run of tosses, exactly half will yield heads. The motivation for this definition is clear: in any finite run, the actual proportion of heads will be less than a half or greater than a half; it will not be *exactly* a half (except accidentally). But the longer the sequence, the closer will be the proportion of heads among tosses of an ideal coin to the number $\frac{1}{2}$. On the other hand, we all know that we cannot just blithely talk about ratios in infinite classes. Corresponding to ratios in finite classes, we have limits in *ordered* infinite classes. Consider Venn's hypothetical infinite class of tosses of a coin. This is an infinite sequence ordered in a natural way: there is the first toss, the second toss,... the *n*th toss,.... After any finite number of tosses the ratio of heads to the total number of tosses is well defined; let us denote it by 'f_n/n', where f_n is the number of tosses that have resulted in heads. Then, corresponding to the infinite sequence of tosses, we have an infinite sequence of well-defined ratios: $f_1/1, f_2/2, f_3/3 \ldots f_n/n, \ldots$. Such a sequence may not have a limit, of course, but if it does, that limit would seem to be just the number that Venn had in mind.

A definition of probability along these lines, as the limit of a relative frequency in an infinite class, was offered by Richard von Mises.[2] Von Mises defined probability as the limit of a relative frequency in a *collective*, where a collective is an ordered infinite class of events (tosses of a coin, measurements of a quantity, trials of an experiment, etc.). The same class of events, ordered in a different way, may constitute a class in which the limiting frequency has a different value, or is not defined at all. For example, consider a sequence of tosses of a coin, in which every odd-numbered toss yields a head. The limiting frequency of heads in the sequence is one half. But we can re-order the sequence as follows: tosses 1 and 3, then toss 2, then tosses 5 and 7, then toss 4, then tosses 9 and 11, etc. The limiting frequency of heads in this new sequence – consisting of just the same tosses as the old one – is $\frac{2}{3}$ rather than $\frac{1}{2}$. In fact, we can rearrange the sequence in such a way that the limiting frequency of heads has any value we wish. Thus the order the elements of a collective is as important as the set of elements itself.

Furthermore, if we had a sequence of tosses of a coin, in which all the odd numbered tosses yielded heads, and all the even numbered tosses yielded tails, we would not, according to von Mises, talk about the probability of heads. Knowing where we are in the sequence, we could predict with perfect certainly whether or not heads would occur on the next toss. He therefore took, as an essential characteristic of a collective, that the events in it occur in an unpredictable, random, way. This is a difficult idea to spell out clearly, but it has been done. The general idea is this: consider an ordered sequence of trials of some experiment which may result either in success or failure. We suppose that the limiting frequency of success exists and is *s*. Now consider the subsequence, consisting of all those trials that immediately follow a success. This subsequence is to have a limiting frequency of success, and the value of this limiting frequency is to be *s*. Now consider the subsequence of trials that consists of all those trials that immediately follow a pair of successive successes. This new subsequence is to have a limiting frequency of success, and the value of this limiting frequency is to be *s*. Now consider the subsequence of trials that consists of all those trials that immediately follow a pair of trials consisting of a success followed by a failure. This new subsequence is to have a limiting frequency of successes, and the value of this limiting frequency is to be *s*. If we can continue in the same way, we say that the original sequence is after-effect free. It is possible to show that such after-effect free sequences exist – indeed, it is possible to construct them according to rule. Furthermore, given any finite or enumerable set of ways of selecting subsequences of an arbitrary collective, it is possible to show that there are collectives in which the original collective and each of this set of subsequences has the same limiting frequency *s* of success.[3]

Von Mises' frequency definition of probability has for many years been regarded as the last word on the subject by many of those statisticians who consider the problem worth taking seriously at all. Philosophers have tended to be less satisfied with it, for a number of reasons. First of all, given an infinite sequence of events, it is an open question whether or not it is a collective. Does the limiting frequency exist? Is the limiting frequency insensitive to place selection (is the sequence after-effect free)? While we can get evidence which will render affirmative answers to these questions plausible, such an effort seems a bit beside the point when what we are concerned about is the likelihood of success in a particular sort of experiment. Second, most of the sequences of events with which we deal in real life are known not to be collectives, because we know perfectly well that they are not even infinite. The tosses of a given coin, for example, cannot by any stretch of the imagination be construed as an infinite sequence. Von Mises, of course, was perfectly well aware of all this. In speaking of limiting frequencies in collectives, he was offering an idealization and reconstruction of a certain body of scientific theory, just as the astronomer, in taking heavenly bodies to be mass points, is offering an idealization. The question, then, is whether some other idealization might be more fruitful and constructive, or at any rate simpler.

A number of alternatives have been offered. The simplest is Neyman's:[4] probability is defined as a ratio in a finite class, called the Fundamental

Probability Set (or FPS). We need not specify the size of this class – it may be as large as we wish – but we certainly have no reason to suppose that it is infinite. Continuous probability distributions require infinite classes of course, but they may be regarded as approximations to the finite truth. H. Cramér[5] takes probability to be a concept characterizing what he calls mass phenomena; we do not have to suppose that the phenomena are ordered, or that the frequencies approach limits in the mathematical sense. The connection between the probability model, and the mass phenomenon, is provided by observed frequencies (and statistical inference); it is no different than the connection between any theoretical model, and the empirical phenomenon of which it is the model.

The most recent empirical interpretation of probability offered by philosophers is the propensity interpretation, given in slightly varying forms by Karl Popper and Ian Hacking.[6] The starting point of this interpretation is the observation that we not only use probability to characterize actual mass phenomena, but also to characterize kinds of events that may occur only a few times, or not at all. We might design an asymmetrical die, for example, in such a way that we could have perfectly good grounds for supposing that if it were tossed a very large number of times, it would tend to fall with the ace up $\frac{1}{4}$ of the time. Popper would regard this as a propensity of the die, which would be manifested under the standard tossing procedures were they to be carried out many times, but which the die *has*, even if it is tossed only a few times or not at all. This propensity is a physical characteristic of the die. Hacking regards the propensity of falling with an ace uppermost a quarter of the time as a propensity, not of the die as a physical object, but of the chance-set-up which consists of that die being thrown by some standard dice-throwing mechanism. Again, of course, to evaluate this propensity, as to evaluate a limiting frequency, or a probability model, or even a class ratio in a Fundamental Probability Set, is to perform an act of statistical investigation. And in any of these interpretations, a probability statement is an empirical statement about the world.

In general, playing on the vagueness of the term 'model', we may say that on most empirical interpretations of probability, the concept of probability plays its role in a model of a physical process. We may then leave open the question, for the time being, of whether probability in the model should be interpreted as a propensity, as a finite frequency, as a limiting frequency, or what. In any event, the statistical problem, in the framework we have set up on the basis of an empirical interpretation of probability, is the problem of ascertaining whether a given model fits the data, or whether some model in a given set of models fits the data, or of finding a model of a certain class that will fit the data, or.... On the one hand there is the real world, in which we notice that certain circumstances yield certain outcomes in an unpredictable (by us) but not entirely chaotic way. These circumstances we call – more or less appropriately – a random experiment. They may consist of performing a measurement under certain standard conditions (e.g., with a certain instrument, with care, etc.), or of tossing a coin or throwing a die under certain conditions (e.g., on a smooth hard surface). Corresponding to this physical real-world experiment, we have an abstract entity, a probability model. In the random experiment, we take account of certain results (e.g., heads, a five uppermost, a reading of 10.3 cm, or 15 °C). In the probability model, there correspond to these results certain simple events, (e.g., H, 5, 10.3, 15). In the real world, some of the results occur with significant (and useful) stability, i.e., with relative frequencies that are fairly stable. One does not have to be an obstetrician for very long before one notices that boys are born roughly as often as girls. In the probability model, we would like to assign to events probabilities that correspond to these observable frequencies. The probability model is thus the abstract counterpart of the real random experiment. Its connection with the real world is neither more nor less intimate than the connection of any other abstract theory with the part of the world of which it is the theory. There is some looseness of fit – the model and the reality modeled need not be perfectly fitted to one another – and the grounds on which we accept one model or another are precisely the grounds for accepting one statistical hypothesis or another.

In a later chapter we shall again take up the question of the empirical interpretation of measure-theoretical statements. We shall provide an explicit semantics for such statements – that is, truth conditions for them – and it will be seen that although various frequency interpretations are possible, something like the propensity interpretation is most fundamental and general. But by then we shall have seen that there are strong arguments against interpreting these statements directly as probability statements.

There are certain things to be observed about empirically interpreted probabilities. First, probabilities are defined for a set of sets or for a set of kinds of events. Since we may construe a *kind* of event as the set of events of that kind, we can simplify things by talking about sets only. The set S of sets for which (in any given model) probability is defined is taken to be a σ-algebra of sets: that is, if A and B belong to S, then their union and intersection belong to S; if A belongs to S, then the complement of A, $\bar{A}$, belongs to S, and if $\{A_i\}$ is a countable set of members of S, then the union of all the A_i, $\bigcup A_i$, also belongs to S. It should be observed that as a representation of our knowledge of the world, this is a slight exaggeration: it is not generally the case if we know a certain probability (say the probability of surviving for one year) in each of two classes (coal miners and ministers) that we also know the probability in their intersection. At the same time, of course, it should be remembered that this is a problem concerned in the *application* of probabilities to the world, and not in any way a problem in either the empirical interpretation of probability, or in the model that we want to apply. As in many other applications of scientific knowledge, we turn out to be ignorant of things that it would be convenient to know.

Second, on the empirical interpretation of probability, the probability axioms are tautologous. In the finite case, the axioms are simply set-theoretical truths; in the infinite and abstract case (where probability is referred to as the limit of a relative frequency, or an abstract property of a model, or a propensity) the axioms are taken as part of the defining characteristics of probability.

Third, probability can be attributed only to kinds of events or to sets, and not (usefully) to individual events or to the members of the sets that belong to S. We cannot properly talk about the probability that heads will occur on the *next* toss of this coin, but only about the probability that heads (in general) will occur on tosses (in general) of this coin. We cannot talk about the probability that John Smith, a 40-year old coal miner will survive for a year, but only about the probability that 40-year old coal miners will survive for a year. This fact has been emphasized by Neyman, by Reichenbach, by Salmon, by von Mises, by Cramér [7] – by nearly every serious writer who has adopted an empirical interpretation of probability: probability is applicable only to sets of events or kinds of events. When it appears that probability is being applied to an individual event, it is being applied elliptically to the *kind* of event of which that individual event is a repeatable instance. This holds for every empirical interpretation of probability, including the propensity interpretations. When we have a finite reference class or fundamental probability set, it is obvious: we define probability only for a certain field of subsets of the FPS, and not all for members of that set. The same thing is of course true under any interpretation of probability which takes it to be explicitly definable in terms of limiting frequencies. Probability, on these views, is a property of classes only and not individuals; indeed, it is applicable only to infinite classes.

We must consider separately two approaches to the interpretation of probability which takes it to be the abstract counterpart of frequencies encountered in the real world. One approach takes the elements of the probability model to correspond to the individuals or individual events of the real world: thus the points or elements of the model correspond to particular throws of the die, or to individual measurements, or to the longevities of individuals such as Samuel Smith. Probabilities are then assigned to sets of these points, and not to the points themselves. The other approach regards the elements of the probability model to be events in a generic sense: kinds of events, e.g., the event of landing with a six uppermost, the event of yielding a measurement of 4.3 centimeters, the event of dying at age 76. These are the points or elements, on this approach, and probability is defined directly for these points or elements. But then the individual events of the world are not elements of the model at all; Samuel Smith is not represented in the model at all – only the generic property of dying at age 76. Under the first approach, Samuel Smith is represented by a point in the model; but probabilities are assigned only to sets of points. Probabilities are assigned to individual points in the second approach, but under that approach, Samuel Smith is not represented in the model at all. Thus whatever the form of the abstract counterpart interpretation of probability, probabilities are not assigned to individual entities.

When, under a propensity interpretation of probability, we assign the probability $\frac{1}{2}$ to getting heads with a particular coin, or to the outcome *heads* on a certain chance set up, it may seem as if we are attributing probability to a particular thing, rather than a class of things. But the propensity attached to the coin or the chance set up characterizes the potential mass behavior of that coin or chance set up, and not any particular instance of tossing or any particular outcome of a trial of the chance set up. We may speak of the probability of getting a head on the next toss of the coin, but only insofar as this represents an *instance* of a kind of chance set up. In the same way, we may interpret probability as a limiting frequency, and speak of the probability of heads on the *next* toss as an elliptical way of talking about the limiting frequency in a sequence of tosses of which that one is an *instance*. Nevertheless the probability is properly and directly attributable to the infinite sequence in the second case, and to the chance set up in the first. Just as the probability to be assigned to heads on a particular coin toss, on a frequency interpretation, depends on which of any number of infinite sequences the toss is to be regarded as an 'instance' of; so on a propensity interpretation, the probability of heads on a particular coin toss will be dependent on what kind of chance set up that particular toss is to be regarded as an 'instance' of.[8] Without a mechanism for choosing the correct reference class or the appropriate 'kind' of chance set up, these formulations must remain elliptical as applied to individual occurrences, and impotent as a guide in life. We shall see later that this problem of choosing an appropriate reference class is precisely the fundamental problem of both the application of probability and the foundations of statistical inference.

One final point must be brought out in this connection: a complex event or individual can no more have a probability on these empirical views than can a simple event or individual. We can no more talk (on this kind of interpretation of probability) of the probability that less than three-quarters of the *next* 1000 coin tosses we perform will result in heads, than we can talk of the probability that the next toss of the coin will result in heads. To put the matter more precisely: in a model for coin-tossing, we have either an infinite number of elements (representing possible tosses of coins) of which a certain subset, H, represents the tosses resulting in heads, and is generally assigned a measure $\frac{1}{2}$; or else we have a model containing two events, H and T, each of which is assigned a probability-measure $\frac{1}{2}$. When we want to talk about sets of tosses, we have to use different models. If we are concerned with pairs of tosses, we consider a model whose individuals (points) are ordered pairs in the single-toss model. The space in which probability is defined is just the cartesian product of the space of the old model with itself. In the first style of interpretation it will be the infinite set consisting of all the possible infinity of ordered pairs of tosses; in the second interpretation, it is the four-membered set consisting of the cross product of $\{H, T\}$ with itself. Consider now sets of n tosses: the behavior of these sets of tosses will be represented by a probability model consisting of a set of *individuals* or *simple events* which is the n-fold cartesian product of the original set with itself. On the usual assumption of independence of tosses, the probability measure to be assigned in the new model depends merely on the measures assigned in the one-toss model; we get an induced measure on the n-toss model. On the basis of that measure, we can compute such things as the probability that of n tosses, less than three-quarters will result in heads. On either of the n-toss models, there will be a set of events (an infinite set in the first case, a finite set in the second) which will have the property: less than three-quarters of the sequence of n-tosses result in heads. Let us take the measure of this set to be $1-\varepsilon$, since on the usual assumptions it will be close to 1. If we consider some definite description, such as 'the next n tosses', we see that in the first model it exists as an individual member of the infinite set for which we have defined the probability measure; but we do not know whether it belongs to the set of individuals (sequences of n tosses) possessing the property in question or not, any more than we know whether 'the next toss' belongs to the set H or not. And whether it does or not is really of no importance, because the probability function P is simply not defined for those individual sequences of n tosses, but only for sets of them. In the second model, the entity described as 'the next n tosses' doesn't even have a counterpart; again it is grammatical nonsense, on this view of probability, to attempt to assign a probability measure to it.

These three points should be kept in mind during the discussion in Chapter Two of the classical approach to statistical inference. First, probabilities are assumed to be defined for a field or σ-field (sigma-field) of sets, so that if $P(A)$ and $P(B)$ are defined in a certain model, so are $P(A \cup B)$, $P(A \cap B)$, and $P(\bar{A})$; and in the case of a σ-field, if $P(A_i)$ is defined for all i, then $P(\bigcup A_i)$ is defined. Even when we have quite complete statistical knowledge about a certain subject matter, it is not generally this complete. Second, the probability axioms are, on empirical interpretations, analytically true. Third, on empirical interpretations of probability it is nonsense to talk about the probability that an individual has a certain property or belongs to a certain set, unless that individual is construed as an 'instance' of a type or kind of individual; and this is true whether the individual is conceptually simple, like the next toss of a particular coin, or conceptually complex, like a particular ten-thousand member random sample of particular population.

On the other hand, it should also be remarked that there is a very close connection between empirical interpretations of probability and logical or subjectivistic interpretations. In each case the typical mathematical framework is the same, with the occasional exception that some writers of logical or subjectivistic persuasion will accept an axiom of finite additivity (if for all i and j, the probability of A_i and A_j is 0, the probability of the union or disjunction or alternation of a finite number of A_i's is the sum of the probabilities of each of the A_i's), while rejecting the countable counterpart. With this exception the probability calculus is the same on each of these theories.

It should also be observed that the interpretation of the items constituting the domain of the probability function is independent of the interpretation given to the probability function itself. Thus one may adopt an empirical interpretation of probability, and suppose that it is defined over a field of sets (this is perhaps most typical); or one may take it to be defined over a field of events; or one may take it to be defined over an algebra of statements. Similarly, one may adopt a logical or subjectivistic view of probability, and (typically) take the domain of the probability function to be an algebra of statements, or, more rarely, a field of events, or least often, a field of sets.

2. We shall now introduce some of the conventional concepts and terminology of probability theory. We shall do so relatively informally, since we merely require to have these ideas at hand for the next six chapters. A more formal development is provided in the Appendix. That formal development will be employed as a basis for the later chapters, from Chapter Seven on.

Consider a certain set S, of individuals. They may be individual events, or trials, or measurements, or insurance policies, or whatever. This set corresponds to Neyman's Fundamental Probability Set. It will not generally be finite. The probability function P is an additive, non-negative, real valued function, defined for a sigma-field of subsets of S, $\mathcal{S}$. This sentence says almost everything, but it says it rather too compactly. Let us break it down. To say that $\mathscr{S}$ is a sigma-field of subsets of S, is to say:

(F-1) $S \in \mathscr{S}$; S belongs to $\mathscr{S}$

(F-2) If A belongs to $\mathscr{S}$, the complement of A in S, $S-A$, belongs to $\mathscr{S}$: $A \in \mathscr{S} \supset S-A \in \mathscr{S}$

(F-3) If $A_1, A_2, \ldots A_n, \ldots$ is a countable set of subsets of S, each of which belongs to $\mathscr{S}$, and if the intersection of each pair to them is empty, then their union, $\bigcup A_i$ belongs to $\mathscr{S}$.

The properties of the probability function defined over all the members of $\mathscr{S}$, are given by the three axioms:

(P-1) $P(S)=1$

(P-2) If $A \in \mathscr{S}$, then $P(A) \geqslant 0$

(P-3) If $A_1, A_2, \ldots A_n, \ldots$ is a finite or countable set of members of $\mathscr{S}$, and if the intersection of each pair of them is empty, then $P(\bigcup A_i) = \sum P(A_i)$. (This is the important property of additivity.)

Within this abstract framework, we may define a number of concepts important in the discussion of statistical inference. These concepts are important also in other theories of statistical inference than those based on an empirical interpretation of probability.

Random Quantity: A random quantity (often referred to as a random variable) is a function from S into R, the set of real numbers or real number magnitudes. For example, if S is the set of tosses of a coin, the function X defined for all members of S whose value is 0 if the member toss results in heads, and 1 if the toss results in tails, is a random quantity. The function, defined on the same set, whose value is 1 for every member of S is also a (constant) random quantity. The function, defined on the same set, whose value is 5 for the members of S that result in heads, and 17 for the other members of S, is a random quantity. If S is a set of ball bearings, the function Y, whose value for a given member of S is the diameter of that member of S, is a random quantity; the function Z whose value for a given member of S is the weight of that member of S is also a random quantity. Obviously any numerical function of a random quantity is also a random quantity. By a simple and obvious generalization we may speak of n-dimensional random quantities; these will be functions from S to R^n.

Borel sets: Consider the random quantity X, a real valued function defined for the individuals of a set S. Among the subsets of S that may hold a special interest for us may be those composed of individuals for whom the function X has a value lying within certain limits: the set of individuals x, belonging to S, such that $X(x)$ is greater than t_1 and less than or equal to t_2. The set of Borel sets associated with a random quantity X comprises just the smallest sigma-field of sets that contains every set of this form. To gain another perspective on the matter, we may approach the set of Borel sets from below, so to speak. Consider the set of all sets of the form $\{x:x\in S \,\&\, t_1<X(x)\leqslant t_2\}$, for any real numbers t_1 and t_2. These sets correspond to the half open intervals on the real line. Let this set of sets be $\mathscr{S}_1$. It is obviously not a σ-field. Now consider $\mathscr{S}_2$, consisting of the members of $\mathscr{S}_1$, complements (in S) of members of $\mathscr{S}_1$, and finite or denumerable unions of members of $\mathscr{S}_1$. $\mathscr{S}_2$ is still not a σ-field, since the complement of a union of members of $\mathscr{S}_1$ need not be in $\mathscr{S}_2$. Let $\mathscr{S}_3$ consist of the members of $\mathscr{S}_2$, complements of members of $\mathscr{S}_2$ and finite.... For similar reasons, $\mathscr{S}_3$ is not a σ-field. Every time we repeat this process, we introduce new sets. The limit of this process, however, is a σ-field. The set of Borel sets is just the set of sets that we approach in the limit if we continue this process without stopping.

Distribution: A distribution is a probability measure defined on the Borel sets corresponding to a random quantity. Let S be our basic set, and let X be a real valued function whose domain is S. Then X is a random quantity. Its distribution is generally given by a distribution function, F_X. F_X is a real valued function, whose domain is the entire set of reals, such that for every real number r, $F_X(r)=P(\{y:y\in S \,\&\, X(y)\leqslant x)$; in other words, the distribution function of the random quantity X over the set S is the numerical function which gives the measure of the subset of S for which the value of the random quantity is less than or equal to r, for any real number r.

We may also consider distributions of two or more random quantities. Just as the set of Borel sets corresponding to a single random quantity X is the smallest sigma-field containing all sets of the form $\{y:y\in S \,\&\, t_1< <X(y)\leqslant t_2\}$, so the set of Borel sets corresponding to a pair of random quantities X, Y is the smallest sigma field containing all sets of the form $\{y:y\in S \,\&\, t_1<X(y)\leqslant t_2 \,\&\, s_1<Y(y)\leqslant s_2\}$, and in general the set of Borel sets corresponding to the random quantities $X_1, X_2, \ldots X_n$ is the smallest

sigma-field containing all sets of the form $\{y{:}y\in S\ \&\ t_{11}<X_1(y)\leqslant \leqslant t_{12}\ \&\ t_{21}<X_2(y)\leqslant t_{22}\ \&\ldots\&\ T_{n1}<X_n(y)\leqslant t_{n2}\}$. The distribution of two random quantities X and Y, will be given by their joint distribution function: $F_{X,Y}(x, y)=\mathrm{P}(\{w{:}w\in S\ \&\ X(w)\leqslant x\ \&\ Y(w)\leqslant y\})$. The distribution of n random quantities may be defined similarly. There are relatively few kinds of distribution encountered frequently in statistics; for future reference, as well as for illustration of the foregoing ideas, I shall mention a few of them.

The Binomial Distribution. Consider an experiment that can have two outcomes, which we may consider success and failure. It is the sort of experiment we may try over and over again – for example it may consist of drawing a ball from an urn, observing whether or not it is black, returning it to the urn, mixing up the contents, drawing a new ball, etc. Let our fundamental probability set S consist of all n-fold repetitions of this experiment. Each element of S thus consists of n-trials of the experiment or n draws from the urn. We define the random quantity B as the number of successes on a given sequence of n trials. Under certain circumstances (which of course must be established empirically) the distribution of the random quantity B is given by:

If $x<0$, $F_B(x)=0$

If $0\leqslant x\leqslant n$, $F_B(x)=\sum_{i=0}^{[x]}\binom{n}{i}p^i(1-p)^{(n-i)}$, where '$[x]$' denotes the largest integer less than or equal to x, and '$\binom{n}{i}$' is the number of ways of picking i things out of n, i.e.,

$$\frac{n(n-1)\,(n-2)\ldots(2)\,(1)}{i(i-1)\,(i-2)\ldots(2)\,(1)\,(n-i)\,(n-i-1)\,(n-i-2)\ldots(1)}$$

If $n<x$, $F_B(x)=1$.

The numbers n and p are the parameters of the distribution. The parameter n represents the number of trials of the experiment being considered, the number of items in a sample, etc. The parameter p represents the measure of the set of successes in the set of trials of experiment. Under circumstances which will insure fairness of the drawings, p will be the proportion of black balls in the urn. Given an appropriate method of tossing, and taking heads to be a success, the random quantity, number of heads on n tosses of a coin, will have a binomial distribution, in which the parameter p is characteristic of the coin (it is $\frac{1}{2}$ for a fair coin), and represents the measure of the set of heads in the set of all tosses made by that method.

The Poisson Distribution: Let S consist of a set of equal intervals (of distances or times or areas or the like), and consider a certain kind of event which may occur in such an interval. The random quantity Y is to be the function, defined over elements of S, which takes as its value for a given time interval, the number of times that an event of the kind in question occured during that interval. Thus Y may take the values $0, 1, 2, \ldots n, \ldots$. It sometimes happens that the distribution function for Y is:

$$\text{If } y<0,\ F_Y(y)=0$$

$$\text{If } 0\leqslant y,\ F_Y(y)=\sum_{i=0}^{[y]} \frac{\lambda^i e^{-\lambda}}{i!},$$

where 'e' denotes the base of the natural logarithms, and $i!=(i)\ (i-1)\ldots(2)\ (1)$, and λ represents the 'average rate' at which the event occurs.

A common quantity which often very nearly has this distribution is the number of telephone calls that arrive at a switchboard during a given time interval. We let the random quantity T be the function defined for the time intervals 12:00 and 12:05, 12:05 to 12:10, etc., whose value is the integer representing the number of calls received at the switchboard during that interval. The measure of the set of intervals during which 0 calls arrive will be $e^{-\lambda}$; the measure of the set of intervals during which 5 or less calls arrive will be

$$e^{-\lambda}+\frac{\lambda e^{-\lambda}}{1}+\frac{\lambda^2 e^{-\lambda}}{2}+\frac{\lambda^3 e^{-\lambda}}{6}+\frac{\lambda^4 e^{-\lambda}}{24}+\frac{\lambda^5 e^{-\lambda}}{120}$$

The distribution function contains the parameter λ, which is characteristic of the particular distribution.

The Normal Distribution: Let S consist of a set of objects or events, and let the random quantity X be a function representing some quantitative attribute possessed by such objects or events. S might consist of the members of a certain segment of the community, and X might be the function whose value for a given individual is his I.Q.; or S might be the set of packages of cereal produced by an automatic packaging machine, and X

might be the function whose value for a given package is the weight of cereal in that package; or S might be the set of events consisting of an egg being hatched, and X might be the function whose value is the number of hours of incubation required to hatch the egg. It often happens that such quantities are distributed *approximately* according to a function of the form:

$$\text{for all real values of } x,\ F_X(x) = \int_{-\infty}^{x} \frac{1}{\sqrt{2\pi}}\, e^{-(y-\mu)^2/2\sigma^2}\, dy$$

Here there are two parameters that characterize the distribution, μ and σ.

The prime importance of the normal distribution lies in the fact that a very great number of empirically significant quantities are distributed approximately normally. A clue to the pervasiveness of the normal distribution lies in the central limit theorem, which asserts that under very general circumstances, the sum of a large number of independent random quantities will be essentially normally distributed, regardless of the distributions of the random quantities that comprise the sum. (The sum of a number of random quantities is the function whose value, for a given individual in the domain S, is the sum of the values of the functions which constitute the individual component random quantities for that individual; the sum of a number of random quantities is thus a random quantity.)

3. There are certain characteristics of distributions that are of central importance in the discussion of statistical inference. Among them are the mean and standard deviation of a random quantity. The mean of the function G, integrable with respect to F_x, of the random quantity X, is generally defined as:

$$E(G(X)) = \int_{-\infty}^{\infty} G(x)\, dF_X(x),$$

where the integral is taken to be a Stieltjes[9] integral. The mean of the quantity X itself is just

$$E(X) = \int_{-\infty}^{\infty} x\, dF_X(x)$$

For the three kinds of distribution described above, we have

Binomial:

$$\mathrm{E}(X)=\sum_{0}^{n} i\binom{n}{i} p^{i}(1-p)^{n-i}=np.$$

Poisson:

$$\mathrm{E}(X)=\sum_{0}^{\infty} \frac{\lambda^{i}\mathrm{e}^{-\lambda}}{i-1}=\lambda.$$

Normal:

$$\mathrm{E}(X)=\int_{-\infty}^{\infty} \frac{x}{\sqrt{2\pi}}\,\mathrm{e}^{-(x-\mu)^2/2\sigma^2}\,\mathrm{d}x=\mu$$

The mean of a random quantity is a kind of 'average' or 'on-the-whole' value. It provides some indication of the character of the distribution of the random quantity. We may also be interested in knowing how closely the values of the random quantity X are clustered about its mean. Thus the random quantity X might have as equiprobable values 5, 6, and 7, and have as its mean 6, while the random quantity Y might have as equiprobable values all the integers from 1 to 11, and also have 6 as its mean. An indication of this degree of dispersion is provided by the *variance* of a random quantity. This is defined as:

$$\mathrm{D}^2(X)=\int_{-\infty}^{\infty} [x-E(X)]^2\,\mathrm{d}F_X(x)$$

In the examples above, we have:

Binomial Distribution:

$$\mathrm{D}^2(X)=\sum_{i=0}^{n} (i-np)^2\binom{n}{i} p^{i}(1-p)^{n-i}=np(1-p)$$

Poisson Distribution:

$$\mathrm{D}^2(X)=\sum_{i=0}^{\infty} (i-\lambda)^2\,\frac{\lambda^{i}e^{-\lambda}}{i!}=\lambda$$

Normal Distribution:

$$\mathrm{D}^2(X)=\int_{-\infty}^{\infty} (1/\sqrt{2\pi})\,(x-\mu)^2\,\mathrm{e}^{-(x-\mu)^2/2\sigma^2}/\sigma^2\,\mathrm{d}x=\sigma^2$$

It should be noted that in these three distributions we can infer the parameters of the distributions from the mean or the mean and variance, and we can infer the mean and variance from the parameters. This need not, however, always be the case.

The standard deviation of a random quantity is simply the positive square root of its variance. Standard deviation plays an important role in statistical inference. An indication of this importance is proved by Tchebycheff's theorem, which says that the probability that a random quantity will differ by more than k standard deviations from its mean is no more than $1/k^2$. For example, the probability that X will differ by more than four of its standard deviations from its mean value is at most $\frac{1}{16}$.

Tchebycheff's Theorem. If X is a random quantity, F_X its distribution function, and both $D^2(X)$ and $E(X)$ exist, then for every positive number k,

$$P(\{y: |X(y) - E(X)| \geqslant kD(X)\}) \leqslant 1/k^2.$$

Proof:

$$D^2(X) = \int_{-\infty}^{\infty} [x - E(X)]^2 \,\mathrm{d}F_X(x) =$$

$$= \int_{-\infty}^{E(X)-kD(X)} [x - E(X)]^2 \,\mathrm{d}F_X(x) +$$

$$+ \int_{E(X)-kD(X)}^{E(X)+kD(X)} [x - E(X)]^2 \,\mathrm{d}F_X(x) +$$

$$+ \int_{E(X)+kD(X)}^{\infty} [x - E(X)]^2 \,\mathrm{d}F_X(x)$$

by a basic property of the integral.

If $x \leqslant E(X) - kD(X)$ or $E(X) + kD(X) \leqslant x$, then $[x - E(X)]^2 \geqslant k^2 D^2(X)$. Therefore,

$$D^2(X) \leqslant k^2 D^2(X) \left[\int_{-\infty}^{E(X)-kD(X)} \mathrm{d}F_X(x) + \int_{E(X)+kD(X)}^{\infty} \mathrm{d}F_X(x) \right]$$

But

$$\int_{-\infty}^{E(X)-kD(X)} \mathrm{d}F_X(x) = P(\{y : X(y) \leqslant E(X) - kD(X)\}).$$

And

$$\int_{E(X)+kD(X)}^{\infty} \mathrm{d}F_X(x) = P(\{y : X(y) \leqslant E(X) + kD(X)\}).$$

Thus $P(\{y : |X(y) \geqslant E(X)| \geqslant kD(X)\}) \leqslant 1/k^2$, as advertised.

4. It will be convenient to have at hand a general notion of a probability system. We shall say that $\mathscr{P}$ is a standard probability system (SPS) if it is an ordered pair $\langle \mathscr{B}, P \rangle$, where $\mathscr{B}$ is a Boolean algebra $\langle B, V, \Lambda, i, j, k \rangle$ and P a probability function defined over B. The Boolean algebra may be construed as a set of subsets of V, in which case Λ is the empty set, i the operation of intersection, j the operation of union, and k that of complementation. This is what we have done earlier. But $\mathscr{B}$ may also be the Lindenbaum-Tarski algebra of a language, so that V is the set of tautologies, Λ the set of contradictions, etc.; or $\mathscr{B}$ may be an algebra of events or of types of events, the constants V and Λ and the operations i, j, k receiving appropriate interpretations. To call P a probability function defined over B, the set of elements of $\mathscr{B}$ is to stipulate that it is real valued, non-negative, that $P(V) = 1$, and that if $a, b \in B$ and $a \; i \; b = \Lambda$ then $P(a \; j \; b) = P(a) + P(b)$.

We shall say that $\mathscr{P}$ is a standard sigma probability system (S σ P S) if it is an ordered pair $\langle \mathscr{B}, P \rangle$ where $\mathscr{B}$ is a σ-algebra, and P a σ-additive probability measure.

Some writers (Keynes, Koopman) have maintained that probabilities do not have the structure of [0, 1], and in particular that it is false that all probabilities are comparable. To accommodate such views it suffices to allow the range of the probability function to be included in a space R with a maximal element 1 and a minimal element 0, in which '+' and '×' have been given appropriate meanings. Then we may say that $\mathscr{P}$ is a nonstandard probability system (N P S) if it is an ordered pair $\langle \mathscr{B}, P \rangle$ where $\mathscr{B}$ is a boolean algebra and P a probability function additive in R.

Similarly, $\mathscr{P}$ will be a nonstandard sigma probability system (N σ P S) if it is an ordered pair $\langle \mathscr{B}, P \rangle$ where B is a σ-algebra and P a probability function σ-additive in R.

NOTES

[1] John Venn, *The Logic of Chance*, London, 1866.

[2] Richard von Mises, *Wahrscheinlichkeit, Statistik, und Wahrheit*, J. Springer, Berlin, 1928. Translated as *Probability, Statistics, and Truth*, 2nd revised English edition, prepared by Hilda Geiringer, London and New York, 1957.

[3] The concept of randomness in a collective has a long and interesting history discussed with clarity and precision by J. A. Coffa in his dissertation, *Foundations of Inductive Explanation*, University of Pittsburgh, 1972.

Among the distinguished contributors to this history are A. Wald, 'Über die Wiederspruchsfreiheit des Kollectivbegriffes', *Ergebnisse eines math. Kolloquiums*, 1947, pp. 38–72; A. Church, 'The Concept of a Random Sequence', *Bull. of American Mathematical Society* **46** (1940), 130–135; and C. P. Schnoor 'A Unified Approach to the Definition of Random Sequence', *Mathematical Systems Theory* **5** (1971). An important recent review article is Per Martin-Löf, 'The Literature on von Mises' Kollektives Revisited', *Theoria* **35** (1969).

[4] J. Neyman, *First Course in Probability and Statistics*, New York, 1950; *Lectures and Conferences on Mathematical Statistics and Probability*, Washington, D.C., 1952.

[5] H. Cramér, *Mathematical Methods of Statistics*, Princeton, 1951. (First edition, Uppsala, 1945.)

[6] K. R. Popper, 'The Propensity Interpretation of Probability', *British Journal for the Philosophy of Science* **10** (1959, 1960), 25–42.

Ian Hacking, *The Logic of Statistical Inference*, Cambridge, 1965. Hacking's 'chance' corresponds roughly to Popper's 'propensity'. For a recent work that makes much of the application of propensities to the single case, see D. H. Mellor, *The Matter of Chance*, Cambridge, 1971. To these references we should now add D. A. Gillies, *An Objective Theory of Probability*, Methuen, 1973. A general review of propensity theory is H. E. Kyburg, 'Probabilities and Propensities', *British Journal for the Philosophy of Science*, forthcoming.

[7] For the views of Neyman, von Mises, and Cramér, see the works cited. Hans Reichenbach's views may be found in *Experience and Prediction*, Chicago 1938 and *The Theory of Probability*, Berkeley and Los Angeles 1949. The latter appeared first in German in 1934. It should be observed that Reichenbach also employs a concept of weight, analogous to (empirical) probability, which is applicable to the single case. Wesley Salmon's views are most completely expressed in *The Foundations of Scientific Inference*, Pittsburgh 1967. Propensity theorists are prone to talk of the probability of the 'single case', but this only amounts to predicating distributively of each member of the class what is – on the frequency view – true of the whole class collectively.

[8] This is not true of the view of Mellor.

[9] These integrals are defined formally in the appendix. In most instances they amount to the familiar Riemann integral of calculus.

CHAPTER TWO

CLASSICAL STATISTICAL THEORY

In an area where there is so much acknowledged turmoil, such outspoken disagreement, and in which so many divergent points of view are so strongly represented as in statistics today, it is almost anomalous to refer to one particular theory as the 'classical' one. It is the stranger to use this term, in that the theory to which it is applied may be said to be no more than fifty years old. But in statistics, as in mathematics, things happen quickly – far more quickly than the layman, or even the professional consumer of statistics or mathematics, realizes. Many of the fundamental ideas of the theory of testing statistical hypotheses were developed in the first quarter of this century by Karl Pearson, Jerzy Neyman, and R. A. Fisher. By the nineteen-forties, the view, primarily associated with the name of Neyman, that the fundamental form of statistical inference was the choice between statistical hypotheses, had come to dominate other views in the English speaking world. It merely dominated other views among statisticians; but it utterly overwhelmed other views among those whose interest in statistics was primarily practical. In medicine, in biology, in psychology, in sociology, in economics – everywhere that statistical hypotheses were formulated as such and tested directly – the vocabulary and formulations of the Neyman approach were so thoroughly ascendent that for many scientists no other approach was even thinkable. 'Significance level', 'One-sided', 'Two-sided', 'Uniformly most powerful', 'Power', 'Error of type I', 'Error of type II' – these terms were part of the vocabulary of scientific inference, and seemed to many who used them to admit of no alternatives. Even the Greek letters alpha and beta seemed to need no more explanation in statistics than pi in geometry. We may call the theoretical framework in which these terms gained currency 'classical', at least in the sense that for a long time there seemed to be no live options. It is only recently that radical alternatives to this framework have begun to be proposed. (Among the alternatives embodying greater or lesser departures from the classical theory are decision theory, Bayesian inference,

Fiducial inference, and the empirical Bayes approach.)

1. The view of statistical inference that we are now exploring is determined in part by a certain view as to the nature of probability, and in part by a certain highly pragmatic perspective on the role of statistical inference in human affairs. Underlying these differences there may be yet more fundamental differences of attitude toward man and knowledge, but for the moment we can ignore both psycho- and philosophical analysis.

From this pragmatic point of view, the need for statistical inference arises from our uncertainty as to how we ought to behave under certain circumstances. We are ignorant concerning the state of the world, and thus, in order to achieve our ends as best we may, we attempt to develop rules of behavior which we may follow, whatever the state of the world, in the expectation that we are following rules whose characteristics are generally (1) desirable, and (2) attainable. It is only in virtue of the fact that desiderata (1) and (2) often conflict that we must delve into the difficulties of statistical inference. The most desirable rule, of course, is one that tells us how to anticipate the future perfectly; the most attainable and simplest rule is to forget about the arithmetic and act as we feel like acting. E. H. Lehman puts it as follows:

> The need for statistical analysis stems from the fact that the distribution of X, and hence some aspect of the situation underlying the mathematical [probabilistic] model is not known. The consequence of such a lack of knowledge is uncertainty as to the best mode of behavior. The observations... provide guidance as to the best decision. The problem is to determine a rule which, for each set of values of the observations, specifies what decision should be taken.[1]

Although one thinks of such actions as accepting a shipment of electric motors, vaccinating a schoolful of children, and the like, the proponents of this point of view have also less purely pragmatic decisions in mind, such as the decision (in agricultural experiments) that variety X yields more bushels of corn per acre than variety Y, or the decision (in medical research) that drug D is without benefit in the treatment of such and such a disease, or even the decision to continue testing or to augment a sample. Some proponents of these views – particularly philosophers who have adopted them – have even proposed that such decisions as to publish or not to publish a certain statistical hypothesis in a learned journal can be treated in the same pragmatic framework. At the other end of the scale of critical reflection are the many practising scientists (particularly in psychology and the social sciences) who automatically apply the statistical tests developed for the purposes of 'inductive behavior' to the evaluation of the significance of experimental results. Indeed, the actions contemplated by Neyman himself are sometimes purely intellectual, for example, "a_1 – Assume the attitude that the observational errors in the studies of the distribution of galaxies in space are adequately represented by the original model, and concentrate...."[2] We cannot therefore (as might be tempting) simply reject this whole point of view out of hand, on the ground that pure scientific research is not pursued with definite practical decisions in mind. As Neyman and others have made perfectly clear, the distinction between 'pure' and 'applied' science is very muddy; and in any event, the engineering applications of probability and statistical inference are as much in need of clarification and foundation as any other applications. Only in the application of probability theory to carefully designed gambling equipment is there no controversy among proponents of various schools of thought. Indeed, this is not surprising, for one might expect that the underlying logical mechanisms of statistical inference, whether they be employed in quality control problems in industry, or in the quest of knowledge for its own sweet sake, are the same. We must therefore see what light is cast by the pragmatic approach to statistical inference on these underlying logical mechanisms.

Hodges and Lehman[3] characterize the mathematical approach to a statistical problem as developing (typically) through four stages: In the first stage, a problem is specified in detail, and our interest focusses on finding some intuitively plausible statistical technique for dealing with it. In the second stage, we become concerned with exploring the properties of that technique – what characteristics of the technique make it a good one? Or a bad one? Third, we can consider several proposed techniques – i.e., general formulas or rules – and explore the question as to which of them is best. Finally, in the fourth stage, we become involved in the design of experiments. This has a practical side (how large should our samples be? Should we use stratification or randomization?) and also a formal side. Neyman describes the formal side this way:

When the one by one examination of performance characteristics becomes difficult, we give it up and, instead, try to solve a problem which, in a sense, is the inverse problem. We specify in advance the properties of the desirable performance characteristics and try to determine the decision function, if such exists, whose performance characteristics possess the indicated properties.[4]

As an illustration of these stages of development, let us consider the problem of estimating the proportion of defective motors in a shipment. In the first stage it seems plausible to take the proportion of motors in a sample as indicating the proportion in the shipment as a whole. But we know perfectly well that the proportion in a small sample can vary from the proportion in the shipment as a whole – in fact it can be shown that for large shipments and even large samples, the chances of having exactly the same proportion in the sample as in the shipment are negligible. What is the relationship, then, between the proportion in the sample and the proportion in the shipment? To explore this relationship mathematically is the business of the second stage of statistical analysis. By ordinary methods of probability theory, given relatively uncontroversial assumptions, given the proportion of defective motors in the shipment, we can calculate the proportion of samples that come close to reflecting the proportion of defective motors in the shipment. We can do this for a number of different assumed proportions in the shipment, and thus get an idea of how our test statistic (proportion in the sample) behaves under various circumstances. In the third stage of statistical analysis we can cast about for other techniques, and compare them with the one we arrived at intuitively. It is often possible to come up with unintuitive techniques that satisfy our criteria better than the intuitive ones. (It is a question, which we shall have to take seriously in later chapters, whether this indicates a defect of our intuitions or a defect of our criteria.) Thus we might propose as an estimate, not d/s, the actual proportion in the sample, but $(d+1)/(s+2)$. Given certain criteria of desirability for estimates, we can examine the behavior of these two estimates (d/s and $(d+1)/(s+2)$) and establish the preferability of one of them. In the fourth stage we can consider abstractly *all possible* estimates, and choose from among them the one that most nearly (in some appropriate sense of 'nearly') satisfies our criteria.

The paradigm case for this style of statistical analysis is the choice between two simple statistical hypotheses, or the test of a simple hypothesis H_0 against a simple alternative H_1. (For example, H_0 might state that a certain quantity was distributed according to a Poisson distribution with parameter 5, and H_1 might state that it was distributed according to a Poisson distribution with parameter 2.) We shall carefully examine a case of this sort below. But beyond this one, there are many applications of the classical statistical analysis, and the concepts developed for this one may be slightly modified for application to other circumstances. The obvious immediate generalization, is to the case of testing between two hypotheses which may not be simple. (For example, H_0 states that the mean of a certain normal population is less than 5.0; H_1 states that it is equal to or greater than 5.0.) This is the form of hypothesis testing most pervasive in the literature of psychology and the social sciences; it is also the form of hypothesis testing most slavishly and naively followed by far too many scientists. As a further generalization, we may consider the classical approach to interval estimation (the parameter p of a certain binomial distribution lies between r_1 and r_2) and to point estimation (the estimate of the parameter λ of a certain Poisson distribution is k). The basic concept involved in all these forms of statistical analysis, looked at from the classical point of view, is *error*; and the criteria in terms of which we evaluate proposed tests and possible tests are themselves formulated in terms of errors. A natural outgrowth of this approach to statistical problems is decision theory, in which we introduce a measure of the importance (or expense) of the various possible errors. Indeed the testing of statistical hypotheses is nowadays often introduced as a special case of decision theory, applicable to problems in which the cost functions applicable to errors have a certain general form. Below I shall present, in as simple a form as possible, a quick review of the treatment offered by the Neyman-Pearson school of the classical forms of hypothesis testing; the general decision theoretic approach will be dealt with separately.

2. The statistical problem that, on the classical view, is the simplest of all is the problem of choosing between two simple statistical hypotheses. Let us suppose, by way of an example, that there are two kinds of tulip bulbs, one of which produces red flowers, and one of which produces yellow flowers. A large shipment of tulip bulbs is received, in which these two kinds are mixed up. Furthermore, the buyer cannot remember whether he ordered 40% yellow tulips and 60% red tulips, or vice versa.

We wish to perform a test in order to find out what the absent-minded buyer bought. An obvious thing to do is to take a sample of bulbs – ten, say – from the shipment, plant them, and to decide what the composition of the shipment is according to the number of tested bulbs that produce red flowers. Intuitively it seems relatively plausible to regard the shipment as composed of 60% red tulips if six or more of the sample turn out to have red flowers.

The first thing we can do is to inquire into the properties of this rule. It is obvious that the rule may well lead us astray: there is no guarantee that by some fluke we won't end up with no yellow tulip bulbs at all in the sample, even if the shipment as a whole contains 60% of them. What the chances are, in fact, can be calculated precisely, subject to certain assumptions. In fact, subject to these assumptions, we can calculate exactly – or to whatever degree of approximation we choose – the probability that a sample will contain six or more red tulips.

What are the assumptions on which such a calculation will be based? There are two: an assumption regarding the so-called randomness of the sampling method, and the assumption of one of the two alternative hypotheses. Randomness of the sample means a special thing, on the interpretation of probability we have at hand here; and correspondingly, the conclusion – the probability statement at which we arrive – must be interpreted in a special way.

Random Sampling Assumption I: The usual assumption concerning the sampling of the bulbs is that it be done by a method which selects each individual with the same probability. As it stands, this doesn't make sense: our probability model assigns probabilities to classes of individuals (the class of red-flowering bulbs and the class of yellow-flowering bulbs) and not to individuals themselves. What is involved in the claim that the sampling is done by a method which would select each individual bulb with the same probability? A new probability model, based on a new fundamental probability set, is involved. Our original hypotheses concerned the measure, in the set of n bulbs in the shipment, of the set of red-flowered bulbs. The Random Sampling Assumption I concerns an *FPS*, *D*, consisting of drawings from the shipment made by a given method; *D* thus consists of events, rather than bulbs. This indefinitely large class of events is partitioned into a finite set of subsets: the set of all those events consisting of the selection of bulb number 1 of the shipment by means of the method; the set of all those events consisting of the selection of bulb number 2 of the shipment by means of the method; ... Random Sampling Assumption I assigns the measure $1/n$ to each of these n subsets of D. This is nothing more or less than the spelling out of the assumption that the method of sampling is such that each individual has the same probability (chance) of being selected. This is the standard condition imposed on the sampling method.

But this condition is inadequate. It will not lead to the conclusions we want. To see this, suppose that the tulip bulbs are numbered, and let the first bulb of our sample be chosen by means of a table of random numbers. Let the subsequent bulbs in the sample be chosen also by means of the table of random numbers, subject to the condition that if the first bulb produced red flowers, the subsequent bulbs of the sample shall also be the bulbs of red-flowered tulips, and if the first bulb produced a yellow bloom, the subsequent bulbs of the sample shall also be bulbs of yellow tulips. Of course this is an awkward rule because we must test the bulbs as we go along, and in general we shall have to test many more than 10 bulbs to get a *sample* of 10 bulbs. But there is nothing about the statement of the randomness assumption which precludes this. Indeed, it is easy to see that under this silly rule of selection I have just spelled out, and subject to the usual assumptions (well justified) about tables of random numbers, the method is such that the measure of the set of events consisting of drawing bulb number i, is the same for every i from 1 to n, and is $1/n$, as required by the Random Sampling Assumption I. Every bulb stands the same chance of being selected as every other one, by the method I have outlined.[5] But while the method satisfies the letter of the Random Sampling Assumption I, it surely does not conform to its intent. What we really need is the following:

Random Sampling Assumption II: The sample is selected by a method under which each possible equinumerous sample has the same probability of being selected. This is the assumption implicitly made in most applications of decision theory and hypothesis testing. It may be explicated in terms of probabilities or measures, just as the first assumption was: we consider a fundamental probability set D_{10}, consisting of selections of 10 individuals from the population of n bulbs, by a given method. This indefinitely large class of events is partitioned into the finite set of possible distinct samples that may be selected from the population by that method. Random Sampling Assumption II corresponds to the assignment of the same measure to each of these $\binom{n}{10}$ sets, namely, $1/\binom{n}{10}$, or in the case of sampling with replacement, the assignment of $1/n^{10}$ to each of the n^{10} possibilities.

The two random sampling assumptions are connected by a relation involving independence. It is sometimes said simply, 'Well, if you have independence, the two assumptions are equivalent'. This is true – obviously in the silly rule of selection I proposed as a counterexample to Assumption I we have dependence, in the sense that the color of the flower produced by the second bulb of the sample *depends* on the color of the flower produced by the first bulb selected. But to spell out what it is to 'have independence' is not as simple as it seems.

The relevant concept of independence is 'numerical' independence.

Intuitively, we say that the selections made by the method of drawing samples are *numerically independent*, if the chance that the ith selection is bulb number n_j, given that the first $i-1$ selections are bulbs number $n_{k_1}, n_{k_2}, \ldots, n_{k_{i-1}}$, is just the same as the unconditional chance that the ith bulb is bulb number n_j. (For sampling without replacement, we stipulate that i not belong to $\{k_1, \ldots, k_{i-1}\}$ and that the chance depend only on how many bulbs have already been removed.) The formal treatment of numerical independence is perfectly straight-forward: we begin with our original FPS D consisting of instances (possible instances) of drawing a bulb by the method in question. The second random sampling assumption entails that we have a probability measure defined on a finite field of subsets of D, $\mathscr{D}$, based on the partition of D into the set of events consisting of the drawing of bulb number one, the set of events consisting of the bulb number two, etc., and that each element of this partition is given the same measure, namely, $1/n$. We interpret this assignment as meaning (roughly) that in the long run, each bulb will be selected by the method about $1/n$th of the time. This is already a hypothesis, and a fairly strong one, concerning our method of sampling; it is sometimes easily justifiable (as when the sampling is done by a table of random numbers); it is more often and less justifiably simply asserted that the difference between this hypothesis and the truth is irrelevant to the situation at hand. These comments apply as well, however, to Random Sampling Assumption II.

Numerical independence of pairs of draws will be an empirical characteristic of the measure we assign to a new FPS consisting of n-tuples of the elements of D. The new partition sets are of the form: bulb i is drawn on the first draw, and bulb j is drawn on the second draw,..., bulb k is drawn on the nth draw. In the new model the measure assigned to the set of events equivalent to drawing bulb i on the first draw will be the same as the measure assigned to the event of drawing bulb i in the first model, $1/n$. Similarly, the measure assigned to the set of events equivalent to drawing bulb j on the second draw will be $1/n$. The measure we assign to the set of events consisting of first drawing bulb i, and then bulb j is not determined, however; consistent with the assignments made so far, it can vary from zero to $1/n$. We say that the first pair of draws is numerically independent if and only if the measure assigned to the set of events equivalent to drawing first bulb i, and then bulb j, is the product of the measure assigned to the sets of events equivalent to drawing bulb i first and the measure assigned to the set of events equivalent to drawing bulb j second, namely, $1/n^2$. In general, we say that a set of draws is numerically independent if the measure assigned to a particular sequence of draws is simply the product of the corresponding measures in the original model.

Parenthetically, it should be observed that numerical independence is inconsistent with sampling without replacement: If we are sampling without replacement, the measure of a set of events in which we get bulb number i on two separate draws is 0. But (*a*) a sense of numerical independence appropriate to sampling without replacement can be defined, and (*b*) in drawing relatively small samples from relatively large populations, there is little difference between sampling with replacement and without.

We may now state the relation between the two random sampling assumptions:

> If Random Sampling Assumption I is satisfied for each of the n draws, and the draws are numerically independent, as specified in the preceding paragraphs, then Random Sampling Assumption II is satisfied as well.

Now given that Random Sampling Assumption II is satisfied, and given one of the two hypotheses (that 40% of the bulbs are red flowering or that 60% of the bulbs are red flowering), we can calculate, through the binomial probability law, or its normal approximation, the measure of the set of sets of ten bulbs of which six or more will turn out to have red flowers. In view of the fact that in such cases as this we would sample by a method which did not involve replacing a bulb once sampled, a more accurate treatment would involve using the hypergeometric distribution in place of the binomial, and modifying the random sampling assumption to yield the measure zero for all those subsets which include the same bulb more than once. The assumption of independence required to get from Random Sampling Assumption I to Random Sampling Assumption II would also have to be modified and indeed changed to a stipulation concerning dependence: we would stipulate, for example in the two member case, that the measure of the set of all events consisting of a draw of bulb i followed by a draw of bulb j should be $(1/n)\ (1/n-1)$ if $i \neq j$, and 0 if $i = j$. But in most practical applications between the usual binomial treatment and this more accurate hypergeometric treatment are slight.

Using the binomial calculation, we are led to the following conclusions: if the proportion of red-flowering bulbs is 0.4, and Random Sampling Assumption II is satisfied, then the measure of the set of sampling events in which we will be led to accept (correctly) the hypothesis that the proportion is 0.4 is 0.8337, and the measure of the set of sampling events in which we will be led to accept (erroneously) the hypothesis that the proportion is 0.6 is 0.1663. On the other hand, if the proportion of red-flowering bulbs is 0.6, and Random Sampling Assumption II is satisfied, then the measure of the set of sampling events which will lead us to accept (correctly) the hypothesis that the proportion is 0.6 is 0.6330, and the measure of the set of sampling events in which we will be led to accept (erroneously) the hypothesis that the proportion is 0.4 is 0.3670.

To make this description of simple hypothesis testing coincide with the classical view, we need to take one hypothesis to be the *hypothesis tested*, or the *null hypothesis*. In the case of testing a simple hypothesis against a simple alternative, choosing the null hypothesis is somewhat arbitrary; it is generally taken to be the least desirable, in some sense.

Here we shall take it to be the hypothesis that only 40% of the bulbs are red-flowering. In this framework, we may speak of two sorts of error.

Error of the First Kind consists in rejecting the hypothesis tested, when it is in fact true. Most classical statistical tests are set up in such a way that this error can be computed; even if we are testing a simple hypothesis against a complex alternative, we will be able to compute the measure of the set of events which will lead us to erroneously reject the hypothesis. In our example, an error of the first kind consists in rejecting the hypothesis that 40% of the bulbs are red-flowering, when it is in fact true that 40% of the bulbs are red-flowering. The measure of this set of events on the assumption that the hypothesis is true, is about 0.1663. Note that this cannot be construed as the probability of error in our results; in the case at hand our results either lead us into error or they do not. There is no question of probability.

Error of the Second Kind consists in erroneously accepting (or failing to reject) the hypothesis tested when it is in fact false. In this particular case, when the hypothesis tested is false, there is another hypothesis (that 60% of the bulbs are red) which is true, and which enables us to calculate the measure of the set of events, under the assumption of the falsity of the hypothesis tested, which will result in our making an error of the second kind. In our illustrative case, the error of the second kind consists in falsely accepting the hypothesis tested, i.e., in accepting the hypothesis that 40% of the bulbs are red-flowering, when in fact it is false, and actually 60% of the bulbs are red-flowering. Above we have calculated this measure as 0.3670. Often we will be unable to calculate a numerical value for the measure of the set of events in which we make an error of the second kind. Thus if we were testing the hypothesis that 40% of the bulbs were red-flowering against the hypothesis that some other proportion of the bulbs were red-flowering, the assumption that the hypothesis tested was false would not give us a basis for calculating the error of the second kind.

Error of the second kind gives us an indication of the ability of the test to discriminate between the hypothesis tested and its alternative. The greater the probability of error of the second kind, the less significant it is if the test *fails to reject* the hypothesis tested. The ability of a test to discriminate between the hypothesis tested and its (possibly complex) alternative is referred to as its *power*. When the alternative hypothesis is simple, as in the case under discussion, the power of a test is taken to be simply one minus the probability of error of the second kind. Thus the power of the test described is roughly $0.63 \doteq 1 - 0.37$.

This is as far as statistical analysis can take us on the classical view; given the proposed test, this is all we can say about it. Indeed this is farther than we can generally be taken by this form of analysis, since it is only in these rare cases in which we have a *simple* hypothesis being tested against a *simple* alternative hypothesis, that we can actually calculate the two types of error. Observe, in particular, that even here in this special case, there is no way of combining the two kinds of error to get an overall 'probability' of error. Furthermore, these two kinds of errors are qualitatively different. "The adoption of hypothesis *H* when it is false is an error qualitatively different from the error consisting of rejecting *H* when it is true."[6] It is thus a blunder from the very beginning even to *attempt* to combine them into a single blanket 'error', as we might be tempted to do, and as Bayesian theorists, for example, do. Probabilities, construed as measures on this view, can only be calculated relative to the relevant hypotheses; and even the probabilities thus calculated do not characterize the *particular* inference or decision made, but only that *kind* of inference or that *kind* of decision. Thus we can say that under the hypothesis in question, given the second randomness assumption, we will in the long run commit an error of the first kind about 17% of the time; and an error of the second kind will be committed about 37% of the time, in the long run, if the hypothesis tested is false. It is these long-run errors that the view of statistical inference now classical in the English speaking world takes to be essential to the characterization of any form of statistical inference.

3. We may follow out the other stages of statistical analysis with the help of this example. We can consider the behavior of alternative rules of decision for this case: e.g., to reject the hypothesis that the proportion of red-flowering bulbs is 0.4 only when there are seven or more red-flowering bulbs in our sample of ten. Given the truth of the hypothesis tested, and the randomness assumption, the measure of the set of cases in which we will falsely reject the hypothesis is 0.0548; this is the 'probability of error of the first kind'. The probability of error of the second kind, or of failing to reject the hypothesis, even though it is false, is 0.6178. Although this

new test gives us more safety with respect to committing an error of the first kind, it does worse with respect to errors of the second kind. In fact, if we adopt this test, it is more likely than not that we'll make an error of the second kind if the hypothesis tested is false.

This will be true in general: more safety with respect to committing an error of the first kind exposes us to more risk of committing an error of the second kind. The convention then is to choose the maximum risk we are willing to run of falling into an error of the first kind, and then to design the test so as to minimize the chance of falling into an error of the second kind. Suppose, for example, we are just willing to reject the hypothesis tested (that the proportion of red bulbs is 40%) a tenth of the time when it is true. That is, we are just barely willing to accept a probability of $\frac{1}{10}$ of committing an error of the first kind. This is called the *significance level* of the test. It is generally denoted by alpha, and in some sciences (psychology, e.g.) it bears a large part of the burden of indicating the inductive support a hypothesis has. The plausibility of using a number as an indication of such a property is one of the things we shall be most concerned to examine later. We can then ask for a test which will satisfy us with respect to its long run probability of committing an error of the first kind, and which will also minimize the probability of making an error of the second kind, i.e., we seek to maximize power. Of the tests we are considering now, the second satisfies the requirement of having a significance level of less than $\frac{1}{10}$, i.e., of leading to a false rejection of the hypothesis tested not more than $\frac{1}{10}$ of the time. In the particular case we are considering, – the testing of a simple hypothesis against a simple alternative – the test determines both the significance level and the power (the long-run probability of *not* making an error of the second kind) simultaneously. One might think that this was an end of the matter.

But in the terms provided by this theory of hypothesis testing, we can do 'better'. By employing a *randomized* test, we can set the significance level at precisely $\frac{1}{10}$, and at the same time increase the power of the test. To see how this is accomplished gives us significantly more insight into the way the theory works, into its standard of 'better', and into its general tenor. Let us call the first test we have described, with significance level 0.1663 and power 0.6330, test I, and the second test, with significance level 0.0548 and power 0.3822, test II. Consider now the following test. We construct a deck of 200 cards, 119 of which are red, and 81 of which are black.

Test R is the following: we shuffle this deck and draw a card at random from it. At random means what was specified above: that in the FPS consisting of draws from the deck, the measure of the set of draws resulting in any given card is the same as the measure of the set of draws resulting in any other card. In the long run, each card will be chosen with equal frequency. If the card thus drawn is red, we perform test II; if the card is black, we perform test I. It is easy to calculate that the significance level of test R is just 0.10, and that its power is 0.484. We have achieved a satisfactory significance level (better than that of test I), and better power than that of test II.

Notice that in fact when we perform this test, we know perfectly well whether we have ended up performing test I or test II – that is, we know whether we reject the hypothesis that 40% of the bulbs are red-flowering when there are six or more in the sample, or whether we reject it when there are seven or more in the sample, so that from a certain point of view it would seem that in the former cases the probability of making an error of the first kind ought to be 0.1663, and in the second case it ought to be .0548. This point of view is based on a principle of conditionality, first explicitly discussed by Alan Birnbaum.[7] We shall return to the principle of conditionality later; here we wish merely to note that this point of view is not the Neyman-Pearson point of view; from this latter point of view, we are performing a single (randomized) test, with significance level 0.1, and power 0.5. If we hired someone to perform the test for us, and all he reported was whether or not the hypothesis was rejected, he would report false rejection a tenth of the time he reports rejection and false acceptance half the time that he reports acceptance. (Not 'of the time' he reports at all; 'of the time' in the long set of tests to which these numbers are relevant.) That is the sense in which the numbers we arrived at characterize the test.

A further step in the analysis of the bulb problem would consist in perhaps considering larger tests – i.e., tests involving more bulbs. In this way we could achieve a higher level of power with the same significance level. But the analysis of a test of any size would proceed along the same lines, and it is possible to show that the test we arrive at in this way is the most powerful test of its size, of the given level of significance. This is shown in general, for a broad class of problems, by the Neyman-Pearson Fundamental Lemma, to which we shall return when we have given a general characterization of hypothesis testing.

We have dwelt on the test of a simple hypothesis against a simple alternative at great length, because it is one which presents the characteristics of the classical approach to statistical inference in a particularly clear way. The generalization of this approach to other kinds of problems is relatively straight-forward; since it does result in the introduction of some new and more general concepts that are involved in other theories as well, it is worth carrying out.

The first step involves the deletion of the adjective 'simple'. This can be done in two steps. We consider first the testing of a simple hypothesis against a complex alternative: for example testing the hypothesis H that exactly 40% of the bulbs are red-flowering against the alternative that more than 40% are red-flowering. The analysis of this form of test is just the same, essentially, as the analysis of the binary case above. A test consists of a rule which directs the rejection of H, on the basis of a randomly selected sample, when the sample has such and such a composition. We can, subject to the randomness hypothesis, and the hypothesis tested, calculate precisely the long run probability that the sample will be such as to lead to the false rejection of H. This probability, is again, the level of significance alpha, or the probability of error of the first kind. But there is no way in which we can calculate the probability of committing an error of the second kind, for this probability – the long-run probability of getting a sample *not* leading, by the rule of the test, to the rejection of H, even though H is in fact false – will depend on precisely what alternative to H is true. There is no way of calculating the measure on the set of these events when we are not given a measure on the simple event of selecting a red bulb. Sometimes, however, it is possible to prove of a given test that it is the most powerful test, of the given level, considered as a test of H against *any* simple alternative making up the complex alternative. Such a test is called 'Uniformly Most Powerful'; a precise formal definition will be given shortly. It should be observed here merely that 'any simple alternative' must be understood as referring to any alternative in a certain set of alternatives we have under consideration: namely, in this case, the set of alternatives corresponding to different values of p in the binomial distribution.

The next development in the process of devising more practicable and useful tests of hypotheses consists of the test of a composite hypothesis (e.g., the proportion of red-flowering bulbs is less than or equal to 0.40) against a composite alternative: (the proportion of red-flowering bulbs is more than 0.40.

Rather than to pursue this simple example further, it seems more profitable to proceed immediately to a general treatment of statistical inference from this classical point of view.

4. Consider a test of the hypothesis that the random quantity X is distributed according to a frequency function $f_X(\theta_1, \theta_2, \ldots \theta_n)$ or according to one of a class of frequency functions $\{f_X(\theta_1 \ldots \theta_n)\}$. Let $\boldsymbol{\theta}$ represent the vector $(\theta_1, \ldots \theta_n)$; then we may say that the distribution of the random quantity X is a member of a class of distributions that depend on a parameter $\boldsymbol{\theta}$, which may take any one of a set of values in the parameter space Ω. Thus Ω is the set of all those values of $\boldsymbol{\theta}$ that correspond to admissible distributions for X, that is, the hypotheses concerning the distribution of X that we are prepared to take seriously. The hypothesis tested corresponds to a certain member of Ω, if the hypothesis tested is simple, or to a certain subset of Ω, Ω_H, if the hypothesis tested is composite.

Generally the test variable, Y, is a random quantity related to the random quantity X. Thus if X is a quantity with a distribution in a certain population, Y may be the average value of that quantity for a subset of the population. But given the distribution of X, and a hypothesis concerning randomness of our sampling, we will be able to derive a distribution for Y. The random quantity Y may have any of a certain set of values; this set S of values is the *sample space*. Thus, subject to the random sampling hypothesis, every point in the parameter space corresponds to a distribution for X, and determines a distribution of the random quantity Y over the sample space. In general, of course, Y will be vector valued.

Given, now, a set of admissible parameter values Ω, a subset Ω_H or a single member corresponding to the hypothesis tested, and a sample space in which the test quantity Y takes its values, we may say that a test of the hypothesis consists of a rule which for each point of the sample space directs us to accept or reject the hypothesis tested, i.e., to accept or reject the hypothesis that $\boldsymbol{\theta}$ belongs to Ω_H. A simple test may thus be construed as a rule or function, defined over the sample space, such that if a point in the region of rejection, or *critical region* S_C, of the sample space is observed, then the hypothesis is rejected, and if a point elsewhere in the sample space (in $S - S_C$) is observed, then the hypothesis is not rejected.

A simple test of a hypothesis is a special case of a randomized test, which may be construed as a function ϕ defined over the whole sample space, such that for every point x in S, $\phi(x)$ is the probability of rejecting H, and $1-\phi(x)$ is the probability of accepting H. Frequently a randomized test procedure consists of a partition of the sample space into three parts: S_0, in which $\phi(x)=0$ and the hypothesis H is never rejected; S_C in which $\phi(x)=1$ and the hypothesis is always rejected; and S_{C^*} in which $\phi(x)=p$ and the hypothesis is rejected or not according to the outcome of some auxiliary experiment which yields the result 'reject' with known probability p.

Another (equivalent) way of looking at randomized tests is to consider an enlarged sample space $S \times R$, in which the elements of R are 'accept' or 'reject' and correspond to the outcomes of an experiment with known stochastic properties; these properties will in general depend on the outcome of the observation in S. The conditional probability of observing 'reject' in R, given the observation x in S, is simply $\phi(x)$. Thus corresponding to the random rejection function ϕ we have the nonrandom rule: reject H if we observe an element (x, reject) in $S \times R$, and otherwise accept H.

Some examples may help to make these concepts clearer.

(1) *Binomial example:* Suppose we have a population P, some members of which have a characteristic K. The parameter space Ω consists of points p, $0 \leqslant p \leqslant 1$, such that the proportion of n-membered subsets of P that have k members with the characteristic K is given by $\binom{n}{k} p^k (1-p)^{n-k}$.

A sample consists of n members of P, and the sample space is thus an n-dimensional vector space, i.e., a set of n-sequences $(x_0, \ldots x_{n-1})$. We suppose that the only relevant characteristics of the sample concern the presence or absence of the characteristic K in the individual members of the sample. Let us take the random quantity $X_i=1$ if the ith member of the sample has the characteristic K, and $X_i=0$ otherwise. Then the sample space consists of n-sequences of zero's and one's. Among the natural hypotheses to test are: (a) $p=0$; (b) $p>\frac{1}{2}$; (c) $p_1 \leqslant p \leqslant p_2$; (d) $p_1 \leqslant p \leqslant 1$; (e) $p=p_1$. Each of these hypotheses singles out a certain subset of Ω, and in effect, asserts that the subject matter of the hypothesis can be characterized by a member of that subset of Ω. Possible selections of critical regions for these hypotheses are (a) the set of vectors $(x_0, \ldots, x_{n-1})$ such that some $x_i \neq 0$, i.e., such that $\sum x_i > 0$; (b) $1/n \sum x_i \leqslant \frac{1}{4}$; (c) $1/n \sum x_i \notin [p_1, p_2]$; (*d*) $1/n \sum x_i \in (0, p_1 - 0.1)$; (e) $|p_1 - 1/n \sum x_i| > 0.015$.

Possible randomized tests are: (f) Reject H with probability 1 if $\sum x_i > 0$; if $\sum x_i = 0$, perform an auxiliary experiment, and reject H with probability $\frac{1}{10}$. (g) Reject H with probability 1 if $1/n \sum x_i < \frac{1}{4}$; reject H with probability 0 if $1/n \sum x_i > \frac{1}{2}$; and if $1/n \sum x_i \in [\frac{1}{4}, \frac{1}{2}]$ reject H with probability $4(\frac{1}{2} - 1/n \sum x_i)$, by means of an appropriate auxiliary experiment.

(2) *Normal example:* Suppose we have a population P, and that there is a real-valued function X defined over the elements of P. We assume that X has a normal distribution, characterized by its mean m and standard deviation σ. The parameter space Ω in this case is two-dimensional, consisting of pairs (a, b), where a may have any value, and b is non-negative. The truth is represented by the (unknown) pair (m, σ). A sample consists as before of n members of P, and the sample space is the n-dimensional vector space $(x_0, \ldots x_{n-1})$, where x_0 is the value of the function X for the first member of our sample, x_1 its value for the second member, etc. Among the hypotheses we might want to test are: (a) $m = 0$, no restriction on σ; (b) $0 < \sigma < 1$, no restriction on m; (c) $m < 5$, $\sigma < 1$. Relatively natural critical regions would be:

(a) the set of vectors $(x_0, \ldots x_{n-1})$ such that $[1/n \sum |x_i|]^2 > (1/n - 1) \sum (x_i - 1/n \sum x_i)^2$

(b) the set of vectors $(x_0, \ldots, x_{n-1})$ such that $1/(n-1) \sum (x_i - 1/n \sum x_i)^2 > 2$

(c) the set of vectors $(x_0, \ldots x_{n-1})$ such that $1/(n-1) \sum (x_i - 1/n \sum x_i)^2 > 2$ or $(\sum x_i)^2 > 5n/(n-1) \sum (x_i - 1/n \sum x_i)^2$.

For the sake of clarity and helpfulness, the examples have been chosen with an eye to their plausibility. The sample space, in an obvious and intuitive sense, is taken to be one that is relevant to the hypotheses being tested. Furthermore, the characteristic of the sample which places it either in or outside of the critical region is a characteristic which in a similarly obvious and intuitive sense, is relevant to the particular hypothesis being tested. The parameter space Ω characterizes a set of hypotheses that belong to a recognizable family; it is in an apparent sense *homogeneous.* Ultimately such relatively arbitrary restrictions must be lifted, or, what comes to the same thing in the final analysis, justified in particular cases.

5. Back in the early days of the development of the theory of testing hypotheses, the distinction was formulated between committing an error of the first kind – rejecting the hypothesis tested even though it is true – and committing an error of the second kind – accepting the hypothesis tested even though it is false. The significance level of a test is the sort of thing that can often be known about a test without too much difficulty. In the example we considered earlier, concerning red-flowering tulip bulbs, we could easily calculate from a given hypothesis concerning the proportion of red-flowering bulbs in the whole population, what the relative frequencies were with which we might expect to find various proportions of red-flowering bulbs in the samples. When we test a specific (simple) hypothesis, such as '40% of the bulbs are red-flowering', it is a straight-forward and simple process to calculate the long-run frequency with which an error of the first kind will be committed. When we test a composite hypothesis, such as 'less than 40% of the bulbs are red-flowering', the calculation of significance level is less straight-forward, but still presents no serious problems. This will be true, in general, for in general the 'hypothesis tested', even if it is composite, will be of fairly simple structure.

This is not the case with respect to probabilities of error of the second kind. The left-over hypotheses – $\Omega - \Omega_H$ – will in general be far less uniform or homogeneous in character than the hypotheses in Ω_H. In some simple cases this is not so – consider our bulb-testing example. But in general the chances of not committing an error of the second kind – i.e., of accepting (or failing to reject) the hypothesis when it is in fact false, depends upon which of the alternative hypotheses is true, and a test which is a powerful test of our hypothesis against one alternative might turn out to be very weak indeed as a means of deciding between our test hypotheses and some other alternative.

More generally, we say that a test of a hypothesis H has the significance level α if the long-run frequency with which the test will reject H, given that H is true, is less than or equal to α. Of course a test with a significance level α is also a test with significance level $\alpha + \varepsilon$. It is thus also convenient to have a term for the infinum of the significance levels; this is called the *size* of the test. Thus the size of the test is the maximum probability, given the truth of the hypothesis, that the rule will lead to its false rejection.

The fundamental condition we impose on a test (now construed generally as a rejection function ϕ defined over the sample space S) is that the test have a certain preassigned size α. Given a hypothesis represented by the parameter point $\boldsymbol{\theta}$, the probability of rejection is simply the mean value of the function ϕ of the random quantity X whose value we observe, evaluated according to the distribution characterized by the parameter $\boldsymbol{\theta}$. Thus

$$E_{\boldsymbol{\theta}}\phi(X) = \int \phi(x)\,\mathrm{d}P_{\boldsymbol{\theta}}(x).$$

The condition we impose on the test ϕ is:

(1) $\quad E_{\boldsymbol{\theta}}\phi(X) \leqslant \alpha$, for all $\boldsymbol{\theta}$ in Ω_H.

Given that this condition is satisfied, and thus that the chances of our committing an error of the first kind are less than or equal to α, we wish to minimize our chances of committing the other sort of error. The long-run frequency with which we avoid committing an error of the second kind, given that the true distribution is $P_{\boldsymbol{\theta}}$, where $\boldsymbol{\theta}$ is in $\Omega - \Omega_H$, is $E_{\boldsymbol{\theta}}\phi(X)$ – the mean value of the rejection function evaluated according to the distribution characterized by the parameter $\boldsymbol{\theta}$. This will be a function of $\boldsymbol{\theta}$, of course, $E_{\boldsymbol{\theta}}(\phi(X))$, and is conventionally denoted by '$\beta_\phi(\boldsymbol{\theta})$'. We want to maximize the power, $\beta_\phi(\boldsymbol{\theta})$, for all $\boldsymbol{\theta}$ in $\Omega_K = \Omega - \Omega_H$. In general, of course, a test which maximizes power against one alternative $\boldsymbol{\theta}_1$ in Ω_K will fail to have maximum power against some other alternative $\boldsymbol{\theta}_2$ in Ω_K. But there are special cases in which the same test has maximum power against all the alternatives in Ω_K.

The most important special case is that in which Ω contains just two points, and we are testing one simple hypothesis against a simple alternative. This special case is established as such in the *Fundamental Lemma of Neyman and Pearson.*

Let P_0 and P_1 be probability distributions possessing densities[8] p_0, p_1 respectively with respect to a measure μ. For testing the hypothesis $H: P_0$ against the alternative $K: P_1$,

(i) Existence: there exists a test ϕ and a constant k such that

(A) $\quad E_0\phi(X) = \alpha$

(B) $$\phi(x) = \begin{matrix} 1, & \text{when} & p_1(x) > kp_0(x) \\ 0, & \text{when} & p_1(x) < kp_0(x) \end{matrix}$$

(ii) Sufficient condition for a most powerful test: If a test satisfies (A) and (B), for some k, then it is most powerful for testing P_0 against P_1 at level α.

(iii) Necessary condition for a most powerful test: If ϕ is most powerful at level α for testing P_0 against P_1, then for some k it satisfies (B), except possibly on a set of μ-measure 0; and satisfies (A) unless there exists a test of size smaller than α and with power 1.

There are many circumstances under which this theorem can be directly applied. Such, for example, was the original tulip bulb situation; we were testing there the hypothesis H, that the proportion of red-flowering bulbs was 0.6, against the hypothesis K, that the proportion of red-flowering bulbs was 0.4. Our test, since it was not randomized, did not yield a test of precisely level alpha. Randomization concerns only that set of elements x of the sample space for which $p_1(x)=kp_0(x)$. Observe that condition (B) can be expressed in terms of the likelihood ratio

$$\frac{p_1(x)}{p_0(x)},$$

a quantity that we shall be much concerned with throughout our investigation. The condition (B) can be expressed: reject H when the likelihood ratio is greater than k; accept H when the likelihood ratio is less than k, and when the likelihood ratio is k, apply the randomization procedure required to ensure that the test is of size alpha.

There is another very broad class of circumstances under which there exist uniformly most powerful tests. An instance was the second version of the tulip bulb example, when we were concerned with testing the hypothesis that the proportion of red bulbs was less than 0.4, against the hypothesis that is was more than 0.4. These circumstances arise when we are concerned with a set of hypotheses characterized by a single real valued parameter θ; Ω is thus just a certain set of real numbers – the possible values of the parameter θ. We are testing a hypothesis of the form $H:\theta\leqslant\theta_0$ against the alternative $\theta>\theta_0$. In general, the most powerful test for testing H against the particular alternative θ_1 will depend on θ_1; but in one special class of cases a uniformly most powerful test exists. That is where the real parameter family of densities $p_\theta(x)$ corresponding to the hypotheses under consideration possesses a *monotone likelihood ratio.* This is defined to be the case when there exists a real valued function $T(x)$, such that:

(a) for any $\theta<\theta'$, the distributions P_θ and P_θ are distinct;

(b) the ratio $p_\theta(x)/p_\theta(x)$ is a non-decreasing function of $T(x)$.

Under such circumstances, we have the following theorem:

Let Ω be a set of real numbers, θ, let the set of statistical hypotheses correspond to probability densities $p_\theta(x)$, and suppose that this family of probability densities has a monotone likelihood ratio in $T(x)$. For testing the hypothesis H_0: $\theta\leqslant\theta_0$ against K:$\theta>\theta_0$,

(i) There exists a Uniformly Most Powerful test which is given by

$$\text{(A)} \qquad \phi(x)=\begin{cases}1 \text{ when } T(x)>C\\ \gamma \text{ when } T(x)=C\\ 0 \text{ when } T(x)<C\end{cases}$$

where C and γ are determined by the condition:

$$\text{(B)} \qquad E_{\theta_0}\phi(X)=\alpha$$

(ii) The power function

$$\beta_\phi(\theta)=E_\theta\phi(X)$$

of this test is strictly increasing for all points θ for which $\beta_\phi(\theta)<1$

(iii) For all θ', the test determined by (A) and (B) is UMP for testing H':

$$\theta\leqslant\theta' \text{ against } K'\text{: } \theta>\theta' \text{ at level } \alpha'=\beta_\phi(\theta')$$

(iv) For any $\theta<\theta_0$ the test minimizes $\beta_\phi(\theta)$ (the probability of an error of the first kind) among all tests satisfying (B).

By reversing the inequalities, we see that there is also a Uniformly Most Powerful test for testing H:$\theta\geqslant\theta_0$ against the alternative K:$\theta<\theta_0$. This theorem also has an obvious bearing on question of confidence-interval inference. We shall therefore return to it. Among the one-parameter families of distributions with monotone likelihood ratios to which this theorem applies, are hypergeometric distributions, binomial distributions, and poisson distributions. The latter two distributions are one-parameter exponential families of distributions – i.e., they can be expressed in the form:

$$p_\theta(x)=C(\theta)\, e^{Q(\theta)T(x)}\, h(x),$$

where Q is strictly monotone. In the binomial case, for example, we have

$$p_\theta(x) = \binom{n}{x} \theta^x (1-\theta)^{n-x} = \binom{n}{x} (1-\theta)^n \exp[(\log(\theta/(1-\theta)))(x)]$$

Another important instance in which there exist uniformly most powerful tests also concerns one-parameter exponential families of distributions. Here we are testing the composite hypothesis $H:\theta \leqslant \theta_1$ or $\theta \geqslant \theta_2$ against the composite alternative $K:\theta_1 < \theta < \theta_2$.

The test has the form,

$$\phi(x) = \begin{cases} 1 \text{ when } C_1 < T(x) < C_2 (C_1 < C_2) \\ \gamma_i \text{ when } T(x) = C_i \\ 0 \text{ when } T(x) < C_1 \text{ or } T(x) > C_2, \end{cases}$$

where the C's and γ's are determined by the requirement that

$$E_{\theta_1}\phi(X) = E_{\theta_2}\phi(X) = \alpha.$$

It is important to observe that although there is a UMP test for testing the Hypothesis $H:\theta \leqslant \theta_1$ or $\theta \geqslant \theta_2$, in an exponential family, against the alternative $K:\theta_1 < \theta < \theta_2$, the dual problem does not have a UMP solution: there exists no UMP test for testing K against H.

Probably the most important testing problems in all of statistics arise in connection with the normal distribution. This is partly because the normal distribution is so pervasive in nature, partly because errors of measurement tend to be distributed normally, and partly because even when a number of independent quantities are not themselves distributed normally, their sum and their average are distributed approximately normally. The parameters of interest here are the standard deviation σ, and the mean, μ. There are four natural one-sided hypotheses we may consider:

H_1	$\sigma \leqslant \sigma_0, -\infty < \mu < \infty$
H_2	$\sigma_0 \leqslant \sigma, -\infty < \mu < \infty$
H_3	$\mu \leqslant \mu_0, 0 \leqslant \sigma < \infty$
H_4	$\mu_0 \leqslant \mu, 0 \leqslant \sigma < \infty.$

For a sample of $n > 1$, there is a UMP test of a hypothesis of the form H_1, against an alternative $K:\sigma_0 > \sigma, -\infty < \mu < \infty$. No UMP test at all exists for testing H_2; and UMP tests for H_3 and H_4 exist only in the completely uninteresting case in which the size of the test is greater than one-half. The UMP test for testing H_1 against its alternative is

$$\sum(x_i - \bar{x})^2 \geqslant C.$$

For some of these problems, and for others not mentioned yet, tests exist meeting slightly weaker criteria. These criteria will be the subject of the next two sections.

5. A test which is uniformly most powerful is one which has maximum power against every alternative in Ω_K; i.e., it is a test such that the long run frequency of making an error of the second kind (failing to reject the hypothesis tested, though it be false) is a minimum, given $\theta \in \Omega_K$, regardless of which $\theta \in \Omega_K$ we consider. Naturally this is a difficult condition to meet; in general a test having maximum power against one alternative $\theta_1 \in \Omega_K$ will not also have maximum power against another alternative $\theta_2 \in \Omega_K$. A natural condition to impose, if we abandon the strong condition that power should be maximized for all alternatives in K, is that at any rate the power should be greater (or at least as great) against all the alternatives in K than it is against alternatives in H. That is, we don't want it to be more likely, under some alternative in K, that we will accept the hypothesis H, than it is that we will accept it under some alternative in H. Expressed in an extreme form: we don't want it to be more likely that we will accept H when it is false (because some alternative in K is true, against which the power is less than $1-\alpha$) than it is that we will accept H when it is true (because the size of the test is α). Put in formal terms, we want the power function of the test ϕ, $\beta_\phi(\theta) = E_\theta \phi(X)$ to satisfy:

$$\beta_\phi(\theta) \leqslant \alpha \quad \text{if} \quad \theta \in \Omega_H$$
$$\beta_\phi(\theta) \geqslant \alpha \quad \text{if} \quad \theta \in \Omega_K$$

Such a test ϕ is said to be *unbiased.*

If we now restrict our attention to unbiased tests, it often turns out that there is among these unbiased tests one which is uniformly most powerful. Such a test is called a Uniformly Most Powerful unbiased test. A UMP unbiased test is thus a test such that the long run frequency of making an error of the second kind is a minimum, not in the class of all possible tests, but in the class of unbiased tests. Here, as elsewhere in statistics, the terminology is more suggestive than it has a right to be. Everybody knows that bias is bad; and *surely* one wouldn't want to use a test which was more likely to fail to reject H when it is false because K is true than it is to fail to reject H when H is true. But it is not difficult to imagine situations in which precisely this is desirable, and therefore in which a biased test would be desirable. Consider specific alternatives H^* in H, and K^* in K. H^* we suppose to be one of the most marginal alternatives in H, and we suppose that the chances of rejecting H^* when it is true are just alpha, the significance level. Similarly, let K^* be close to the edge of the K's – i.e., be very similar in content to the alternatives in H. Under these circumstances we might not care very much if we made an error of the second kind when K^* was true, and we might be willing to accept a test ϕ whose power against K^* was less than $1-\alpha$, so long as its power against alternatives that differed a good bit from the alternatives in H were high. Thus we might well be willing to accept a test that was biased with respect to some alternatives not so very different from those we lump together under H, so long as it did a very good job of discriminating against alternatives in K that are very different from those in H.

There is another set of circumstances which would quite properly lead us to prefer a biased test to an unbiased one. Suppose that we have some means of evaluating the relative likelihood of the alternatives in H and K, and suppose that there is an alternative in K, K^* which is significantly different from the alternatives in H, but which is very much less likely than any other alternative in K. A test which was very powerful in discriminating against the alternatives in K that are very likely might be desirable, even if it were biased in respect to K^*.

These two considerations are taken care of by two developments of this classical theory of testing hypotheses; the first, decision theory, takes account of the seriousness of various sorts of mistakes by representing the mistakes as losses. The second, Bayesian inference, explicitly considers the prior likelihoods of various hypotheses and takes these likelihoods into account. We shall consider both of these approaches in due course.

There are many important testing problems in which there exist UMP unbiased tests, even though there is no generally UMP test.

Let Ω be a set of real numbers, and let the set of statistical hypotheses we are considering correspond to probability densities belonging to a one-parameter exponential family

$$p_\theta(x) = C(\theta)\, e^{\theta T(x)}\, h(x)$$

Let H be the composite hypothesis $\theta_1 \leqslant \theta \leqslant \theta_2$, and let K be the composite hypothesis $\theta < \theta_1$ or $\theta_2 > \theta$. Then there exists a UMP unbiased test of H against K, given by:

$$\phi(x) = \begin{cases} 1 \text{ when } T(x) < C_1 \text{ or } T(x) > C_2 \\ \gamma_i \text{ when } T(x) = C_i, \text{ for } i=1 \text{ or } i=2 \\ 0 \text{ when } C_1 < T(x) < C_2 \end{cases}$$

where C_1, C_2, γ_1, and γ_2 are determined by the condition:

$$E_{\theta_1}\phi(X) = E_{\theta_2}\phi(X) = \alpha.$$

Another situation, in which there exists a UMP unbiased test, but no generally UMP test, concerns the same set of hypotheses, but consists of a test of $\theta = \theta_0$ against $K: \theta \neq \theta_0$. This test has the same form, but the constants are determined by the conditions:

$$E_{\theta_0}\phi(X) = \alpha$$
$$E_{\theta_0}[T(X)\,\phi(X)] = E_{\theta_0}[T(X)]\,\alpha$$

Uniformly Most Powerful unbiased tests exist for various tests concerning normal distributions, or for the comparison of two normal distributions. But there are also many circumstances, including practical ones, in which UMP unbiased tests do not exist, for example, in the presence of nuisance parameters. We may however achieve new results by considering only tests that satisfy another criterion, invariance, that also has some intuitive plausibility to recommend it. Roughly speaking we say that a test is invariant if it is independent of the particular coordinate system chosen for the sample space in which the data are expressed. Among the restricted class of invariant tests appropriate to a particular problem, we may find a UMP invariant test.

Lehman writes:

> The principles of unbiasedness and invariance complement each other in that each is successful in cases where the other is not. For example, there exist UMP unbiased tests for the comparison of two binomial or poisson distributions, problems to which invariance considerations are not applicable.... Conversely there exist UMP invariant tests of hypotheses specifying the values of more than one parameter but for which the class of unbiased tests has no UMP member.

Furthermore,

> The principles of unbiasedness and invariance can be used to supplement each other in cases where neither principle alone leads to a solution but where they do so when applied in conjunction.[9]

6. The theory of confidence bounds and limits is also a part of the classical tradition. Let $\underline{\theta}=\underline{\theta}(X)$ be a function of the observable random quantity X, and suppose that the distribution of X depends on a single unknown parameter θ. We consider in particular functions $\underline{\theta}$ such that $P_\theta(\{x:\underline{\theta}(X(x))\leqslant\theta\})\geqslant 1-\alpha$, for all θ, i.e., such that no matter what the true value of θ may be, the measure of the set of sample points x such that $\underline{\theta}(X(x))\leqslant\theta$ is at least $1-\alpha$. Such a function is called a *lower confidence bound* for θ. In case the distribution depends also on other parameters than θ, say $\theta_1,\ldots,\theta_n$, we stipulate that $P_{\theta,\theta_1,\theta_2,\ldots\theta_n}(\{x:\theta(X(x))\leqslant\theta\})\geqslant 1-\alpha$ for all θ, $\theta_1,\ldots,\theta_n$. The definition of upper confidence bounds is perfectly analogous. A *confidence interval* for θ at confidence level $1-\alpha$ is defined as a random interval with endpoints $\underline{\theta}(X)$, $\bar{\theta}(X)$ such that $P_{\theta,\theta_1,\theta_2,\theta_3,\ldots\theta_n}$ $(\{x:\underline{\theta}(X))\leqslant\theta\leqslant\bar{\theta}(X)\})\geqslant 1-\alpha$.

We want $\underline{\theta}(X)$ to be as informative as possible; therefore we would like it to be as close to θ as possible, while still maintaining the confidence level $1-\alpha$. One way of expressing this condition is to stipulate that for all $\theta'<\theta$, and for all θ, $P_\theta(\{x:\underline{\theta}\,(X(x))\leqslant\theta'\})$ is to be a minimum. Such a function $\underline{\theta}$ does not generally exist in the presence of nuisance parameters $\theta_1\ldots\theta_n$. When it does exist, it is a *uniformly most accurate lower confidence bound for* θ. The existence of such a function corresponds to the existence of a UMP test for testing $\theta=\theta_0$ against alternatives $\theta\neq\theta_0$. We may define a uniformly most accurate upper confidence bound for θ similarly.

When $\underline{\theta}$ and $\bar{\theta}$ are uniformly most accurate lower and upper confidence bounds with confidence coefficients $1-\alpha_1$ and $1-\alpha_2$, for θ, and for every x, $\underline{\theta}(x)<\bar{\theta}(x)$, the intervals $(\underline{\theta},\bar{\theta})$ are confidence intervals for θ, with confidence coefficient $1-\alpha_1-\alpha_2$. They contain the true parameter value with probability $1-\alpha_1-\alpha_2$ since

$$P_\theta(\{x:\underline{\theta}(X(x))\leqslant\theta\leqslant\bar{\theta}(X(x)))=1-\alpha_1-\alpha_2,\text{ for all }\theta,$$

and furthermore, for given values of α_1 and α_2, they minimize any natural loss function of the errors, e.g.,

$$L(\theta,\underline{\theta},\bar{\theta})=\begin{cases}\bar{\theta}-\underline{\theta} & \text{if } \underline{\theta}\leqslant\theta\leqslant\bar{\theta}\\ \bar{\theta}-\theta & \text{if } \theta<\underline{\theta}\\ \theta-\underline{\theta} & \text{if } \bar{\theta}>\theta.\end{cases}$$

Since in the presence of nuisance parameters uniformly most accurate confidence bounds do not generally exist we are led to look for a less demanding condition. In analogy with UMP unbiasedness, we find that often uniformly most accurate *unbiased* confidence intervals and bounds exist. The interval function $(\underline{\theta}, \bar{\theta})$ is unbiased if $P_{\theta, \theta_1, \ldots \theta_n}(\{x : \underline{\theta}(X(x)) \leqslant \leqslant \theta' \leqslant \bar{\theta}(X(x))\}) \leqslant 1-\alpha$ for all θ', θ and all $\theta_1 \ldots \theta_n$ and $\underline{\theta}$ is unbiased if $P_{\theta, \theta_1 \ldots \theta_n}(\{x : \underline{\theta}(X(x)) \leqslant \theta'\}) \leqslant 1-\alpha$ for all $\theta' < \theta$.

In general, we may consider confidence *sets*, $S(X)$, rather than bounds or intervals; $S(X)$ is a function of the random quantity whose values are the points of the sample space; its value is a set of points (not necessarily an interval) in the parameter space. The confidence set $S(X)$ is of level $1-\alpha$, or has confidence coefficient $1-\alpha$, when

$P_{\theta, \theta_1 \ldots \theta_n}(\{x : \theta' \in S(X(x))\}) \geqslant 1-\alpha$, for all $\theta \in H(\theta')$ and all $\theta_1 \ldots \theta_n$,

where $H(\theta')$ is a set of hypotheses being tested. If $H(\theta')$ is $\theta = \theta'$ and $S(X)$ is an interval, we have the usual confidence interval; if $H(\theta')$ is $\theta \leqslant \theta'$ and $S(X)$ is $\{\underline{\theta} : \underline{\theta}(X) \leqslant \theta\}$ we have the lower bound case. The set of confidence sets of level $1-\alpha$ is unbiased if

$P_{\theta, \theta_1, \ldots \theta_n}(\{x : \theta' \in S(X(x))\}) \leqslant 1-\alpha$ for all θ' such that $\theta \notin H(\theta')$ and for all $\theta, \theta_1, \ldots \theta_n$.

The set of confidence sets $S(X)$ is uniformly most accurate unbiased at confidence level $1-\alpha$ if it minimizes the probabilities

$P_{\theta, \theta_1, \ldots \theta_n}(\{x : \theta' \in S(X(x))\})$ for all θ' such that $\theta \notin H(\theta')$ and for all $\theta, \theta_1 \ldots \theta_n$.

The conditions under which uniformly most accurate unbiased confidence sets exist are roughly those under which uniformly most powerful unbiased tests exist.

We may illustrate these ideas with the help of a binomial example. Suppose we wish to find a 0.95 equal tail confidence interval for the unknown parameter p of a binomial distribution from a sample of 50. The sample space may be regarded as the set of integers 0 through 50 – the possible numbers of items in the sample with the property in question. The probability that we will get less than k items in the sample, given that p is the true probability is

$$\sum_{x=0}^{x=k} \frac{50!}{x!(50-x)!} p^x (1-p)^{50-x}.$$

If we set this equal to 0.025 we have a condition on x and p that will be satisfied 0.025 of the time, or with probability 0.025. Suppose we have observed $x=20$. We compute p^* to make this sum no greater than 0.025. Since p is not a random quantity, it makes no sense to speak of the *probability* that p is greater than p^*; it either is or is not greater than p^*. Nevertheless, we can say that whatever p may be, *if* it is greater than p^*, then the probability is less than 0.025 that we *would* observe 20 or less items in the sample with the property in question. If we assert that p is less than p^*, we can justify our assertion by the fact that the probability of making mistaken assertions in general accord with our rule is less than 0.025. From the tables of the cumulative binomial distribution, we can observe that p^* is 0.55. Similarly, we may obtain a lower bound p_* for p which is such that if p is less than p_* the probability is less than 0.25 that we will observe 20 or more items in our sample with the property in question; p_* turns out to be 0.26.

We may combine these results, and observe that the probability is at least 0.95 that an interval constructed in accordance with the techniques we have illustrated will include the true value of the unknown parameter p. We cannot, of course, say anything about the probability of the case at hand; but we can in perfect propriety associate the number 0.95 with the interval (0.26, 0.55). This number is called a *confidence coefficient*, not a probability. This indicates that the *methods* we have used are such as to lead to error not more than 5% of the time, *whatever* value p may have.

7. The justification of techniques of estimation and inference is by their long run properties; surely to be told that whatever the state of the world, a certain technique will lead us astray with a probability no greater than 0.05 is to be told something interesting and important. And according to the classical theory we are expounding, it is *only* these long run properties that can be called upon as justification for our statistical methods.

A recent development along these lines, which has been hailed by

Neyman himself as the greatest breakthrough in statistics in the past fifty years, is Herbert Robbins' *Empirical Bayes* or *Compounding* approach to statistical inference.[10] This approach stems from the discovery that if problems which have been traditionally treated separately are treated jointly, improvements in the long run properties of our decision, inference, and estimation techniques can be demonstrably assured. For example, if data of the same general sort from astronomy, agronomy, genetics, and economics are combined, the decision procedures it is possible to employ will have improved long run operating characteristics. One begins to suspect that something has gone wrong here, especially when a statistician points out (as a *reductio ad absurdum*) that in accordance with the compounding approach it should be possible to go through the *Handbook of Physics* and improve the accuracy with which physical constants are known by compounding the disjoint inferences on which they were originally based; one would not need to dirty one's hands in a laboratory.

From the new point of view, however, this is not a *reductio ad absurdum*, but an application of the compounding approach which is not only legitimate and interesting, but even quite useful. If it is a *reductio ad absurdum*, it is difficult to know where its bite takes hold: at what point, or under what conditions, do we want to say that long-run properties no longer interest us? What other properties are we going to cite as relevant in their place? Or are we to distinguish some long-run properties from others, and if so, how?

It is already sufficiently clear that the general theory of statistical inference that emerges from studies along the lines laid down by Neyman, Pearson, and the other giants of the British-American school is not a theory that the logician can admire. Though it has moments of sparkling incisiveness (exemplified, for example, by the Neyman Pearson Lemma), though it is often elegant in the highest sense of the word, there remain at its foundations only a constantly increasing set of awkward, *ad hoc*, limited principles. Of course the theory must be founded on such principles as we have available; the work must get on. If this were all that could be said against this point of view, we could let the matter drop; logical infelicity in the foundations might be found in a far more grating form in any of half a dozen other disciplines. But there are more penetrating objections that can be raised to this whole approach.

One type of problem is introduced directly by means of the concern with long-run relative frequencies. It is illustrated in an example formulated by D. R. Cox.[11] Consider two normal populations, N_1 and N_2, where N_1 has a mean of θ and a standard deviation σ_1, N_2 has mean θ and standard deviation σ_2, and $\sigma_1 \gg \sigma_2$. We wish to test the simple hypothesis $H_0 : \theta = 0$ against the simple alternative: $\theta = \theta' \cong \sigma_1$.

This situation is not artificial: consider a factory producing parts on two complicated machines. Only one machine is operated on a given day; on the average each machine is used as often as the other. One machine being more accurate than the other, the standard deviation σ_1 of a critical measurement of the parts produced by the first machine is far greater than the standard deviation σ_2 applicable to the parts produced by the second machine. The machines are either correctly set, and the mean of this measurement is 0; or, if either machine goes out of adjustment, it produces parts with measurements having a mean about equal (coincidentally) to σ_1. The statistician arrives at the factory to check the adjustment of the machine that happens to be operating that day. He knows, of course, which machine it is. But he has been obliged to design his test in advance.

The test consists of drawing a sample member of N_1 with probability $\frac{1}{2}$ or a sample member of N_2 with probability $\frac{1}{2}$. The populations are perfectly distinct, so that we know which population we have sampled. Suppose we want a test with size (and therefore in this case significance level) 0.05; by the Neyman Pearson Lemma there is a uniformly most powerful test, the details of which depend on the actual values of θ', σ_1, and σ_2, which is approximately this:

ϕ: Take as the critical region for rejection H_0

$x > 1.28\ \sigma_1$ if the first population has been sampled,
$x > 5\ \ \sigma_2$ if the second population has been sampled.

Although this means our erroneous rejections of $H_0 (\theta = 0)$ will occur almost 10% of the time that we sample the first population, we will be sampling that population only half the time; and the other half of the time we will practically never make a mistake. The long-run probability of rejecting H_0 when it is true is 0.05, and the power (in this case, 1 − the long run probability of failing to reject H_0 when it is false) is a maximum.

The test ϕ is thus demonstrably better than another test of the same size that might be proposed:

ϕ': $x > 1.64\ \sigma_1$, if the first population has been sampled,
$x > 1.64\ \sigma_2$, if the second population has been sampled.

The test ϕ has greater power than ϕ'; the probability of accepting H_0 when it is false is greater than the corresponding probability under ϕ.

Nevertheless it seems clear that any rational person who is concerned to keep the probability of error of the first kind down to 0.05 should prefer ϕ' to ϕ. The 'long-run' here constitutes not a clarification, but a obfuscation; what happens in the long-run is a red herring, since we always know which of two parts of the long-run we are in, and what happens in either of those two parts of the long-run is very different from what happens in the long-run as a whole.

The principle which Cox formulates to justify the obvious choice of the second test is a general heuristic one: "[the sample space] should be taken to consist, so far as possible, of observations similar to the observed set S_1 in all respects which do not give a basis for discrimination between the possible values of the unknown parameter θ of interest."[12]

An analogous principle is formulated by Alan Birnbaum as the principle of conditionality: "...the evidential meaning of any outcome (E_h, x_h) of any experiment having a mixture structure is the same as the evidential meaning of the corresponding outcome x_h, of the corresponding component experiment E_h, ignoring otherwise the over-all structure of the original experiment E."[13]

Similar difficulties with the 'long-run relative frequency' approach (or as Birnbaum calls it, the 'error frequency' approach) are given by Pratt in his generally highly laudatory review of Lehman's book. One of the counterexamples (in an intuitive sense of the word) that Pratt gives has nearly the same point and the same cure as the preceding one. Consider an experiment whose outcome may be registered on either of two meters: one meter gives twenty-six readings: *a*, *b*, *c*,... *z*; the other meter gives only two: *d*, not-*d*. If the result of the experiment is *d*, it will register as *d* on either meter, and in that event it shouldn't make any difference to the import of the experiment which meter '*d*' happened to be read on. But

... the result of a significance test would ordinarily depend on which meter you were reading. A direct continuation of this argument shows that an inference or decision should depend on the probability under the possible hypotheses of the outcome observed only... [and not on] the probabilities of other outcomes also....[14]

Another controversial consequence of the classical approach, illustrative of the fact that in order to apply the approach we must include the specification of results that might have been but were not obtained, is the fact that the 99% confidence interval obtained from the examination of fifty items, of which 20 turn out to have a certain property, is different from the 99% confidence interval obtained from drawing items until 20 have been obtained with the property in question, when a total of fifty are drawn. In the first instance p^* and p_* are determined by:

p^* = the least value of p such that

$$\sum_{\nu \geqslant 20} \binom{50}{\nu} p^{\nu}(1-p)^{50-\nu} \leqslant 0.005$$

p_* = the greatest value of p such that

$$\sum_{\nu \leqslant 20} \binom{50}{\nu} p^{\nu}(1-p)^{50-\nu} \leqslant 0.005$$

In the second instance p^* and p_* are determined by:

p^* = the least value of p such that

$$\sum_{\nu \geqslant 50} \binom{\nu}{19} p^{20}(1-p)^{\nu-20} \leqslant 0.005$$

p_* = the greatest value of p such that

$$\sum_{\nu \leqslant 50} \binom{\nu}{19} p^{20}(1-p)^{\nu-20} \leqslant 0.005$$

Still another sort of point is made by the discussion of examples derived from the following experiment: let U and V be the smallest and largest of a sample n from a uniform distribution between θ and $\theta+1$. We can define a uniformly most accurate upper confidence bound for θ, under which it may be known with certainty that θ is below this confidence bound, 'belying the confidence level $1-\alpha$'. (For example, if $V-U=1$.) Similarly, we can define uniformly most accurate unbiased, and uniformly most accurate invariant confidence intervals for θ, possessing the long run property that they will cover the true value of θ with probability $1-\alpha$, but which are such that in a particular case, given knowledge of

U and V, we can sometimes know for *sure* that θ does not fall in the interval; and we can sometimes know for *sure* that θ does fall in the interval!

Thus Pratt comments that "the theory is arbitrary be it however 'objective', and the problems it solves, however precisely it may solve them, are not even simplified theoretical counterparts of the real problems to which it is applied."[15] To be sure, the theory is not always applied to the problems which it is *supposed* (for example, by Neyman) to be applied. As Birnbaum points out,

> Neyman himself has been foremost in insisting that standard statistical methods such as estimation and testing cannot appropriately be interpreted as methods of inference.... Nevertheless, the latter interpretation is taken as basic, implicitly if not explicitly, in a major part of the exposition and application of the Neyman-Pearson theory.[16]

Pratt says,

> What has made the confidence interval popular is 'indicating what information is available.' Decision problems seem beside the point; a confidence interval probably contains the parameter, and the confidence level measures how probably. But does it? By the formal definition, it no longer does, once we insert numerical values for the endpoints.... We think, and would like to say, it 'probably' does; we can invent something else to say, but nothing else to think.[17]

An instance of the degree to which one can get carried away by one's concern for long-run properties is to be observed in Lehman's own book. Suppose we are interested in the effectiveness of a certain treatment, and perform an experiment on $m+n$ patients, treating n of them and not treating m. We assess the effectiveness of the treatment by measuring some quantity X of the patients. We suppose that the value of the measurement X_i for the ith patient is the sum of two components, U_i and V_i, where U_i is a random quantity representing the initial state of the patient and V_i is a random quantity representing the additive impact of the treatment on the ith individual. Each of the V_i's is assumed to be normally distributed with standard deviation σ, and mean η if the treatment has been applied, and mean $\xi(\leqslant\eta)$ if the treatment has not been applied. Clearly the treatment is effective, under these assumptions, if and only if $\eta\neq\xi$. Thus the null hypothesis to be tested is $\eta=\xi$. if we assume, in addition to the other assumptions, that the patients are selected in such a way that the U's are a random sample from a normal distribution with mean μ and variance σ_1^2, then the classical machinery applies, and yields a uniformly most powerful unbiased test, the t-test.

Alas, it is not generally – perhaps not ever – possible to ensure that the sample of patients is random in the required sense. Even if the quantities U_i have a normal distribution of the required character, we cannot select patients from the whole class of such patients, but only individuals in a particular hospital, or a particular group of hospitals, only individuals at certain stages of the disease, only individuals having the disease at those stages at this particular time in history, etc. Thus on the classical view, the values of the functions U_i, that are exhibited in the sample that we *have* are not values of random quantities, but constants. Thus the distribution of the measurements is given by an expression that contains $m+n$ arbitrary constants, and "... it is clearly impossible to distinguish between $H:\eta=\xi$ and the alternatives $K:\eta>\xi$. The most powerful level α test for testing H against any simple alternative specifying ξ, η, σ, and the u's rejects H with probability α regardless of the observations."[18]

All is not lost, however. Focussing carefully on the long run properties of the tests we apply, and forgetting whatever is immediately under our noses, we can *obtain data through the device of randomization* (Lehman's phrase!) which will allow us after all to test hypothesis $H:\eta=\xi$. We assign the treatments to the patients in such a way that each of the $(m+n)!$ possible assignments has the same probability of being chosen. By this remarkable device a totally uninformative experiment becomes perhaps highly informative: though in point of fact it may consist of *exactly* the same patients, treatments, and observations as before. Let $y_1 \ldots y_n$ represent the observed X-values for the treated patients, and $x_1 \ldots x_m$ the observed X-values for the untreated patients.

Without randomization, a set of y's which is large relative to the x-values could be explained entirely in terms of the unit effects u_i. However, if these are assigned to the y's at random, they will on the average balance those assigned to the x's. As a consequence [sic!], a marked superiority of the second sample becomes very unlikely under the hypothesis and must therefore be put down to the effectiveness of the treatment.[19]

Recalling that we know precisely as much about the u_i values without randomization as with randomization, this sounds almost like a *reductio ad absurdum* of the classical theory.

It is at this point, I think, that our objections become directed back toward the very foundation of the theory: the interpretation of probability. It is already easy to see, without even entering into the theory, that on an empirical-objectivist or relative frequency interpretation of probability, we will never be able to use such locutions as 'the probability is 0.95 that the parameter lies in the interval I'. We are prohibited from using such locutions no less than, and just as obviously as, we are prohibited from making probability statements about the weather *tomorrow*, the next toss of the coin, the frequency of heads on the *next 100* tosses, the life expectancy of *Mr. Jones*, etc. It is, as all frequentists have pointed out, quite beside the point to argue that people do in fact use these locutions. People often use 'velocity' for speed, rather than for a speed-direction vector, but that does not prevent velocity from being a far more fruitful concept in mechanics than speed. But it is to the point that frequentists talk this way too, in ordinary life – most of them; and that the occasions when probability talk is appropriate become narrowed down to the point where 'properly speaking' the only context in which it is appropriate to talk of probabilities is that of a statistical hypothesis. Something must fill the gap, and does. It is sometimes 'significance level'; sometimes 'confidence level'. The most careful of statisticians (e.g., Neyman) may succeed in never using these concepts to characterize *particular* experiments (for it is the experimental test *in general*, not the particular test applied to a particular sample, to which these phrases may be appropriately applied), but we all know perfectly well that practically every statistician, except when he is teaching classes, uses these concepts in much the way that ordinary people use 'probability'.

That a sharp distinction must nevertheless be drawn between confidence and probability, on this view, is clear from the fact that when both concepts apply they may very well conflict. There are occasions, on this classical view, when a statistical parameter may appropriately be regarded as a random quantity: for example, when one first draws a bag of black and white balls from an urn containing a large number of bags, representing various possible ratios of black to white balls, and then draws a sample from the bag with a view to making an estimate of the ratio in that particular bag. Under such conditions, we may perfectly well apply Bayes' theorem, and thereby obtain a posterior distribution for p, the probability of drawing a black ball from that bag. We may, for example, arrive at a posterior distribution according to which the probability that p lies between 0.55 and 0.65 is 0.9. But the confidence method is justified by its long run properties *whatever* be p;

therefore we may simultaneously correctly assign a confidence coefficient of 0.9 to the interval (0.75, 0.95). What are we to believe?

No one would maintain that we should believe that p lies in the interval (0.75, 0.95); everyone agrees that when we have prior probabilities at our disposal (at any rate legitimate frequency ones) we should apply a Bayesian analysis. What is unclear is why we should use our knowledge of the prior probability distribution, rather than the confidence analysis. The confidence method gives an upper limit for error; the Bayesian analysis gives an exact value. But for a given parameter, 0.9, say, this would suggest we use the confidence method. In general it may be that the confidence interval is wider than the corresponding narrowest posterior probability interval, but this is not intrinsic to the problem: it is easy enough to construct examples where the opposite relationship holds (e.g., where the sample frequency is close to 1, and the prior density is concentrated toward 0).

Birnbaum calls attention to a similar anomaly. When the random quantity determined by the sample can take only a finite number of values, we cannot determine exact $p\%$ confidence intervals, except by introducing an auxiliary continuous random variable that has nothing to do with the parameter in question; that is, we can only obtain exact $p\%$ confidence intervals with the help of randomization. The introduction of the irrelevant variable conflicts with Birnbaum's *Principle of Sufficiency* (about which more later), and he points out that in fact, the "usual practice in applied statistics,... is to restrict consideration to non-randomized confidence region estimators... [thereby] sacrificing a degree of efficiency and exactness in terms of error-probabilities...." [p. 125].

As the examples mentioned above show, it is not always the case that a high confidence level reflects a high degree of evidential support; it is not always the case that a high significance level represents a high degree of significance. If it were the case, we could perhaps acquiesce in the rather extensive and inconvenient change of vocabulary that is recommended. But although many statisticians, and essentially all psychologists, sociologists, political scientists, economists, biologists, ecologists, bacteriologists, pathologists, physicians, toxicologists, astronomers, anthropologists, etc., cite significance levels (the smaller the more proudly) and confidence levels (the higher the more joyfully) as though they reflected a level of evidential support applicable to the instance at hand, we know that in general this cannot be the case. In general the level of evidential support, the degree to which the evidence supports one hypothesis as against another, depends on more than some convenient and calculable long run frequency.

The theory of statistical inference we have been considering in this chapter not only gives no indication as to what factors there are that this degree of support depends on, but often specifically counsels (as in its rejection of the principle of conditionality) *against* taking account of factors other than long-run frequencies. There is much that is important on *any* theory of statistical inference contained in the work of those who write from this particular point of view, just as there is in the work of Bayesian theorists. It is, however, clear that the last word on the foundations of statistical inference is not to be found in the classical British-American School.

NOTES

[1] E. L. Lehman, *Testing Statistical Hypotheses*, John Wiley and Sons, New York 1959, p. 1.

[2] Jerzy Neyman, 'Inductive Behavior as the Basic Concept of the Philosophy of Science', *Review of the I.S.I.* **25** (1957), 18.

[3] Hodges and Lehman, *Basic Concepts of Probability and Statistics*, San Francisco 1964, p. 211.

[4] *Op. cit.*, p. 16.

[5] *Proof:* We show that the chance of bulb i being in the sample is $10/n$. This is the sum of two probabilities: the probability that the sample is all red and bulb i is red and bulb i is in the sample, and the probability that the sample is all yellow and bulb i is in the sample. Case I: Assume that 40% of the n bulbs are red. Then the first probability is $0.4 \times 0.4 \times 10/0.4n$ and the second is $0.6 \times 0.6 \times 10/0.6n$; the sum is $10/n$. Case II is similar.

[6] Neyman, *First Course in Probability and Statistics*, p. 261.

[7] See, for example, Alan Birnbaum, 'On the Foundations of Statistical Inference', *Journal of the American Statistical Association* **57** (1962), 269–306.

[8] A definition and discussion of probability densities will be found in the appendix.

[9] E. L. Lehman, *Testing Statistical Hypotheses*, New York and London 1959, pp. 228–229.

[10] Herbert Robbins and Ester Samuel, 'Testing Statistical Hypotheses – the 'Compound' Approach', in R. E. Machol *et al.* (eds.) *Recent Developments in Information and Decision Processes*, New York 1962, pp. 63–70.

[11] D. R. Cox, 'Some Problems Connected with Statistical Inference', *Annals of Mathematical Statistics* **29** (1958), 357–363.

[12] *Ibid.*, p. 10.

[13] *Op. cit.*, p. 11.

[14] John W. Pratt, Review of Lehman, *Testing Statistical Hypotheses*, in *Journal of the American Statistical Association* **56** (1961), 166. The voltmeter example is refined and expanded in Pratt's, 'Comments on A. Birnbaum's, 'On the Foundations of Statistical Inference'', *Journal of the American Statistical Association* **57** (1962), 314–315.

[15] Review of Lehman, p. 164.

[16] Alan Birnbaum, 'The Anomalous Concept of Statistical Evidence', *Hectograph*, 1966, p. 10.
[17] Review of Lehman, p. 165.
[18] Lehman, *op. cit.*, p. 190.
[19] *Ibid.*, p. 191.
[20] Alan Birnbaum, 'Concepts of Statistical Evidence' in *Philosophy, Science, and Method* (Morgenbesser *et al.*, eds.), New York 1969, p. 125.

CHAPTER THREE

R. A. FISHER: LIKELIHOOD AND FIDUCIAL INFERENCE

R. A. Fisher, although he too claimed to adopt an empirical conception of probability, saw clearly some of the problems alluded to in the last chapter. According to him, "'The Theory of Testing Hypotheses' [the ironical quotes are Fisher's] was a later attempt, ... to reinterpret [tests of significance] in terms of an imagined process of acceptance sampling...."[1] In designing a test for accepting batches of manufactured items, it is indeed the long run properties of the test that are of interest; if we are willing to reject acceptable samples 5% of the time, it may well be that the only way in which we can achieve just that level of false rejection is to employ a randomized test, and it may make good sense to want to achieve exactly that level of false rejection. It is, after all, the indefinitely long run of repetitions of the test that concerns us. As Fisher clearly saw, testing a scientific hypothesis is another matter altogether: "...such processes have a logical basis very different from those of a scientist engaged in gaining from his observations an improved understanding of reality."[2]

1. Fisher's own views, which are based upon a concern for scientific inference as opposed to acceptance sampling, are therefore worth examining. Fisher's approach leads to a unified theory; lacking both the fragmentary and *ad hoc* character of many of the principles mentioned in the last chapter, and the unrealistically excessive concern with the 'long run'. It has, however, two shortcomings: the range of problems for which it provides usable answers is rather narrow; and the principles on which its unity rests are very obscure. The difference between Fisher's views and those of the Neyman-Pearson tradition comes out most clearly in connection with significance tests. The tests and the form of the results are nearly the same for each group; yet the underlying philosophy and the interpretation of the results is profoundly different.

Fisher calls attention[3] to the difference between a technician performing a mechanical test over and over again for the Royal Navy or a manufacturing concern, and a scientist attempting to evaluate a scientific hypothesis. While it is obviously appropriate for the test that is designed to be repeated over and over again to be designed with an eye to, and evaluated in terms of, its long run characteristics, it seems not at all obvious that the same standards should be applied to scientific inference. In testing a scientific hypothesis, we want to take account of everything we know; we *should* take account of everything we know.

An acceptance procedure is devised for a whole class of cases. No particular thought is given to each case as it arises nor is the tester's capacity for learning exercised. A test of significance on the other hand is intended to aid the process of learning by observational experience.... To regard the test as one of a series is artificial; the examples given have shown how far this unrealistic attitude is capable of deflecting attention from the vital matter of the weight of the evidence actually supplied by the observations on some possible theoretical view, to, what is really irrelevant, the frequency of events in an endless series of repeated trials which will never take place.[4]

As a consequence, randomized tests of significance, designed, as we saw in the last chapter to yield tests in which the error of the first kind has an exactly specified value, strike Fisher as absurd; "...planned randomization...has no useful part to play in the formation of opinion, and consequently in the tests of significance designed to aid the formation of opinion in the Natural Sciences."[5]

These criticisms, similar to those mentioned in the last chapter, point up the very great difference in approach and attitude of Fisher and Neyman. Fisher's problem is close to the traditional problem of induction; he is concerned with certain aspects of the procedure by which scientists, using everything available to them in the way of data and ingenuity, draw inferences concerning the structure of the world. Neyman's problem is modeled on the acceptance sampling problem; he is concerned with establishing general procedures, which can be followed with a predictable number of mistakes of a certain sort in the long run. The distinctions in practice may often be small. But there are distinctions in principle which become obvious in certain special cases – for example, in the question of using randomized tests. Fisher, being concerned with the individual instance of inference, cannot but regard randomization in tests (not, of course, in material) as a complete red herring. Given certain data, to decide whether or not to accept a certain scientific hypothesis H according to whether a red card is drawn from a certain deck is surely absurd. If we are thinking of scientific inference, we should clearly form our judgment as best we may on the basis of the relevant evidence. Given that we know what the turn of the card was, and given that the color of the card is irrelevant to the truth of our hypothesis (as it surely is in most cases) our opinion of the hypothesis is best indicated by the test that the card directed us to apply. From the point of view of acceptance sampling, however, the analysis must be quite different. If we have decided that we are willing to be wrong 10% of the time, and if it takes a randomized test to yield a 10% level of significance, then we must follow a general rule with the property that it will not result in our being wrong more than 10% of the time. That we can distinguish the times when we are more likely to be wrong and the times when we are less likely to be wrong is a fact that is irrelevant to the general characterization of our procedure. In fact this point of view may very well be appropriate if we are going to relegate our testing to lackies, and if there is a fair degree of homogeneity in the applications of the tests. But where there is no homogeneity in the populations being sampled, the stipulation that the level of significance should be uniformly 10% doesn't make much sense.

Fisher claims that it "... has been frequently stated, as though it were a characteristic property of Confidence Limits, that the interval between them will in repeated samples cover the true value with the exact frequency corresponding with the level of significance chosen. E.g., that in 98% of trials the true value would be found to lie between the two 1% points."[6] Fisher remarks that in general the true value will lie between the assigned limits in more than 98% of such trials and no exact statement of probability can be inferred. It is true, at least in the case of discrete distributions such as the binomial, that the interval will cover the true value in perhaps more than the canonical portion of the cases; but this clearly is not a serious matter: we are hardly losing cogency in our argument if we must admit that we are saying that the probability is 'at least' 0.98 that we are right, instead of 'exactly' 0.98.

What is more to the point is that the long run probabilities mentioned in the confidence interval statement are assumed to hold regardless of background knowledge, and this may be (it is a bit hard to tell) what Fisher has in mind. As we shall see shortly, Fisher makes it very clear that the argument by means of which we arrive at a fiducial[7] distribution for an unknown parameter (corresponding, in a superficial way, to a confidence distribution) requires as one of its most crucial premises that we have no prior knowledge of the distribution of the parameter. This premise is neither needed nor used in the conventional statement of the confidence interval argument. Rather it is the conclusion of the argument which is given a strict interpretation: Before drawing the sample, the probability is (say) 0.95 that the computed interval will cover the true value of the parameter. This is true for all values of the parameter, and therefore for the one at hand, regardless of what we know about the parameter. Of course, if we know something special about our method of sampling this probability may not be *relevant* for us, but that is another matter. Even if we knew exactly what the parameter was, and therefore *knew* whether the interval we got from our sample included it or not, the usual interpretation of confidence intervals would not be affected: we would still say, before we took our sample, that the probability was 0.95 that the interval would cover the true value. The corresponding fiducial statement, on the other hand, would no longer apply.

2. There are two circumstances, according to Fisher, when the result of a scientific inference can be expressed as a *bona fide* probability statement. The first, about which there is relatively little argument, is that in which we start with a prior distribution for the parameter we are interested in, and proceed to apply Bayes' theorem. But, of course, in most applications of statistical inference, we are hard put to find prior distributions for the parameters we are seeking to evaluate, unless we adopt a subjectivistic viewpoint, which is another ballgame altogether. The second circumstance, according to Fisher, is when the fiducial argument is available.

A typical problem in which the fiducial argument works is that of estimating the mean of a normal distribution of unknown variance. The treatment is perfectly analogous, *formally*, to the treatment in terms of confidence intervals. Let X be normally distributed with mean μ and variance σ^2. The quantity $t=\sqrt{N-1}\,(\bar{x}-\mu)/s$ (where $\bar{x}$ is the average of N observations, and s is the standard deviation of those observations) has a distribution which is independent of both μ and σ. Such a quantity, which relates a population parameter (or in general the quantity under investigation) to a sample characteristic, and which has a distribution independent of the (unknown) population parameters, is a *pivotal quantity*. In the case at hand, the distribution of the quantity t is extensively tabulated, and provides explicit numerical values for the probabilities concerning the unknown μ. If we have looked at a sample of 10 x's, and found their mean to be 3.0 and their standard deviation to be 3.0, then (from the tables for Student's distribution),

the probability that μ is more than 2 is 0.83
the probability that μ is less than 4 is 0.83
the probability that μ lies between 2.7 and 3.3 is 0.24
the probability that μ lies between 0.74 and 5.26 is 0.95

For a while Fisher was not sure that this fiducial probability was the same as ordinary probabilities. But as early as 1936 he wrote: "This distinctive terminology is not intended to suggest that fiducial probability is not in the strictest sense a mathematical probability, like that of any other to which the term ought to be applied."[8] The indefinite article may suggest the possibility that there is more than one mathematical probability. But this possibility is rejected in 1956: "Probabilities obtained by a fiducial argument are objectively verifiable in exactly the same sense as are the probabilities assigned in games of chance."[9] He also says that fiducial probability is 'entirely identical'[10] with classical probability.

Although Fisher's conception of probability is exceedingly interesting, both in what it takes account of and what it fails to take account of, we shall postpone discussion of it for a while. Let us first examine carefully the conditions of knowledge and ignorance that must be met in order to apply the fiducial argument in this case.

(1) We must know that X is normally distributed.

(2) We must not know anything about μ; in particular we must not know a prior distribution for μ.

(3) We must know that the sample from which we compute $\bar{x}$ is a sample of ten values of X drawn from the population.

(4) We must not know anything about this sample which is relevant to the distribution of the pivotal quantity t; for example we must have no prior knowledge regarding σ, or a distribution for σ.

(5) The sample must embody all of our knowledge concerning the value of μ.

We require that we know that X is normally distributed, because, lacking this knowledge, we would not have knowledge of the distribution of the pivotal quantity. In fiducial inference we always require knowledge of the distribution of the pivotal quantity, and this in general entails that we have some knowledge about the family of distributions to which the distribution of the quantity we are learning about belongs. "We suppose... it to be agreed that the general hypothesis is acceptable, and that what remains is only to make such probability statements about the value of θ as the data allow."[11]

On the basis of this knowledge, it is assumed that it is possible to find a pivotal quantity t with known distribution. Fisher is careful to keep track of the reference sets: "The probability statement... had as reference set all the values of T which might have occurred in unselected samples for a particular value of [the parameter] θ."[12] More precisely: in virtue of our knowledge of the general class to which the distribution in question belongs, we will also know the exact distribution of the random quantity t in samples drawn from any population belonging to the family in question, for a given value of the parameter θ (or vector parameter $\boldsymbol{\theta}$). This distribution, however, has been established

> ...for all values of θ, and so is applicable to the enlarged reference set of all pairs of values (T, θ) obtained from all values of θ. The particular pair of values of θ and T appropriate to a particular experiment certainly belongs to this enlarged set [our condition (3)], and within this set the proportion of cases satisfying [the requisite inequality] is certainly equal to the chosen probability P. It might, however, have been true... that in some recognizable subset, to which his case belongs, the proportion of cases in which the inequality was satisfied should have some value other than P. It is the stipulated absence of knowledge a priori of the distribution of θ, together with the exhaustive character of the statistic T, that makes the recognition of any such subset impossible, and so guarantees that... the general probability is applicable.[13]

The condition of the exhaustiveness of the statistic T corresponds here to our conditions (4) and (5). It amounts essentially to the stipulation that that statistic, in that sample, embodies all of the information provided by the data regarding the parameter μ.

Condition (2) is somewhat different: "Had knowledge a priori been available, the argument... would have been precluded by the consideration that some of the relevant data had been omitted."[14] But, as Fisher goes on to point out,

> ...had there been knowledge a priori, the argument of Bayes could have been developed, which would have utilized all the data, and which would in general have led to a distribution a posteriori different from that to which the fiducial argument leads.... Consequently it is essential to introduce the absence of knowledge a priori as a distinctive datum in order

to demonstrate completely the applicability of the fiducial method of reasoning... [to particular cases].[15]

One of the knottiest problems in developing the fiducial argument for particular cases, even if we assume that the five conditions listed above are met, is the problem of discovering an appropriate pivotal quantity. In certain cases the conditions of the problem are such as to render the fiducial distribution arrived at unique, but this is not always the case. When there is more than one pivotal quantity, the fiducial distribution we arrive at is dependent on which pivotal quantity we choose to work with. We must therefore have principles which will allow us to select the 'right' pivotal quantity where we have a choice. The difficulty of finding such principles is clearly outlined by Dempster.[16] He shows that in certain cases even Fisher's own careful step by step construction of pivotal quantities can lead to a variety of solutions to the same problem. He concludes that the only safe thing is to limit the term 'fiducial' to just those arguments that Fisher himself has approved.

Fisher seems to assume that if there is prior knowledge, then we will simply be able to adopt a straight-forward Bayesian technique. But in the framework provided by a frequency conception of probability, this is not at all trivial. As Barnard writes

It is evident that a fiducial distribution... is a frail thing, easily upset by additional information. In this respect, however, it does not differ very much from more familiar probability distributions. If we are playing poker against an invisible opponent, or one with a poker face, the hand he holds has a perfectly definite distribution, randomly chosen from the set of cards other than those we hold; but if he has not a poker face, and we notice a twinkle in his eye, our probability distribution is upset, in a way it is difficult to specify precisely.[17]

Lindley has noted[18] that the fiducial argument is flatly inconsistent with any prior distribution but one of a special class; Fisher himself goes further and asserts that the fiducial argument is inapplicable if there is *any* prior distribution. (For Lindley, of course, there is *always* a prior distribution, since he accepts a subjectivistic interpretation of probability.)

3. The circumstances under which scientific inference can result in a genuine probability (those circumstances when either Bayes' Theorem or fiducial inference can be applied) are relatively rare. Fisher distinguishes between two other kinds of scientific inference: that in which we have a

particular hypothesis in mind, and merely wish to ascertain, in the light of evidence, whether or not it should be rejected; and that in which we wish to choose the best of a finite number or a continuum of hypotheses. Significance tests are appropriate to the former case; a likelihood analysis to the latter.

Fisher himself offers little in the way of interpretation for statistical significance tests. The basic idea behind significance tests, according to him is "the elementary one of excluding, at an assigned level of significance, hypotheses, or views of the causal background, which could only by a more or less implausible coincidence have led to what had been observed."[19] Savage remarks that "any datum recorded with accuracy is highly improbable on any hypothesis."[20] Other writers are just as baffled that there are those whom small probabilities impress so little as Savage is that there are those to whom they seem somehow so mystically significant. We seem to be confronted here with an out-and-out conflict of intuitions, for which no obvious resolution is at hand.

Level of significance, for Fisher, seems simply to describe a reluctance to accept, all things considered. Here, as elsewhere, Fisher is prone to emphasize the principle that we should not disregard evidence, without being very helpful in spelling out how that plausible principle is to be applied. Fisher agrees that low probabilities can also have the effect of producing a 'reluctance to accept', though of course a low probability and a high significance level are not at all the same thing on his view. Yet it is curious that the same epistemological attitude seems to correspond to two such different things as probability and significance level.

On the testing-hypotheses view (which Fisher refers to in ironical inverted commas) there is a connection between significance level and probability, though it is not altogether straightforward, and can be expressed only within a framework involving an abstract discussion of the operating characteristics of statistical tests, in which alternative hypotheses are considered explicitly. In due course, we may succeed in resolving some of the puzzles surrounding significance tests. Our attitude toward a poker game in which we are dealt a hand containing four aces is not at all the same as our attitude toward a theory according to which an event has probability $\frac{1}{13}$, after we have performed an experiment five times in which that event has occurred four times. We shall attempt to elucidate the difference in a later chapter.

4. The third sort of uncertain inference to which Fisher addresses himself is that involving the choice of one hypothesis from a continuum of possibilities, or from a finite number. He claims that the likelihood function provides the best indication of what evidence an experiment offers concerning a class of hypotheses. The first condition for the application of this mode of inference (if it be called such) is that the sample must be drawn from a population in which the distribution of the random quantity X is known except for a parameter θ, say. For example we may have an asymmetrical coin and tossing apparatus, such that we are convinced that x_n, the number of heads on n throws, is a quantity with a binomial distribution characterized by a parameter p, unknown to us: $P(X_n=m)=\binom{n}{m} p^m(1-p)^{n-m}$. In order to obtain information about the value of p, we perform some experiments, and note that $X_{10}=1$, say. In general, we have a density function for X, $f_X(x, p)$, which gives the probability of observing x, given the value p of the parameter. The likelihood function is simply the same function, or any constant multiple thereof, where p is considered the 'unknown' and x the 'known' variable.

The *likelihood* of hypothesis H given evidence E is simply the conditional probability of E on H. The likelihood ratio of H_1 to H_2 on E is just the ratio of the corresponding conditional probabilities. Since we need be concerned only with the likelihood ratios, we may restrict our attention in the example to the function $p^1(1-p)^9$. It is easy to see that $p=0.1$ maximizes this function, and indeed 0.1 is the maximum likelihood estimate of the parameter p, $\hat{p}$. The likelihood ratio that characterizes a value q for the parameter is the ratio of $q^1(1-q)^9$ to $\hat{p}^1(1-\hat{p})^9$. We may tabulate or plot likelihood ratios for a variety of values of q, and thus obtain a clear picture of what the data tell us. For example, on the data, it appears from the table below that $p=0.1$ is ten times as likely as $p=0.4$ and 40 times as likely as $p=0.5$ (Table I). Figure 1 indicates vividly that values of p not lying between 0.001 and 0.500 are pretty incredible.

It should be observed that for the likelihood function to have this epistemological import, we must have, as in the case of the fiducial inference, *no prior information.* It is obvious that if the prior probability of a value of p equal to 0.5 were 0.99, that value after the experiment would still be the most probable, rather than the value 0.1. The surprising

TABLE I[21]
Relative likelihoods of p to $\hat{p}$
$(R(p/\hat{p}))$

p	likelihood ratio
0	0
0.001	0.0256
0.01	0.236
0.05	0.814
0.10	1.000
0.15	0.897
0.20	0.683
0.25	0.486
0.30	0.313
0.35	0.187
0.40	0.105
0.45	0.0535
0.50	0.0255
0.55	0.0107
0.60	0.00406
0.65	0.00133
0.70	0.000346

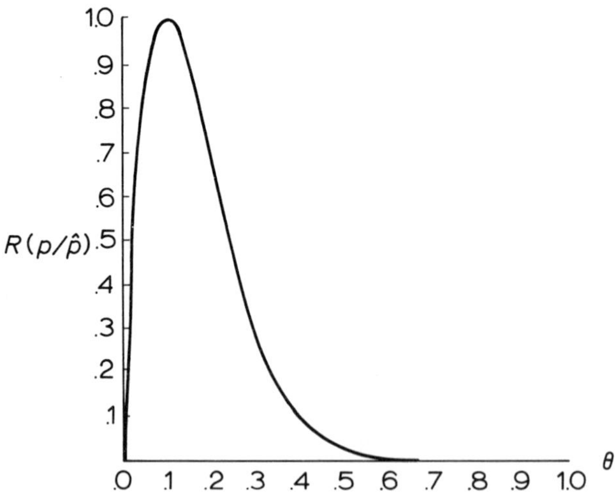

Fig. 1.

outcome of the experiment would reduce the probability that p is 0.5 in accordance with Bayes' theorem. We shall have more to say about this, and about the general implications of the thesis that the evidential content of the data are embodied in its likelihood function, when we come to talk about the relations among the various approaches to statistical inference.

Finally, the likelihood ratio technique only works when we have a specific family of hypotheses to deal with. The likelihood ratio cannot be used as a significance test, and cannot be used to test the validity of a hypothesis except in relation to a class of related hypotheses.

There is one very special case in which we can find a relation between the confidence methods of the last chapter and the likelihood approach discussed here. This concerns the choice between two simple hypotheses H_1 and H_2. If e is the evidence, the $100p\%$ confidence set includes H_1 alone if $P(e/H_1)/P(e/H_2)>1/p$, H_2 alone if $P(e/H_1)/P(e/H_2)<p$, and otherwise includes both H_1 and H_2. The proof, due to Alan Birnbaum, is elementary.

Suppose that H_2 is true. Then the measure of the set of evidence points leading to the rejection of H_2 is $P(\{e:P(e/H_1)/P(e/H_2)>1/p\})= =\sum P(e/H_2)$, where the summation is extended over all sample points e such that $P(e/H_1)/P(e/H_2)>1/p$. But $P(e/H_1)/P(e/H_2)>1/p$ if and only if $P(e/H_2)<p\ P(e/H_1)$. The sum of $P(e/H_2)$ for these sample points e, must be less than the sum of $p\cdot P(e/H_1)$, where the summation is made over points e such that $P(e/H_2)<p\cdot P(e/H_1)$. But obviously the sum of $p\cdot P(e/H_1)$ over anything at all is going to be less than p. Thus

$$P\left(\left\{e:\frac{P(e/H_1)}{P(e/H_2)}>\frac{1}{p}\right\}\right)<p.$$

Similarly, if H_1 is true, we have

$$P\left(\left\{e:\frac{P(e/H_1)}{P(e/H_2)}<p\right\}\right)<p.$$

Thus whichever hypothesis is true, the probability of the confidence set encompassing the true hypothesis is at least $1-p$. Thus the likelihood ratio yields a confidence interval in a natural way for this very special case.[22]

In a later chapter we will examine the claim that the likelihood function embodies the evidential import of an experimental result, and the connection between this principle (Birnbaum calls it the *Likelihood Principle*) and various other principles claimed to underlie statistical inference.

5. Fisher's concept of probability, though he refers constantly to frequencies and 'objective tests' and the like, is not the classical empirical concept. In discussing the meaning of probability, he refers to knowledge and ignorance: "The *knowledge* required for such a [probability] statement refers to a well-defined aggregate, or population of possibilities within which the limiting frequency ratio must be exactly known. The necessary *ignorance* is specified by our inability to discriminate any of the different sub-aggregates having different limiting frequency ratios, such as must always exist."[23] Later he writes, "The subject of a statement of probability must not only belong to a measurable set, of which a known fraction fulfills a certain condition, but every subset to which it belongs, and which is characterized by a different fraction, must be unrecognizable."[24]

In virtue of the repeated references to recognizability, knowledge, and the like, it is clear that Fisher's is not the classical frequency conception of probability. We leave to one side, as unimportant in this context, the serious technical question of how one speaks about ratios in infinite sets – indeed, sets which, to judge by his own examples, may have the cardinality of the continuum. However we choose to handle these problems, the fact remains that a perfectly straightforward probability in a well defined aggregate is what it is, not only in virtue of the existence of a certain number characterizing that aggregate, but *also* in virtue of the fact that *we* don't know how to recognize subsets of that aggregate which would be appropriately characterized by different numbers.

Although the intuitions underlying his notion of probability seem reasonable enough, there are serious objections to Fisher's suggestions.

(1) Surely we need not know *precisely* the value of the probability parameter in an aggregate before we make a probability statement. I don't need to know that this coin is really *perfectly* balanced, nor that my tossing technique is perfectly symmetrical, before I can say that the probability of heads, when I toss the coin, is a half.

(2) It is perfectly possible that we know that the subject of a probability

statement belongs to two or more aggregates, say $A_1 \ldots A_n$, in each of which we know a parameter characterizing the frequency of occurrence of a certain event (say $p_1 \ldots p_n$), without knowing in any particular case, of a subset of the aggregate in which the frequency would be different – that is, without knowing, for any i, of a subset B_i of A_i in which the frequency of occurrence of the event is different from p_i. This entails that we do not know the frequency of occurrence of the event in any of the intersections of our original classes – but this case, so far from being inconceivable, happens all the time in actuarial inference. Surely we are not to take all these numbers as 'the' probability that 'the' subject of our probability statement will turn out to be an 'event'; nor should we say that we know nothing of the probability of the event. But Fisher's formula gives us no means of adjudicating between them or of combining them.

(3) Given any aggregate A, any object a belonging to it, any property P whose frequency of occurrence is characterized by a parameter p, there will exist arbitrarily many subsets of A, to which a is *known* to belong, in which the frequency of occurrence of P is characterized by a parameter different from p. For example: in $\{x:x \in A \& x \text{ has } P \text{ or } x=a\}$, the frequency of P must be essentially 1. We know that a belongs to this set. We can recognize the elements of this set. Similarly for $\{s:s \in A \& x \text{ does not have } P \text{ or } x=a\}$. To get different parameter values we must be able to order the elements of A, or to pick out elements with the help of other properties with known frequencies of occurrence; but there is no difficulty in most cases, given an object a, a real number p, a property Q, and a (large enough) aggregate A, of finding a subaggregate of A to which a is *known* to belong and in which the frequency of Q is *known* to be very nearly p.

6. Statistical inference is very different on Fisher's view, from what it is according to the British American school. Fisher distinguishes several forms of uncertainty in scientific inference, only one of which corresponds at all to a probability. First, the most general form of inference is provided by significance tests. They are applicable in the absence of the conditions that render other forms of inference possible. A hypothesis can be subjected to a significance test all by itself, without having at hand a limited and well-defined set of alternative hypotheses.

These tests are regarded as *sui generis*, and do not receive any sort of probabilistic justification within a general theory of inference. "A test of significance does not authorize us to make any statement about the hypothesis in question in terms of mathematical probability."[25]

Second, likelihood ratio inference may be applicable when Bayesian or fiducial inferences are not. The inference, again, is not characterized by a mathematical probability, but by another quantity, the likelihood ratio. Here we do require a limited and well-defined set of alternative hypotheses. The likelihood ratio exhibits the evidential import of our data with respect to these hypotheses.

Both of these approaches must be profoundly modified in the presence of prior information. Both yield to a form of Bayesian inference if the prior information amounts to a prior distribution of probability for the hypotheses in question.

Third, Bayesian inference does yield a genuine mathematical probability. However the basic condition of Bayesian inference, the knowledge of a prior probability distribution over the alternative hypotheses, is a condition that Fisher regards as being satisfied extremely rarely, and, one gathers, almost never in the case of genuine *scientific* hypotheses, as opposed to hypotheses concerning production lines in factories.

Fourth, there is fiducial inference. Fiducial inference, like Bayesian inference, yields conclusions which can be characterized by genuine mathematical probabilities. But the conditions of ignorance under which fiducial inference is possible are easier to achieve than the conditions of knowledge required for Bayesian inference. Fiducial inference is still a relatively rare possibility, however. It applies unambiguously only in very special cases, and it is highly demanding in what it requires of us in the way of both background knowledge and background ignorance.

Underlying this approach to statistical inference is an interpretation of probability. It would be unfair to criticize 'Fisher's interpretation of probability', because he never actually gives one. To the extent that one can put together an interpretation of probability out of what he does say, it is clearly inadequate. Nevertheless, the principles he offers are very appealing intuitively, when they are applied in a natural way. What better reason, for example, could we offer for not taking a measure in a certain set as determining our degree of rational belief, than that we know that the object in question belongs to some subset of the first set in which the measure is different? But even here two difficulties loom immediately: In general there are all kinds of subsets (perhaps 'artificial' ones) to which we know that a given object belongs, and in which we know the measure of the property in question. And when we abandon the unrealistic supposition that we either know a statistical parameter precisely, or know nothing at all about it, then it becomes somewhat obscure when one parameter is to be regarded as 'different' from another in our body of knowledge.

NOTES

[1] R. A. Fisher, *Statistical Methods and Scientific Inference*, New York 1956, p. 4.
[2] *Ibid.*, p. 5.
[3] *Ibid.*, p. 76ff.
[4] *Ibid.*, p. 99.
[5] *Ibid.*, p. 98.
[6] *Ibid.*, p. 65.
[7] Fiducial distributions and fiducial statements of probability will be defined in due course; they amount to certain special cases of posterior distributions and statements of posterior probability.
[8] 'Uncertain Inference', *Proceedings of the American Academy of Arts and Sciences*, **71** (1936), 253.
[9] *Statistical Methods and Scientific Inference*, p. 59.
[10] *Ibid.*, p. 51n.
[11] *Ibid.*, p. 54.
[12] *Ibid.*, p. 55.
[13] *Ibid.*, p. 55.
[14] *Ibid.*, p. 55.
[15] *Ibid.*, p. 56.
[16] A. P. Dempster, 'On the Difficulties Inherent in Fisher's Fiducial Argument', mimeographed and presented to the 34th session of the International Statistical Institute, 1962.
[17] G. A. Barnard, 'Logical Aspects of the Fiducial Argument', *Bulletin of the International Statistical Institute* **40** (1963), 880.
[18] D. V. Lindley, 'Fiducial Distributions and Bayes' Theorem', *Journal Royal Statistical Society, (B)* **20** (1958), 102–107.
[19] *Statistical Methods and Scientific Inference*, p. 4.
[20] L. J. Savage, 'Discussion', *Bulletin of the International Statistical Institute* **40** (1963), 927.
[21] The table is taken from D. A. Spratt and J. G. Kalbfleisch, 'The Use of the Likelihood Function in Inference', mimeographed, n.d.
[22] Alan Birnbaum, 'Concepts of Statistical Evidence', in S. Morgenbesser, P. Suppes and M. White (eds.), *Philosophy, Science, and Method*, New York 1969, pp. 112–143.
[23] *Statistical Methods and Scientific Inference*, p. 33.
[24] *Ibid.*, p. 57.
[25] *Ibid.*, p. 35.

CHAPTER FOUR

DECISION THEORY

1. Fisher's response to the observation that the classical techniques of testing statistical hypotheses are most appropriate where a hypothesis must be tested over and over again, and acted upon each time, was to denigrate this approach to statistical inference as 'statistics for shopkeepers', to claim that it had nearly nothing to do with the interesting problems of testing *scientific* hypotheses, and to develop a quite different approach to these interesting statistical problems. (To be sure his approach started as early as the Neyman-Pearson approach, and on chronological grounds would have as much right to the honorific 'classical' as that one.) Another response, and one which is more in line with the development of the classical theory, is to agree that this is indeed the statistics of shopkeepers, but to argue that that is just what statistics *ought* to be. If we were psychologists or metaphysicians, the argument might go, then we might want to be interested in what does go on or what ought to go on in the heads of scientists who are doing purely theoretical research; but in point of fact what concerns us is the practical difference it will make to accept one hypothesis or another, and, as Peirce or Dewey or James might have argued, a practical difference can only reveal itself in action. If it makes no difference what hypothesis we accept except when it is a question of acting on that hypothesis, then what we need in statistics is a method for evaluating, in the light of evidence, the actions that depend on hypotheses – and this situation is quite analogous to that of the shopkeeper or the Royal Navy deciding whether or not to accept a shipment of manufactured items. Furthermore, the limitation to the evaluation of courses of action is not a serious one: submitting a result for publication is an action, deciding to believe a hypothesis is an action, even 'accepting' a hypothesis is an action. One ingredient, of course, in any theory of action is a set of utilities of the various possible outcomes. In the case of actions of the sort last mentioned, it may be a bit difficult to come up with an appropriate utility function. This is a problem to which we shall return (it may be the reason that most

scientists still work within the classical framework), but in any event it is a problem that concerns the application of the theory, and not the internal correctness of the theory.

Decision theory does not purport to be a theory of inference at all; it is rather a theory about behaving rationally in the face of uncertainty. Nevertheless many of the situations that are considered by decision theorists are just the same as the situations that are considered by statisticians of other persuasions. Whether a manufacturer is interested in the hypothesis that too many items in a certain lot are defective, or in coming to a decision as to whether or not to accept the lot, is simply a matter of how you look at the situation. Furthermore, decision theory does provide general principles of very broad scope which can in principle (though the details have been worked out only in fairly simple cases) take us much further than the piecemeal and *ad hoc* principles offered by classical statistics.

Many of the most common applications of statistical inference are to problems or processes in which there are a number of choices to make, and in which erroneous choices, in various circumstances, lead to fairly definite and predictable losses. Acceptance sampling in industry provides a paradigm example. There is a certain loss involved in rejecting a shipment of items that is unsatisfactory, a rather larger loss perhaps in rejecting a shipment of items that is, after all, satisfactory, and a still larger loss involved in accepting a shipment which is unsatisfactory. And there are further alternatives we might consider, such as taking another sample, performing a 100% inspection, reprocessing the lot, and so on, each alternative having associated with it certain determinate costs. Furthermore, in the simplest case, in which we are choosing between two alternatives, the structure of the problem is that of choosing between two (possibly complex) hypotheses, and the statistical theory of such choices, which is the simplest and best understood part of statistics, is directly applicable. Thus in many circumstances decision theory is simply a fairly direct generalization of the best understood parts of the theory of testing statistical hypotheses.

The general structure of the decision theoretic problem is the following: there are a number of alternative actions, a_i, usually a finite number (sometimes just two), but perhaps a denumerable or even an uncountable number, as in estimating a magnitude. Similarly, there are a number of states of nature, s_j. One of the conditions of the problem is that we are *given* a loss function, $L(a_i, s_j)$, which gives us the loss of adopting action a_i if the state of nature is in fact s_j. Although it is true that this requirement is often difficult to satisfy realistically, fairly moderate degress of realism will often suffice, for the action to be chosen under given circumstances may be quite insensitive to mild changes in the loss function. At a given stage in our deliberations we have evidence e_k. A decision function is a function whose domain is E, the sample space, or the set of possible evidences e_k, and whose range is S, the set of possible actions a_i. The connection with classical statistics is provided by the assumption that every state of nature, s_j, determines a probability distribution over E, the sample space. Thus given a state of nature, we can compute the *expected* loss of any decision function d. We can do this because each state of nature s_j determines a probability density $dP_j(e)$ over E, the possible set of evidences, and we assume that the action we perform, a_i, and the evidence, are independent. (If we wished, we could stipulate that a state of nature together with an action determined a probability distribution for the evidence.) In a particular state of nature s_j, and for a given set of data e_k, the loss led to by our decision rule d is $L(d(e_k), s_j)$. To obtain the expected loss under the state of nature s_j, we need merely sum or integrate over all the possible evidences: $\int_E L(d(e), s_j)\, dP_j(e)$. We can thus compute, for each possible decision rule d, its expected loss under each state of nature. We then need some way of evaluating these rules.

2. Before turning to the problem of evaluation, let us illustrate these ideas by an example that corresponds precisely to a test of a simple hypothesis against a simple alternative. Let one state of nature, s_0, be that in which 50% of a batch of machines are defective; and let the other state of nature, s_1, be that in which 20% of them are defective. Let a_0 be the action of returning the batch of machines to the manufacturer, and a_1 the action of accepting them. The evidence on which we will base our decision is to be provided by the examination of four of the machines from the batch. To keep things simple, we suppose that we may regard the sample as drawn from a binomial distribution, though in fact the

population is finite and we would not make our drawings with replacement.

Given any state of nature, we can compute the expected loss to be incurred from adopting any given decision function, if we know the loses involved in performing the various actions in each of the various states of nature. To fix our ideas, let us take $L(a_0, s_1)=L(a_1, s_0)=1$ and $L(a_0, s_0)=L(a_1, s_1)=0$ and look at the properties of the following four decision functions:

Expected Losses:

	s_0 (50% defective)	s_1 (20% defective)
a_0 [Return batch]	0	1
a_1 [Accept batch]	1	0

d_1: perform a_0 whatever the evidence.
d_2: perform a_0 if and only if exactly 2 items in the sample are defective.
d_3: perform a_0 if and only if exactly 1 item in the sample is defective.
d_4: perform a_1 whatever the evidence.

Let $E_{s_i}(d_j)=\int_E L(d_j(e), s_i)\, dPs_i(e)$ be the expected loss of the following rule d_j when the actual state of nature is s_i. We have:

$E_{s_0}(d_1)=0$ $\quad$ $E_{s_1}(d_1)=1$
$E_{s_0}(d_2)=0.625$ $\quad$ $E_{s_1}(d_2)=0.1536$
$E_{s_0}(d_3)=0.750$ $\quad$ $E_{s_1}(d_3)=0.4096$
$E_{s_0}(d_4)=1$ $\quad$ $E_{s_1}(d_4)=0$

There are of course many other rules that could be proposed; for example, to perform a_0 just in case the first and third items in the sample are defective. And just as we considered randomized rejection rules for testing statistical hypotheses, so we may consider randomized decision functions or strategies. Thus we might also consider the rule d_5: perform a_1 if no more than one item in the sample is defective, a_0 if more than two are defective, and if exactly two are defective, perform an auxiliary experiment and adopt a_1 if an event of probability $\frac{1}{3}$ takes place. The expected loss under this rule may be computed as follows: Given s_0, the probability that we will end up performing a_1 is

$$\binom{4}{0}(0.5)^4+\binom{4}{1}(0.5)^4+\binom{4}{2}(0.5)^4\left(\tfrac{1}{3}\right)=0.4375$$

The probability of performing a_1 is thus 0.4375 and the expected loss if s_0 is true is 0.4375. Similarly, if s_1 is true, the probability of performing a_1 is 0.8704, and the expected loss under these circumstances is 0.1296.

$$E_{s_0}(d_5)=0.4375 \qquad E_{s_1}(d_5)=0.1296.$$

The idea no doubt already presents itself that we should not merely assess the losses under a helter skelter selection of decision functions that happen to occur to us, but rather should somehow survey the whole set of possible strategies. The consideration of randomized strategies is one device which makes this survey easier; the other is the concept of dominance.

One strategy d_i is said to *dominate* another, d_j, when there is no state of nature in which the expected loss of adopting d_i is greater than the expected loss of adopting d_j, and there is some state of nature in which the expected loss of d_i is less than that of d_j. In the example, d_5 dominates d_2 and d_3.

A strategy or decision function which is not dominated by another is called *admissible*. We may clearly restrict our survey to admissible strategies, provided that we are concerned only with *expected* losses.

In virtue of the fact that we have taken the loss of an incorrect decision to be unity and the loss of a correct decision to be zero, the expected loss of a rule d_i under the statement of nature s_0 is just the probability of doing action a_1, i.e., of doing the inappropriate thing, i.e., of rejecting the hypothesis that the state of nature is s_0 when it is in fact true. This is just the probability of committing an error of the first kind, if we construe d_i as a test of the hypothesis s_0 against the alternative s_1. Thus if we construe d_i as a test in this sense, its significance level is $E_{s0}(d_i)$, and its power against s_1 is $1 - E_{s_1}(d_i)$.

If we allow randomized strategies, then between any two strategies d_i and d_j, with expected losses $E_{s0}(d_i)$, $E_{s_1}(d_i)$, $E_{s0}(d_j)$, and $E_{s_1}(d_j)$, we can interpose a continuum of others $d(p)$ with losses $E_{s0}(d(p)) = pE_{s0}(d_i) + (1-p)\, E_{s0}(d_j)$, $E_{s_1}(d(p)) = pE_{s_1}(d_i) + (1-p)\, E_{s_1}(d_j)$. Thus in general any decision function whose expected losses in each state of nature are a function $pE_{s_i}(d_j) + (1-p)\, E_{s_i}(d_k)$ of the losses of two other strategies will also be a strategy that we might employ. Therefore the set of decision functions is a convex set, that is, a set which contains every linear combination of any two points belonging to it. The set of admissible strategies is a part of the boundary of this convex set.

One enlightening way of looking at this fact is geometrically. In the simple two-dimensional case at hand, we can survey all the possible strategies and plot their expected losses under each of the two hypotheses. Even in such a simple example as this, the possible non-randomized strategies are too numerous to mention. But it is easy to see that the particular order of defective items in our sample cannot serve to differentiate between the two states of nature, on either of the hypotheses. More precisely, the expectation of loss under s_1 of a rule that directs action a_0 exactly when the first and third items are defective is simply proportional (whatever the loss function) to the probability under the state s_1 of getting a sample in which the first and third objects are defective. But under the hypothesis, the probability that the first and third objects are defective is just the same as the probability that the first and second are defective. Since there are six ways in which two of the four objects can be chosen, and since under either hypothesis each of these are equally probable, the expected losses under one of these rules are just the same as the expected losses under a randomized rule that stipulates that action a_0 be performed a sixth of the time in which exactly two defective items are found in the sample, and otherwise to perform a_1. We need not consider these strategies separately, in view of the fact that we are going to consider randomized strategies.

Let d- be the strategy of doing a_1 in any case, d_0 be the strategy of doing a_0 when exactly 0 of the items in the sample are defective, $d_{1,3}$ be the strategy of doing a_0 when either 1 or three of the items in the sample are defective, $d_{0,1,2,3,4}$ be the strategy of doing a_0 in any case, etc. Thus $d_{0,1,2}$ is the strategy of doing a_0 when two or less of the items in the sample are defective. There are in all 32 such strategies (the sum for all i from 0 to 5 of choosing i integers from the set $\{0, 1, 2, 3, 4\}$, or $\sum_{i=0}^{5}\binom{5}{i}=2^5$). The expected loss under each of these strategies, for each of the two states of nature is tabulated in Table II. These pairs of numbers are plotted in Figure 2. A randomized strategy that is a combination of d_i with

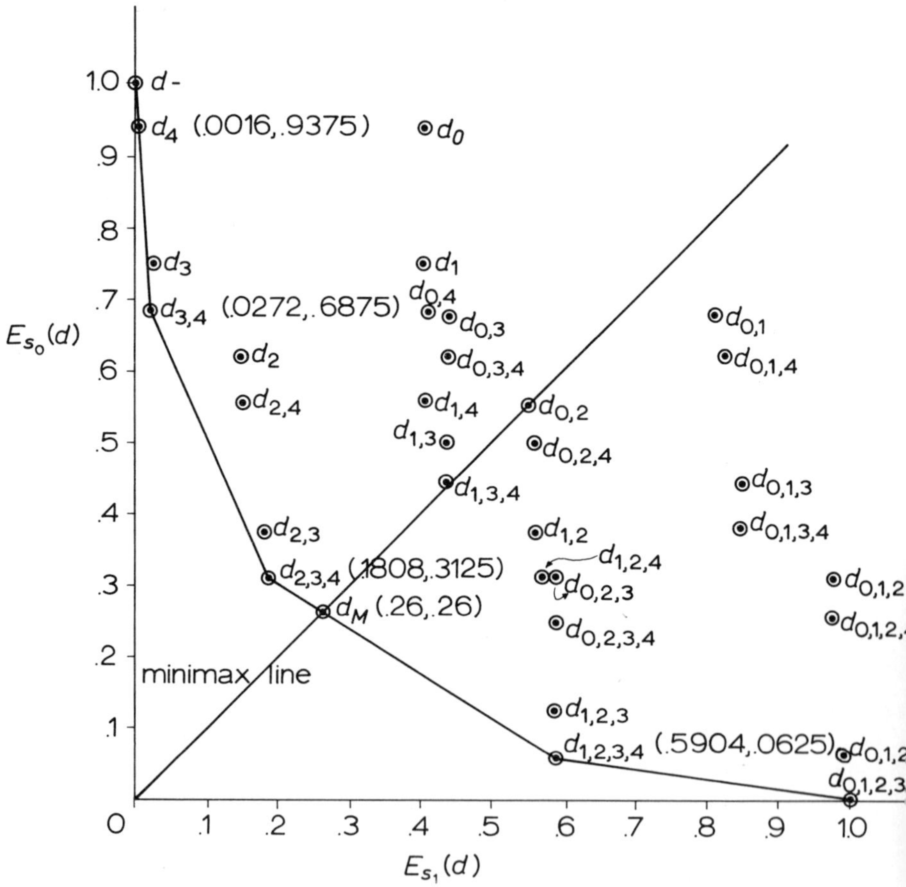

Fig. 2.

TABLE II

	$s_0:p=0.5$ $E_{s_0}(L(d_i))$ (Error of 1st kind)	$s_1:p=0.2$ $E_{s_1}(L(d_i))$ (Error of 2nd kind)
d_0	0.9375	0.4096
d_1	0.7500	0.4096
d_2	0.6250	0.1536
d_3	0.7500	0.0256
d_4	0.9375	0.0016
$d_{0,1}$	0.6875	0.8192
$d_{0,2}$	0.5625	0.5632
$d_{0,3}$	0.6875	0.4352
$d_{0,4}$	0.6750	0.4112
$d_{1,2}$	0.3750	0.5632
$d_{1,3}$	0.5000	0.4352
$d_{1,4}$	0.5625	0.4112
$d_{2,3}$	0.3750	0.1792
$d_{2,4}$	0.5625	0.1552
$d_{3,4}$	0.6875	0.0272
$d_{0,1,2}$	0.3125	0.9728
$d_{0,1,3}$	0.4375	0.8448
$d_{0,1,4}$	0.6250	0.8208
$d_{0,2,3}$	0.3125	0.5888
$d_{0,2,4}$	0.5000	0.5648
$d_{0,3,4}$	0.6250	0.4368
$d_{1,2,3}$	0.1250	0.5888
$d_{1,2,4}$	0.3125	0.5648
$d_{1,3,4}$	0.4375	0.4368
$d_{2,3,}{}^{4}$	0.3125	0.1808
$d_{0,1,2,3}$	0.0625	0.9984
$d_{0,1,2,4}$	0.2500	0.9744
$d_{0,1,3,4}$	0.3750	0.8464
$d_{0,2,3,4}$	0.2500	0.5904
$d_{1,2,3,4}$	0.0625	0.5904
$d_{0,1,2,3,4}$	0	1
d-	1	0

probability p, and d_j with probability $(1-p)$ can be represented by a point on the straight line connecting the point d_i with the point d_j that divides the line in the ratio $p:(1-p)$.

Not all of the pure strategies are admissible. d_1, for example, dominates d_0, so d_0 is inadmissible; but there are still a vast number that cannot be ruled out if we restrict our attention to pure strategies. However, if we look at randomized or mixed decision functions as well as pure ones, it is not hard to see that we need concern ourselves only with d-, d_4, $d_{3,4}$, $d_{2,3,4}$, $d_{1,2,3,4}$, and $d_{0,1,2,3,4}$. Letting X be the random quantity corresponding to the number of defectives in the sample, we need consider only rules of the form: perform a_0 if and only if $X \leq x$, for integral values of x between 0 and 4, inclusive.

For example, consider the rule $d_{1,2,3}$ which is dominated by no pure strategy. The mixtures of $d_{2,3,4}$ and $d_{1,2,3,4}$ may be represented by the set of points on the line connecting the points corresponding to these two strategies. Consider the mixed strategy consisting of applying $d_{2,3,4}p$ of the time and $d_{1,2,3,4}(1-p)$ of the time. Under the state s_0, the loss incurred by this mixed strategy is clearly $pE_{s_0}(d_{2,3,4})+(1-p)$ $E_{s_0}(d_{1,2,3,4}) = 0.3125\,p+(1-p)\,(0.0625)$. Similarly, the expected loss of this mixed strategy under s_1 is $0.1808\,p+0.5904(1-p)$. To find a mixed strategy that dominates $d_{1,2,3}$, we may take the loss under s_1 of $d_{1,2,3}$, 0.5888, and find a mixed strategy among the ones we are considering with just this loss: i.e., a value of p such that $0.1808\,p+0.5904(1-p)=$ $=0.5888$. We find $p=0.0039$. A mixed strategy that uses $d_{2,3,4}$ 0.0039 of the time and $d_{1,2,3,4}$ 0.9961 of the time will therefore have the same loss as $d_{1,2,3}$ under s_1. But it will have the loss 0.0634 under s_0 – much less than the s_0 loss of 0.125 of $d_{1,2,3}$. The mixed strategy thus dominates the pure one.

3. So far we have merely set up the problem. A solution would be a principle telling the manufacturer which decision function to adopt. In view of the way in which we have assigned values to the possible losses, the expected loss, given that s_0 is true, is just the probability that action a_1 will be taken. If we construe action a_1 as rejection of the hypothesis s_0, the expected s_0-loss of a decision function d_i is just the level of significance that would be associated with the corresponding test of the hypothesis s_0. Thus one principle might be to choose a level of significance α, and to adopt that (mixed) decision function whose expected loss is just α when s_0 is the true state of nature. Another principle, to which we shall turn in the next chapter, would apply if there were a way of assigning prior (not necessarily *a priori*) probabilities to the states of the world. Thus if we know that $\frac{3}{4}$ of the lots received by the manufacturer correspond to s_0 and $\frac{1}{4}$ to s_1, then we can compute the expected loss under each decision function d_i, $E(d_i)=\frac{3}{4}E_{s_0}(d_i)+\frac{1}{4}E_{s_0}(d_i)$. Then we

could adopt that decision function for which $E(d_i)$ is a minimum. (In the next chapter we shall see that if it is unique, it is a pure one.)

The most famous principle which is peculiar to the decision theoretic approach to statistics is the minimax principle. The *Minimax Principle* states that we should choose that decision function which will minimize the maximum loss we may expect to suffer under the worst possible state of nature. The principle can be illustrated most easily on a matrix of the expected losses sustained under the hypothetical states of nature S_1, S_2, and S_3, when the decision taken is, say, D_1, D_2, or D_3:

	S_1	S_2	S_3
D_1	0	3	5
D_2	1	2	1
D_3	3	1	0

The minimax principle selects D_2, despite the fact that if S_1 happens to be the case, D_1 would have been preferable, and if S_2 happens to be the case, D_3 would have been preferable. These considerations are irrelevant, since it is a condition of the problem that we don't know which of S_1, S_2, and S_3 actually is the case. Lacking this knowledge, a conservative thing to do is to choose D_2, since the most we can lose is 2 units.

In our example, $d_{2,3,4}$ is the minimax (pure) decision, since in case s_0 is true our expected loss under that rule is 0.3125, and no other pure strategy has such a low maximum expected loss. But we need not limit ourselves to pure strategies; we may note that by allowing the expected loss under s_0 to increase, we may reduce the expected loss under s_1, and conversely. Randomized strategies give us a way of doing this, and in fact there is a randomized strategy (a mixture of $d_{2,3,4}$ and $d_{1,2,3,4}$ mixed about five parts to one) which will give the same loss under *either* state of nature, namely, about 0.26. It is clear that this is the minimax strategy: any point not on the lower boundary of the convex set is inadmissible, any point to the right of the point at which the two expected losses are equal will yield a lower loss if s_0 is true, but a higher loss if s_1 is true; *mutatis mutandis* for any point to the left of the equal loss point. The minimax decision function for any such problem is thus easily seen to be the point at which a 45° line intersects the boundary of the convex set of possible decision functions. For more dimensions, it is the point at which the line of points equidistant from each axis intersects the boundary of the convex set of decision functions. (If there is no intersection, the decision will be that one nearest the equal loss line.)

4. The minimax principle is not always plausible, even where it seems to apply. For example, consider the case of our manufacturer if he loses just as much by sending back a defective lot as he does by sending back a sound lot, but loses even more – say 3 units – if he accepts a defective lot for manufacture. Under these conditions the minimax principle will direct him to employ procedure $d_{0,1,2,3,4}$ – to send back the lot – perform action a_1 – whatever the evidence indicates about it. This will be the minimax decision regardless of how much or little extra he loses by accepting a defective lot for manufacture.

Under these circumstances one might argue that what counts isn't really the total loss, but rather the *difference* between the actual loss and the minimum possible loss. Consider the matrix:

	S_1	S_2	S_3
D_1	1	2	6
D_2	3	2	5
D_3	2	3	4

In S_1 we lose 1 unit, at least, whatever decision we make. In S_2 we lose at least two units, and in S_3 we lose at least 4 units. These losses are not under our control – there is simply nothing we can do about them. What we can do something about is deciding how to get as close to these minimum losses as possible; what concerns us is the *difference* between the actual loss and the minimum loss. Since these are the quantities that concern us, let us construct a matrix that reflects these quantities only; by subtracting the minimum loss in each column from the rest of the entries in that column. This gives us the following table of *regrets* or losses engendered by each decision under each of the three states of nature:

	S_1	S_2	S_3
D_1	0	0	2
D_2	1	0	1
D_3	2	1	0

To this table we now apply the minimax principle, in order to minimize the most we can *lose*, over and above the minimum the state of nature charges us, by our decision. In this case it is clear that D_2 is the safest policy.

We may apply the same technique to our manufacturer's problem. Since as originally formulated the cost of sending back a defective shipment was 0, the minimax regret and minimax strategies are the same. Let us however introduce new losses as mentioned above:

$L(a_i, s_j)$	S_0 (50% bad)	S_1 (20% bad)
a_0 reject	1	1
a_1 accept	3	0

We don't have to go through a whole new computation to obtain a new table or a new graph. The values of $E_{s_0}(d_i)$ will remain just as they were before; the new values of $E^*_{s_1}(d_i)$ will be related to the old values by the linear equation $E^*_{s_1}(d_i) = 2E_{s_1}(d_i) + 1$. All we have done is to change the location and scale of the abscissa. Furthermore, it is easy to see why $d_{1,2,3,4,0}$ is the minimax solution to the new problem: the 45° line touches the convex set of decision functions only at that point. To arrive at the minimax regret solution, we change the location of the abscissa again: we shift the 0 point to the left (this corresponds to subtracting 1 from each of the s_1 losses, and therefore from their means). The result is graphed in Figure 3. It is easy to see that the mixed strategy recommended by the principle of minimax regret consists of a mixture of $d_{2,3,4}$ and $d_{1,2,3,4}$ in proportions of approximately 55 to 45. Under the circumstances envisaged this seems more reasonable than the minimax solution.

5. There are any number of schemes like minimax and minimax loss that might be used to choose between decision functions. One could be optimistic and opt for that decision which offered the absolute minimum possible loss – a minimin policy. Or one could split the difference: average maximum loss and minimum loss in each state of nature, and then choose the decision function for which this value is least. Indeed

it does not seem at all implausible to suppose that given any decision problem whatsoever, given any *admissible* decision function whatsoever in that problem, an ingenious person could invent an almost reasonable policy which would lead to the choice of that decision function. I do not mean a reasonable policy that depends on prior probabilities having

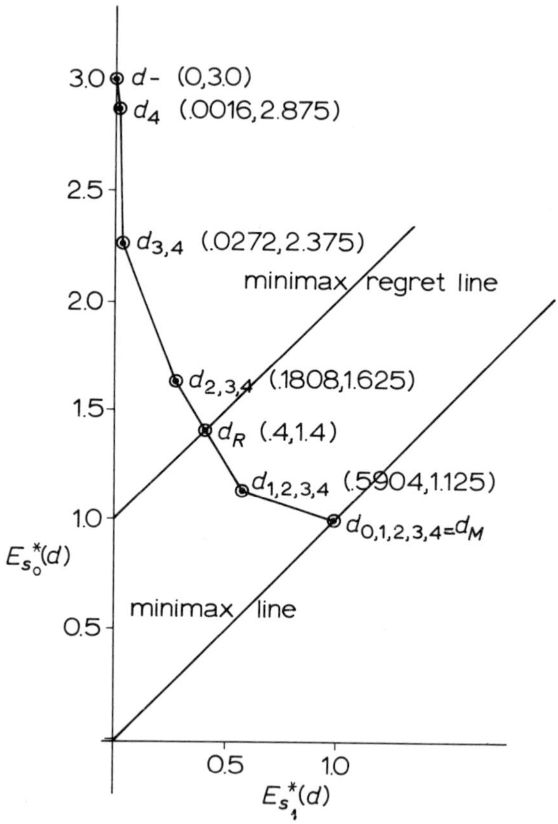

Fig. 3.

certain values or relative values. That a Bayesian decision policy can lead to any admissible decision, we shall see in the next chapter. What is here claimed is that there is a non-Bayesian decision policy, of the same general character as the minimax and minimax regret policies which will lead in the given situation to the decision function d_i.

Of course it may be regarded as typical philosophical pessimism to feel that there is always somebody around ingenious enough to make the better cause appear the worse. And it is possible on the other hand that someone could come up with a general super-policy that corresponded with minimax, minimax regret, etc., in those circumstances where these policies seem indubitably correct.

In general the strategy to which we will be led by a minimax or minimax regret policy will be a mixed one. But there is something epistemologically anomalous about a mixed strategy construed as a way of deciding between statistical hypotheses. Our mixed decision function has certain characteristics. In a certain state of nature, for example, it has a certain expected loss. But when we actually apply that decision function to the data we have in a given set of circumstances, we may separate two stages of the application: one stage consists in (say) performing an auxiliary experiment, the second consists in applying a *pure* strategy determined by the outcome of that auxiliary experiment. In view of the fact that under these circumstances, we *know* which of the pure strategies we applied, the *relevant* expected losses in that particular case should be the expected losses of the particular pure strategy we employed as a result of the outcome of the auxiliary experiment, and not the expected losses of the mixed strategy.

For example, the minimax (and minimax regret) policy for the first version of our manufacturer's problem was a mixture of five parts of $d_{2,4,3}$ to one part of $d_{1,2,3,4}$. What this means in effect is that we should construct a deck of 60 cards; ten of them should be black and 50 of them red. We draw a card at random from this deck; if it is red, we reject the lot if there are two or three or four defective items in our sample; if it is black we also reject the lot if there is only one defective item in the sample. The expected loss under this mixed decision function is 0.26 in each of the two possible states of nature. But when we actually perform the experiment we know which card we got from the deck; we therefore *know* which pure strategy we are using; and therefore we *know* that our expected loss, under each state of nature, is the expected loss of that particular *pure* strategy under each state of nature, which in no case is 0.26.

We shall see in the next chapter that Bayes strategies never require randomization in this sense. Some statisticians (for example, Blackwell and Girshick) regard this as a defect of Bayesian principles, although from the foregoing argument it is difficult to see what could conceivably be offered as grounds for this belief. "A further objection to Bayes principles is that they never require randomization; many statisticians consider that random sampling, which is a form of randomization, is useful."[1] But as Fisher has observed, randomization in the sample is a completely different thing than randomization in one's decision rule.

In applications of this approach, even if we decide to employ a mixed strategy, we will in any particular case end up applying a pure strategy, perhaps determined by our auxiliary experiment. This strategy is characterized by its expectation of loss under each possible state of nature. It is these expectations that motivate the decision maker (the scientist, or, more appropriately, the engineer). Now suppose that the decision maker has some opinions about the relative likelihood that each of the states of nature obtains in the case at hand. We need not suppose that these are empirical probabilities and that the state of nature comes from some super-population characterized by certain relative frequencies; we need not suppose that these opinions are well grounded in evidence; we need not inquire into their natural history at all. Furthermore, we need not suppose that they are altogether precise. All we are supposing is that, for example, he would be (roughly) indifferent to choosing between winning a two dollar prize if s_0 were the case and winning a one dollar prize if s_1 were the case. If the decision-maker has opinions of this sort, then he must also be of the opinion that he would prefer decision function d_i to d_j if and only if the sum of the losses sustained under d_i, weighted by his assessments of relative likelihood, is less than the sum of the losses sustained under d_j, weighted in the same way. If this is so, then there will be one (or at most two) decision functions that are preferable, for him, to any others. What, then, could possibly *motivate* him to adopt the minimax strategy or the minimax regret strategy, unless it coincides with his most preferred function? We can only persuade him to follow d_M, if he can be persuaded to disregard his degrees of belief in s_0 and s_1. The natural approach to doing this would be to argue that it is *irrational* of him to have these degrees of belief. But this is just the sort of thing, this discussion of rational and irrational degrees of belief, that decision theory thrusts out of its jurisdiction.

To illustrate this argument, let us look again at the original problem. Suppose the manufacturer, consulting his records, his intuition, or his crystal ball, decides that he is of the opinion that it is roughly three times as likely that his batch of machines contains 20% defective items as it is that it contains 50% defective items. This is merely to say that he would be indifferent between staking a possible gain of $1.00 on the event that 20% turn out to be defective and staking a possible gain of $3.00 on the event that 50% turn out to be defective. His overall expectation of loss, in objective terms, if he follows the minimax strategy, is 0.26, as we observed. In view of the fact that this strategy is designed to yield the same loss under s_1 and under s_0, the relative likelihoods are irrelevant. His overall (psychologically) expected loss under this mixed rule, d_M is

$$E(d_M) = (\tfrac{3}{4})\, E_{s_1}(d_M) + (\tfrac{1}{4})\, E_{s_0}(d_M) = E_{s_1}(d_M) = 0.26.$$

But in applying d_M he will end up applying $d_{2,3,4}$ (with probability 0.802) or applying $d_{1,2,3,4}$ (with probability 0.198). If he applies the latter, he will know it, and his opinion regarding his expected loss must be that it is

$$E(d_{1,2,3,4}) = (\tfrac{3}{4})\,(0.5904) + (\tfrac{1}{4})\,(0.0625) = 0.459.$$

If he applies the former, he will know it, and his opinion regarding his expected losses must be that it is:

$$E(d_{2,3,4}) = (\tfrac{3}{4})\,(0.1808) + (\tfrac{1}{4})\,(0.3125) = 0.214.$$

Under these circumstances he might be rather inclined to cut the deck again if he got a black card the first time!

Furthermore, his expectations (as evidenced by his willingness to stake possible gains on s_0 and s_1) are such that he must regard his expected loss under $d_{3,4}$ to be:

$$E(d_{3,4}) = 0.6875(0.25) + 0.0272(0.75) = 0.189.$$

If his opinions are what we have stated them to be, his (psychological) expectation of loss must be least for the pure strategy $d_{3,4}$.

In what way, then, might he be motivated to follow the minimax policy, or indeed any strategy other than $d_{3,4}$? It is difficult to see any way at all, if we let him keep his opinions. We might succeed in convincing him that his opinions are irrational, and that he should not hold them. But if we convince him that his opinions should not be such as to preclude the use of d_M, we must at the very same time be convincing him that he has no basis for being indifferent between a $3.00 gain if s_0 is true and a $1.00 gain if s_1 is true.

If there is a probability distribution over the states of nature, then that is quite another matter. The decision problem is a different one, and it does not call for a randomized solution. This is perfectly clearly understood, and perfectly well represented within the classical decision theoretic framework. No problem. Where a prior probability distribution over the states of nature is not given as one of the conditions of the problem, however, it is difficult to see how to handle the prior opinions of the decision-maker, if he has any. The difficulty is exacerbated if, as may be the case, the opinions are not merely whimsical, but can in fact be defended as rational, though not in terms of a prior frequency distribution. To say that in such cases the agent does or should act *as if* there were a prior distribution of such and such a sort is to explain the obscure by the incomprehensible. The counterfactual, far from throwing light on the grounds of choice among possible decision functions, throws a curtain over the whole affair: *any* decision at all can be regarded as rational for an agent acting *as if* ... so and so. Even inadmissible strategies can be explained away this way: the agent rejects the lot if and only if the third item examined is defective; he is acting as if the third item had some special role to play – and perhaps it does! Such an agent might, in point of fact, over a long series of trials, suffer an average loss per trial of 0.10. Surely if this were so, there would be something calling for explanation.

A final difficulty with the decision theoretic approach to statistical inference is the same as one of the difficulties in the classical testing approach: all of the motivations of agents and characterizations of decision functions are provided by long run characteristics – in particular, the means of the losses incurred through various decision functions under various states of nature. These means only apply to our decisions under certain conditions of randomness. These conditions are hypothetically and generally stipulated in the decision theoretic statement thetically and generally stipulated in the decision theoretic statement of the problem: thus, for example, it will be stipulated that our sample of four is a *random* sample of the lot received. This means that it is to be selected by a method which will produce each four-membered subset of the lot with equal frequencies in the long run, as we discovered in Chapter II. But this is not to say that we may not have clues as to the likelihood that that *particular* sample we in fact draw will be a good guide, under any decision function, to the quality of the lot. For example, it might be that the items in the top of the packing box tend to be defective more often than those in the bottom of the box, and that we know this. A random sample of four must at least some of the time come from the top of the box. If so, then the probability that one of these should be defective is greater than the proportion of defective items in the box. To be sure, this could be handled – at least if we could formulate an acceptable statistical hypothesis corresponding to my vague supposition – within the decision theoretic framework. Nevertheless it is not always obvious how to do it in particular cases, and decision theory itself offers us no guidance as to when and how to do it. The problem is the classical problem of all empirical conceptions of probability: the problem of choosing the appropriate reference class in a specific situation.

NOTE

[1] David Blackwell and M. A. Girshick, *Theory of Games and Statistical Decisions*, New York 1954, p. 16.

CHAPTER FIVE

SUBJECTIVE AND LOGICAL APPROACHES

We shall now consider that style of statistical inference called subjectivistic or personalistic or (in a special sense) Bayesian. It is based on a conception of probability altogether different from that underlying the investigations of the previous chapters. It was suggested at one point in the last chapter that there might be difficulty in motivating an agent to follow the recommendations of a minimax policy if he had certain convictions concerning the relative likelihood of the various alternative hypotheses. The subjectivistic approach takes this likelihood, and not long-run frequency or propensity, to be the subject matter of probability. Adopting this technique has profound consequences for the theory of statistical inference.

1. Let us begin by considering the subjectivistic or personalistic interpretation of probability. Probabilities, on this view, are (slightly idealized) actual degrees of belief, where 'degree of belief' is understood behavioristically. If a person would rather stake a possible gain on the occurrence *A* than on the occurrence *B*, then we take him to have a higher degree of belief in *A* than in *B*. Given some fairly innocuous sounding postulates concerning preferences, (some of which we shall examine in due course) it follows that degrees of belief satisfy the axioms of the probability calculus. There is a normative element in the postulates concerning preference: it is assumed, for example, that one should not have preferences which violate the normal transitivity of preference: that if *A* is preferred to *B*, and *B*, to *C*, then *A* is preferred to *C*.

Ramsey [1] argued that one's degree of belief in a statement might be measured by one's willingness to bet on the truth of the statement. He showed that having degrees of belief that satisfied the axioms of the probability calculus was a sufficient condition for avoiding having a book made against one. In 1955 Shimony,[2] Lehman,[3] and Kemeny [4] independently showed that the conditions imposed by the probability calculus were necessary and sufficient to avoid sure losses. In statistics, De Finetti [5] adopted this notion of probability for his innovative approach to statistical inference (around 1930); with the publication of Savage's book[6] in 1954, this approach to statistics began to increase rapidly in popularity, and it is now a serious rival to the classical British-American approach.

A few remarks should be made about the underlying argument for the claim that one's degree of belief should satisfy the axioms of the probability calculus, or, in short, should be *coherent*. The argument that your degree of belief in S should be one minus your degree of belief in not-S, for example, goes as follows: suppose your degree of belief in S is p, and your degree of belief in not-S is q, and that $p \neq 1-q$. Suppose $p<1-q$. Since your degree of belief in S is p, you are willing to bet $D(1-p)$ dollars against Dp dollars on the truth of S. Similarly, since your degree of belief in not-S is q, you are willing to bet $D(1-q)$ dollars against Dq dollars on not-S. If S in fact turns out to be true, the net result of both these bets is $Dp-D(1-q)=-D(1-q-p)<0$. On the other hand, if S turns out to be false, the net result for you is $Dq-D(1-p)=-D[(1-q)-p]<0$. You are bound to lose in either case: you have had a book made against you, a Dutch book. (To have a Dutch book made against you is to commit yourself to a set of bets, such that no matter what happens, you lose.) Therefore, your degree of belief in S must not be less than one minus your degree of belief in not-S.

The implication is that if your belief in S is less than one minus your degree of belief in not-S, you will have a book made against you. But this is simply false. That I will not allow a book to be made against me is entailed by my preference ranking for money (that I'd prefer not to suffer a certain loss), and has nothing to do with my degree of belief. Having accepted a bet at odds of $(1-p)$: p on S, *of course* I won't accept a bet at odds of $(1-q)$: q on not-S. This is not to say that if I had *not* accepted the first bet, I wouldn't accept the second bet. The hidden assumption required is that the behavioral counterpart of a distribution of beliefs over an algebra of sentences is a willingness to make book on them: to accept bets of *any* size, on *either* side, with respect to *any* combination of sentences. This is clearly far more powerful (and correspondingly less plausible) than the assumption that the behavioral counterpart of having a degree of belief p in S is the same as the willingness to offer no more than odds of $(1-p)$: p on the truth of S. It puts the person whose beliefs we are discussing in the position of the bookmaker; the arguments of Ramsey, Shimony, and the others do indeed show that a bookmaker had better be jolly sure that the odds he posts satisfy the axioms of the probability calculus.

One exposition of this point of view adopts this framework explicitly; the situation described is that in which there is a bookmaker and a bettor who is perfectly free to make whatever bets he wishes, all of which *must* be covered by the bookmaker. Then the axioms of the probability calculus are shown to control the odds posted by the bookmaker.[7]

A further complication in this theory arises in connection with statistical inference. If we envisage the scientist as offering to bet against Nature, we need require that his degree of (scientific) belief be coherent only if we are willing to assume that Nature is going to be diabolical enough to make bets in such a way as to exploit any incoherence in his degrees of belief. Such an assumption in such a context seems gratuitous. But as we shall see, there is no need to make much of this line of argument.

2. The basic instrument of inference on this approach is Bayes' Theorem. In its simplest form, on this interpretation, Bayes' Theorem simply states that if one's degrees of belief are coherent – that is, obey the axioms of the probability calculus – then one's degree of belief in a hypothesis H, given certain evidence E, must be equal to the product of ones' degree of belief in E, given the hypothesis (the likelihood), multiplied by the ratio of one's prior belief in H to one's prior belief in E:

$$P(H \mid E) = \frac{P(H) \cdot P(E \mid H)}{P(E)}.$$

Two problems immediately present themselves. If H is a statistical hypothesis, the calculation of the probability of E given H, under fairly straightforward assumptions, presents no problem. But the calculation of $P(H)$ and of $P(E)$ do present problems. Let us look at a simple case in which the problems can be solved.

Let us again consider the manufacturer discussed in the last chapter. He is confronted with a lot of items, of which either 20% or 50% are defective. For the time being we shall consider this as a problem in inference, rather than a problem in choosing between two courses of action. Thus on the basis of a sample, as drawn before, he wants to make an inference concerning, say, the statistical hypothesis that 20% of the items are defective. On the usual assumptions concerning sampling, it is reasonable to argue that the probability that the sample of four items has a certain number r of defectives is (just about) the binomial probability $\binom{n}{r} p^r (1-p)^{n-r}$, given that the sample is drawn from a large population containing a proportion p of defective items. One interpretation of this probability on the subjectivistic view is straighforward: it is just the ratio of the degree of my belief in H and E to the degree of my belief in H. The difficulty is that it is not by evaluating these degrees of belief that I would arrive at the ratio. I arrive at it rather by a hypothetical argument of this nature: *if* I knew that H was true (e.g., that 20% of the items were defective), and that the sampling was to be done in such and such a way (the way it will in fact be done), *then* I *would* have a degree of belief equal to $\binom{n}{r} p^r (1-p)^{n-r}$ in the proposition that r of the n items in the sample would turn out to be defective. This argument seems at first sight to be rather perilous: it seems to assert something about my *dispositions* to bet *were* I to be in a state of knowledge which in fact I'm not in. Such inferences in real life are notoriously insecure. But in point of fact we can prove (if I am coherent), that my degree of belief in E conditional on H, is just the ratio it ought to be.

My degree of belief in E, conditional on H, is indicated (just as in the case of unconditional beliefs) by the least odds that I would offer in a bet on E, with the understanding that the bet is to be called off if H fails to occur. Suppose my degree of belief in E, conditional on H, is $a/(a+b)$. Suppose my belief in degree of belief in H is c. I shall show that my degree of belief in H and E, divided by my degree of belief in H, is exactly $a/(a+b)$. Within the framework of the subjective theory of probability, to say that the least odds at which I would bet on E, conditionally on H, is $a:b$, is to say that I would be equally willing to take either side of such a bet. Consider the following pair of bets: (1) a bet at odds of ca to $b+(1-c)a$ on E-and-H; and (2) a bet at odds of $(1-c)\,a$ to ca on not-H. The outcome of the pair of bets, in every state of affairs ($E\,\&H$, $E\;\&\;\bar{H}$, $\bar{E}\;\&\;H$ and $\bar{E}\;\&\;\bar{H}$), is the same as the outcome of a bet on E conditioned on H, at odds of $a:b$. I am willing to take either side of bet (2), since it conforms to my degree of belief in H. I am willing to take either side of the pair of bets, since the pair is just the same as the single conditional bet. It follows that I am willing to take either side of the first bet of the pair, and therefore that my degree of belief in E and H is $ca/[b+(1-c)\,a+ca]$. But, as a little computation will show, $(ca/[b+(1-c)\,a+ca])/c=a/(a+b)$. Therefore the conditional probability (i.e., my degree of belief in E, conditioned on H) is just the same as the ratio of my degree of belief in E and H to my degree of belief in H.

Of course, this does not entail at all that our belief in r defective items in the sample of n should be just $\binom{n}{r}\,0.2^r\,0.8^{n-r}$. It is conventional to regard $P(E \text{ and } H)$ to be $P(H)$ times this number. The reasoning behind the convention is presumably something like this: it is a set theoretical fact that if 20% of L are defective, $\binom{n}{r}\,0.2^2\,0.8^{n-r}$ is the proportion of members of L^n (the n-fold cartesian product of L with itself) that contain exactly r defectives. It is assumed to be a matter of knowledge that we have no grounds for distinguishing the sample reported in E from any other member of L^n, with respect to having a certain number of defective items. But, of course, there are any number of states of knowledge that we *might* be in, in which this would not be the case. Under these circumstances, it would still be the case that $P(E \mid H)=P(H \,\&\, E)/P(H)$; but it would *not* be the case that $P(E \mid H)=\binom{n}{r}\,0.2^r\,0.8^{n-r}$.

To continue our example, let us suppose that the body of knowledge of the manufacturer is such that $P(E \mid H)=\binom{n}{r}\,0.2^r\,0.8^{n-r}$. We must now deal with two rather more difficult problems. Let us consider, first of all, $P(H)$. One natural way to evaluate $P(H)$ would be by means of a history of shipments, say, from the same concern. It might well be known that a quarter of these shipments, on the average, have turned out to contain 50% defectives. A natural value for $P(H)$ then would be 0.75. But it is not always this easy, of course. (Indeed, there would be no conflict between the subjectivistic theory and the classical theory in such a case.) According to the subjectivistic theory there are any number of grounds on which one might assign a value to $P(H)$, including sheer intuition. What the subjective theory does demand, is that $P(H)$ be consonant with (that is, coherent with) the other assignments of probability that are made. Thus past experience, the intuitive assessment of the effect of new management in the supplying company, seasonal variations, hunches, reading tea leaves, might all enter in. The one thing that is clear and definite on this view that if $P(H)$ is taken to be a, that means that $1-a:a$ is taken to be the fair odds to bet on H, and conversely. In fact, one good way to find out what probability a person assigns to H, is to find out (directly or indirectly) the least odds at which he is willing to bet on H.

At least we should be able to find out what probability the manufacturer attributes to H. (One rather circular way would be to look at his policy for rejecting shipments.) But we must also deal with more general statistical hypotheses. And here we encounter a new sort of difficulty. The probability that a person assigns to the hypothesis that half the tosses of a certain odd coin will land heads will be 0.5 if and only if he is willing to bet at even money that half the tosses will land heads. But this is a strange sort of bet: how is it to be settled? The hypothesis concerns an indefinitely long run of tosses; surely it cannot be settled simply by looking at that indefinitely long run of tosses. We could decide that it would be settled by a very long run investigation with certain results – say a series of a million tosses of which $\frac{1}{2}$ plus or minus ε land heads. Of course, if we do this, we run a finite risk of settling the bet erroneously: that is, of getting the specified result when the hypothesis is false, or of not getting that result, even though the hypothesis is true. Thus it might be claimed that even after observing the million tosses, the bet isn't *really* settled. A more constructive answer might be that although we *talk* about indefinitely long runs, all we 'really' mean is some very large finite run. If that is the case, and we specify the length of the large finite run generally and in advance, then it is perfectly possible in principle to settle the bet. We establish this possibility by explaining away talk of 'indefinitely long runs' as hyperbole. But then we have given up the project of assigning probabilities to general scientific hypotheses.

The problem connected with the probability of the evidence, $P(E)$ is rather different. E is the sort of thing that can easily be settled in a finite period of time (draw the sample of n), and this is the sort of thing it is easy to bet on. There should be no problem in assessing someone's degree of belief in E. Nevertheless, in practice, it is very often done through the agency of hypotheses: thus we construe Bayes' theorem in the form:

$$P(H \mid E) = \frac{P(H) \cdot P(E \mid H)}{\sum P(H_i)\, P(E \mid H_i)}$$

where the summation is extended over all possible hypotheses. This makes good sense initially because to begin with we can think up quite a list of hypotheses, like the one that 20% of the items are defective, which yield easily computable, very popular, values for $P(E \mid H_i)$. In the case at hand, the typical way of computing the probability of the evidence would be by the formula:

$$P(E) = P(H_1)\, P(E \mid H_1) + P(H_2)\, P(E \mid H_2)$$

where H_1 is the hypothesis that 20% of the items are defective and H_2 the hypothesis that 50% of the items are defective. But in general we should often encounter difficulty in finding an exhaustive class of hypotheses to consider in a specific situation. One of the hypotheses will often be of the form 'Or else something we haven't thought of is true', and this kind of hypothesis provides a poor basis for the evaluation of a conditional probability.

Let us consider now a more general and more typical case. Suppose we have a population which we take to be arbitrarily large. We are interested in the proportion of A's in that population. We suppose that our sampling procedure is such that the frequency of A's in a sample of n can be represented as a variable with a binomial distribution, with an unknown parameter p equal to the proportion in the population. There is no problem then with $P(E \mid p)$. If we assume a continuum of p-values is possible, the probability, $P(p)$ must be 0. Nevertheless, we can quite reasonably speak of the probability that p falls between any two limits – as properly at least as we can speak of the probability of any general statistical statement. In particular, we can talk of the probability that p is less than or equal to some number x: $P(p \leq x)$. But this is just the form of a cumulative distribution function, and yields a density function $dP(p \leq x)$. This is the prior density for p. In terms of this we can evaluate the probability of the evidence E: $P(E) = \int_{-\infty}^{+\infty} P(E \mid p)\, dP(p \leq x)$. We can now compute the conditional probability that p is less than or equal to x, given the evidence E, as follows (simply from Bayes' theorem):

$$P(p \leq x \mid E) = \frac{P(p \leq x)\, P(E \mid p \leq x)}{P(E)}$$

from which it follows that the density of the posterior distribution of p is

$$dP(p \leq x \mid E) = \frac{P(E \mid x)}{P(E)} \, dP(p \leq x).$$

From this posterior distribution, we can make whatever statements we want about the parameter p: for example we can give the probability that it lies within any limits we choose. This is just the sort of thing that people have always wanted to obtain from statistical inference.

The stumbling block, of course, is the prior distribution of p. There are circumstances, of course, where a prior distribution can be justified in a strong frequency sense. But in most situations we are more or less ignorant of the distribution of p. There are even circumstances in which we may claim to be 'totally' ignorant of p. It has been argued by some writers that in these circumstances one value of p is no more probably than another, and that the prior probability distribution for p should be uniform between 0 and 1. This was LaPlace's solution to the problem. Other writers have suggested that the best prior distribution is a uniform distribution for $\log(p/1-p)$. Others have suggested a Beta distribution for p, which introduces two useful free parameters a and b. Whatever has been suggested, though, the final judge is the way the customer feels: what goes into the Bayes' theorem, for me, is *my* prior distribution, and if what I regard as total ignorance is a uniform distribution between 0 and 1, that does not at all prohibit you from representing *your* total ignorance by a uniform distribution of $\log(p/(1-p))$. That we happen to disagree, from this point of view, is simply a fact of life.

The generalization of this approach to statistical inference to multi-dimensional problems – that is to dealing with statistical hypotheses having several unknown parameters – is perfectly straightforward. Let K be the space in which the parameters take their values, so that each $k \in H$ is (say) an n-dimensional vector. Given the value of k, the probability of the evidence E can (we assume) be computed: $P(E \mid k)$. Given a probability density for k defined over K, dP_k, we have as above,

$$dP(k \mid E) = \frac{P(E \mid k)}{\int_K P(E \mid k) \, dP_k} \, dP_k.$$

The result is a conditional density for k; to compute the probability that k belongs to any particular region of the parameter space, we need merely integrate $dP(k \mid E)$ over that region. In principle. The details may be arbitrarily hard, depending on the distributions we have chosen.

3. Much writing on statistical inference that embodies the subjectivistic point of view is concerned with prior probabilities of hypotheses. Although no statistician can tell a customer what his belief distribution over a class of statistical hypotheses ought to be (it simply is what it is), yet often a suggestion will be very welcome. Harold Jeffreys offered a large number of such suggestions, based in part on invariance principles, and other writers have followed him in this respect.[8]

One observation should be made here on the appropriateness of the word 'inference'. In deduction, an inference allows us to pass from premises to a conclusion; the result is that if we can accept the premises, we can come to accept the conclusion. In the classical approach to statistical inference something analogous occurs: given certain evidence as premises, we may come to reject a certain hypothesis. In either case we arrive at a categorical opinion. There are differences: in classical statistical inference, as opposed to deductive argument, we must take account of *all* the evidence we have: additional premises may undermine an argument that without those additional premises would be all right. Furthermore, the rejection of an hypothesis, for the same reason, can be only temporary: further evidence may lead us to rescind that rejection. Many writers, particularly Neyman, have claimed that to reject a hypothesis is not the same as to accept its denial. Nevertheless, to pass from suspension of judgement to rejection seems an epistemological act not unlike that of passing from suspension of judgement to acceptance.

On the Bayesian view we neither accept nor reject hypotheses (except when they are either deductively inconsistent with, or implied by, the evidence). The result of inference is neither the rejection nor acceptance of a hypothesis, but a change in our *degree of belief* in that hypothesis – i.e., a change in its probability. To be sure we may describe our response, by means of a test of significance, for example, as 'rejection at the α level', but on the Bayesian point of view this is just a shorthand way of describing the import of some evidence when the full description of a resulting posterior distribution of probability is too complicated to be bothered with. In the more general case where the result of our inference is knowledge about the parameter θ, the form of that result is a posterior distribution for θ. Thus the logic of statistical inference, from a subjectivist point of view, is the logic of changing probability distributions, or changing distributions of degrees of belief, and not a logic of the acceptance or rejection of propositions.

This seems strange, *prima facie*, since one tends to think of scientists and investigators as building up a body of knowledge, albeit tentative and fallible knowledge. This was clearly Fisher's view, though it was not the view of the decision theorists we considered in the last chapter. In this respect, the subjectivists are quite close to the classical decision theorists. Knowledge as such is of no importance to them; knowledge counts only insofar as it helps us to make practical decisions. The role of statistical inference is that of assisting us in making decisions in the frace of uncertainty. Indeed, this approach to statistics is particularly natural for the subjectivists in view of the fact that considerations of utility are involved in the very definition of probability.[9]

4. The subjectivistic or Bayesian approach to decision in the face of uncertainty is striking in its simplicity. It is based on one principle, and that a principle one can hardly dispute: choose that act which will maximize your expected utility. Since both probabilities and utilities may be defined (as they are for Savage, for example) in terms of preferences between acts, the principle of mathematical expectation is built into the very framework of the theory. Nevertheless, as independent argument may also be given for the principle of maximizing utility.

Suppose that there is a probability function P known to apply to the states of the world relevant to the choice between the actions A_1 and A_2. We need not specify anything about the nature of this probability function, except that it provides criteria for fair bets. An empirical probability concept does this, of course, as well as a subjective concept, or for that matter, as we shall see, a logical concept. In terms of this probability, we may compute the expected utilities of A_1 and A_2: $E(A_1)$ and $E(A_2)$. We suppose (what does not seem a large supposition, and indeed what would appear as a demonstrable truth in most circumstances) that there are propositions or events (depending on the domain of the probability function) S_1 and S_2, such that the expected utility of a bet on S_1 at odds of a_1 to b_1 (with total stake a_1+b_1) is $E(A_1)$, and the expected utility of a bet on S_2 at odds of a_2 to b_2 (with total stake a_2+b_2) is $E(A_2)$.

We now show that action A_1 is to be preferred to action A_2 if and only if $E(A_1) > E(A_2)$. Let the probability of S_1 be $P(S_1)$ and that of S_2 be $P(S_2)$. These probabilities must exist since the expectations of bets on S_1 and S_2 were assumed to exist. A bet on S_1 at odds of a_1 to b_1 is a bet at odds of $P(\bar{S}_1)\,[a_1/P(\bar{S}_1)]$ to $P(S_1)\,[a_1/P(\bar{S}_1)] + b_1 - P(S_1)\,[a_1/P(\bar{S}_1)]$. Let $d_1 = b_1 - P(S_1)\,[a_1/P(\bar{S}_1)]$. Then a bet on S_1 at odds of a_1 to b_1 is precisely equivalent to a *fair* bet on S_1 with a total stake $P(\bar{S}_1)\,[a_1/P(\bar{S}_1)] + P(S_1)\,[a_1/P(\bar{S}_1)]$, and with a prize of d_1 conditional on S_1. But a prize of d_1 conditional on S_1 is just the same as an unconditional prize in the amount of $P(S_1)\,d_1$. Note that the principle of mathematical expectation is being applied here only to a prize conditional on an event or on the truth of a statement. The expectation of a bet on S_1 at odds of a_1 to b_1 with total stake $a_1 + b_1$ is thus $p(S_1)\,d_1$, since the expectation of a fair bet is 0. Similarly the expectation of a bet on S_2 at odds of a_2 to b_2, with total stake $a_2 + b_2$ is $P(S_2)\,d_2$, where $d_2 = b_2 - P(S_2)\,[a_2/P(\bar{S}_2)]$. By hypothesis the expectation of A_1 is the same as that of the bet on S_1 and that of A_2 is the same as that of the bet on S_2. Since the expectation of the bet about S_1 is equivalent to an unconditional prize of $P(S_1)\,d_1$, and that of the bet about S_2 is equivalent to an unconditional prize of $P(S_2)\,d_2$, A_1 will be preferred to A_2 if and only if $P(S_1)\,d_1 > P(S_2)\,d_2$, which will hold if and only if $E(A_1) > E(A_2)$. The prizes may of course be negative.

The only principle regarding expectations we have used is the principle that two options yielding the same expectation will fall at the same place in our preference ranking. Putative counterexamples to this principle depend on the bird in the hand principle: the principle that a dollar in the pocket is worth more than one chance in a thousand of a thousand dollars. But it is precisely the nature of utilities, as opposed to dollars, that the bird-in-the-hand principle does *not* apply to them. The situation described is classically and properly taken as indicating that the utility of money is not linear, but the very nature of utilities is to be such that equal expected utilities occupy the same place in our preference rankings.

Given the maxim that one should maximize one's expected utility, and given the subjectivistic postulate that probability measures and utility functions always exist, all decision problems are in principle solved, barring mathematical difficulties, and supposing that the utility functions are bounded.

Let the state of the world be reflected by a value of the vector θ, insofar as the state of the world is relevant to the outcome of the decisions that are available to us. Let the set of actions open to us be A. (Acquiring further evidence may be among the actions.) Given the state of the world θ, we may compute the expected utility of an action a, $E_\theta(a)$. Suppose that we acquire a body of evidence e. As described earlier, the evidence yields a posterior probability distribution over the states of the world:

$$dP(\theta \mid e).$$

For each action, we may then compute the expectation of that action, given our evidence: $E(a) = \int_\theta E_\theta(a)\, dP(\theta \mid e)$. In view of the fact that the utility functions are all bounded $\{E(a): a \in A\}$ has an upper bound. There may, of course, be no action that yields the *least* upper bound of this set but there are bound to be actions that come arbitrarily close to it. Thus given a small positive ε, we should be able to find an action a (in general there may be more than one) such that the difference between the expected utility of that action and the least upper bound of $\{E(a): a \in A\}$ is no greater than ε. And that is a reasonable action to take, given the evidence e.

5. By way of illustration, let us turn once more to (and not for the last time) our manufacturer. To make the problem more realistic, let us assign a new utility function to him. Under this new utility function, it costs him 100 dollars in trouble and expenses if he accepts a lot of items containing 50% defectives, (action a_1); it costs him twenty dollars to send it back (action a_0). It costs him thirty dollars (counting the loss of good will of his supplier) if he sends back a lot containing only 20% defectives, while to accept the 20% defective lot he counts as par for the course, or zero loss. We suppose the probability he assigns to getting a 20% defective lot is 0.75, and the probability he assigns to getting a 50% defective lot is 0.25. It is a straightforward matter to compute his expected loss from applying any particular decision function, mixed or otherwise. For example, his expected loss from following $d_{3,4}$ a third of the time and $d_{2,3,4}$ two thirds of the time is just a third of the expected loss under the first decision function plus two thirds of the expected loss under the second decision function. The expected loss under $d_{3,4}$ is just 0.75 times the expected loss under $d_{3,4}$ given that s_1 is true, plus 0.25 times the expected loss under this rule given that s_0 is true:

Expected loss under the mixed rule described =
$\frac{1}{3}[0.75 \times 0.0272 \times 30 + 0.25 \times (0.6875 \times 100 + 0.3125 \times 20)] +$
$\frac{2}{3}[0.75 \times 0.1808 \times 30 + 0.25 \times$
$\times (0.3125 \times 100 \times 0.6875 \times 20)] \cong 39.1.$
(The probabilities come directly from Table II, where the losses tabulated are equal to the probabilities of error, since the loss due to error is taken to be 1.)

In general the expected loss under any rule will be the prior probability of s_0, multiplied by the expected loss under the hypothesis s_0 plus the prior probability of s_1, multiplied by the expected loss under the hypothesis s_1. On a diagram (see Figure 4) on which the expected loss under hypothesis s_0 is plotted against the expected loss under hypothesis s_1, the states of constant loss L are represented by a line $L = P(s_0)\, E_{s_0} + P(s_1)\, E_{s_1}$, a line of slope $-P(s_0)/P(s_1)$. The set of decision functions can be represented, as before, by convex set in the E_{s_0}, E_{s_1} plane. Any decision function lying in the lowest line of slope $-P(s_0)/P(s_1)$ will be optimal. This will be a unique pure decision function (corresponding to a vertex of the convex set of decision functions) or a pair of pure decision functions, together with the set of all their mixtures. In the case at hand, it is $d_{2,3,4}$.

We may achieve the same result by transforming the coordinates so that the loss of making each wrong decision is normalized to 1, and that of making each right decision is normalized to 0 in each state of nature. Thus the lower boundary of the convex set of decision functions, in these coordinates, will be located within the unit square, and it will be easy to see the connection between Bayesian and alternative approaches. The equation of lines of constant loss in the original coordinates was $L = P(s_0)\, E_{s_0} + P(s_1)\, E_{s_1}$. The new coordinates, E'_{s_0} and E'_{s_1} are related to the old ones by the equations:

$$(1) \qquad \begin{aligned} E_{s_0} &= [L(a_1, s_0) - L(a_0, s_0)]\, E'_{s_0} + L(a_0, s_0) \\ E_{s_1} &= [L(a_0, s_1) - L(a_1, s_1)]\, E'_{s_1} + L(a_1, s_1). \end{aligned}$$

In the new coordinates the expected loss of a wrong decision under a given hypothesis, is just the probability of making that wrong decision.

The equations of the new lines of constant loss will therefore be,

$$\begin{aligned} L = P(s_0)\, &[(L(a_1, s_0) - L(a_0, s_0))\, E'_{s_0} + L(a_0, s_0)] + \\ &P(s_1)\, [L(a_0, s_1) - L(a_1, s_1))\, E'_{s_1} + L(a_1, s_1)] \end{aligned}$$

or

$$L' = P(s_0)\,(L(a_1, s_0) - L(a_0, s_0))\,E'_{s_0} + \\ + P(s_1)\,(L(a_0, s_1) - L(a_1, s_1))\,E'_s$$

or of slope:

$$-\frac{P(s_1)\,(L(a_0, s_1) - L(a_1, s_1))}{P(s_0)\,(L(a_1, s_0) - L(a_0, s_0))} = -\frac{0.75(30-0)}{0.25(100-20)} = -\frac{9}{8}.$$

Precisely the same decision function(s) will be selected by means of the

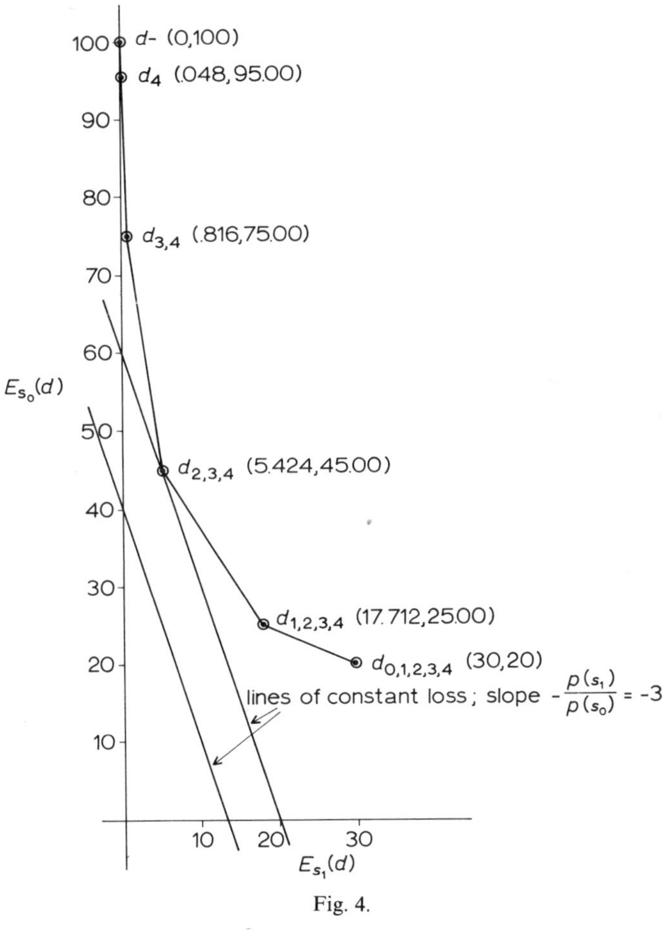

Fig. 4.

new diagram as by means of the old; the corresponding new supporting line will be the line corresponding to the old one, and the point or points at which it touches the set of decision functions will correspond precisely to the point or points touched in the old one. To find expected loss under the selected decision function, of course, we must reconvert to our original loss units, using equations (1). (See Figure 5.)

The same trick can be applied in any finite number of dimensions. Suppose that we have n actions to consider, and q hypotheses. If n is greater than q, there are some actions whose expected loss under any state of the world is greater than that for some other action. Thus, we need consider no more than m actions, where m is less than or equal to q.

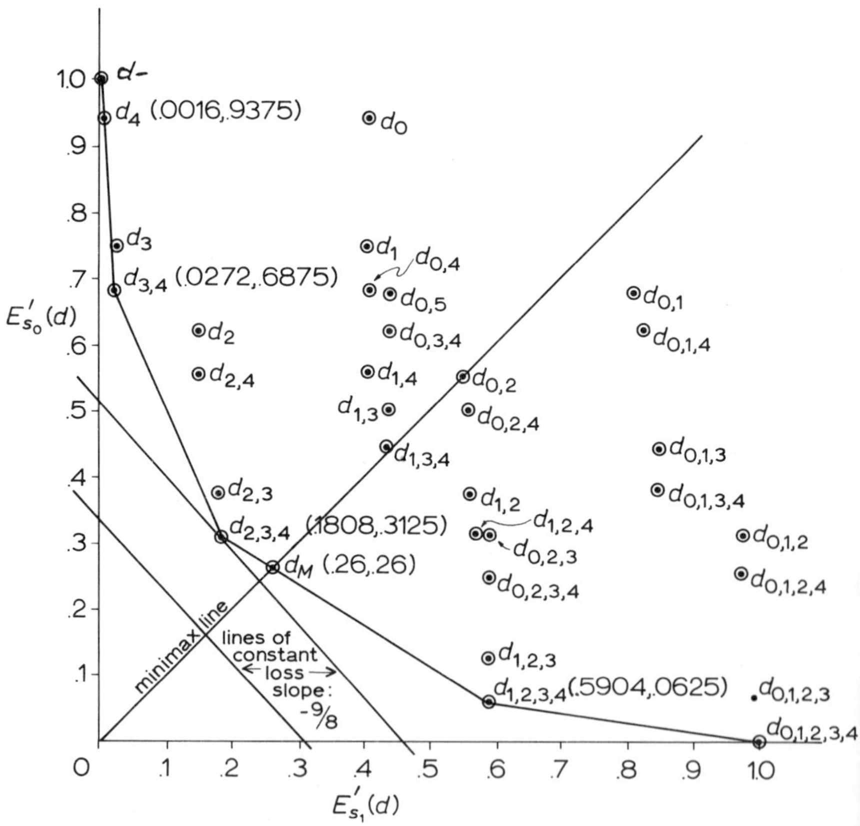

Fig. 5.

On the other hand, if q is greater than m, there must be (at least) two hypotheses, h_i and h_j, under which the least loss is offered by the same action a_q. There is thus no need to consider these hypotheses separately; we may replace h_i and h_j by $h_{ij}=(h_i \vee h_j)$. Since we may suppose that the hypotheses are exclusive in the first place (if they aren't, we may replace them by a set of hypotheses that are), we also have $P(h_{ij})=P(h_i)+P(h_j)$. Thus we may suppose without loss of generality that we are dealing with m possible actions and m exclusive and exhaustive hypotheses. As before, we may now normalize the losses under each hypothesis to the interval between 0 and 1. The equations of the old hyperplanes of constant loss were of the form:

$$L=P(h_0)\,E_{h_0}+P(h_1)\,E_{h_1}+\cdots+P(h_m)\,E_{h_m}.$$

The equations of the new hyperplanes of constant loss will be of the form:

$$L'=P(h_0)\,k_0E'_{h_0}+P(h_1)\,k_1E'_{h_1}+\cdots+P(h_m)\,k_mE'_{h_m}$$

where k_i is determined by the particular change of coordinates. As before, the hyperplane supporting the m-dimensional convex set of decision functions will support it at the same decision function or functions as in the untranslated case.

An example of an inference regarding a continuously distributed random quantity may also be of interest. Suppose that we are conducting a breeding experiment on beef cattle, and we wish to know the distribution of weights to be expected in a certain strain of cattle after a certain feeding program. On the basis of a vast amount of experience with this general breed of cattle, we are very sure that this distribution will be very close to normal, with a mean m, characteristic of the strain, and a standard deviation of 25 pounds. We examine four steers (chosen by lot), and weigh them, obtaining a total weight of 2800 pounds. (We regard the variance of the weighing procedure to be negligible in relation to the variance exhibited by the steers.)

Given a prior distribution for the weight of steers of this strain, we could get to work. But what can we take as this prior distribution? Well, let us suppose that we can regard the prior distribution as roughly normal, and see what happens. This is not implausible; the prior distribution is a reflection of our opinions, after all, and to the extent that our opinions are not terribly articulate we are perfectly free to move them around a bit to make them look normal. Suppose, just looking at the steers, we feel that they are just as likely to weigh less than 600 pounds as more than 600 pounds each. Furthermore, we feel that we would be no more surprised if someone told us that in fact they weigh less than $600-k$ pounds than if someone told us that in fact they weighed more than $600+k$ pounds. Furthermore we feel willing to bet at odds of about 100:1 that the average weight is not less than 475. Supposing that our prior distribution for m is thus normal with a mean of 600 and a standard deviation of 50, and that the weights of these steers are in turn distributed normally with a mean m and a standard deviation of twenty-five, it follows that the posterior distribution of m, given the observations, is normally distributed with mean m^* and variance σ^*:[10]

$$m^* = \frac{2800/25^2 + \dfrac{600}{50^2}}{4/25^2 + 1/50^2} = 695$$

$$\frac{1}{\sigma^{*2}} = \frac{4}{25^2} + \frac{1}{50^2} = 0.0068; \quad \sigma^* = 12.1.$$

The general formulas (where $\bar{x}$ is the mean of n observations, σ^2 is the variance of the observations, m_0 and σ_0^2 are the mean and variance of the prior distribution of m) are:

$$m_n = \frac{n\bar{x}/\sigma^2 + m_0/\sigma_0^2}{n/\sigma^2 + 1/\sigma_0^2}$$

$$\frac{1}{\sigma_n^2} = \frac{n}{\sigma^2} + \frac{1}{\sigma_0^2}.$$

It is easy to see that the larger the sample, the less important the accuracy of the prior distribution.

The posterior distribution allows us to make whatever probability assertions we wish about the mean of the hypothetical population of steers of this strain:

$$P(m \leq 650) = 0.9999$$
$$P(675 < m \leq 715) = 0.90$$
$$P(m > 705) = 0.79.$$

Note how insensitive the final distribution of our beliefs is to minor changes in our initial distribution of beliefs. The evidential impact of even a sample of four tends to swamp our prior distribution. We emerge having a distribution of belief that is normal and centered quite closely about the observed mean weight.

6. The immediate and obvious objection to this approach to statistical inference is the one that its title begs for: it is arbitrary and subjective just in that area in which we are seeking objective and warranted grounds for making inferences and decisions. For example, it is clear that if my prior belief in a hypothesis H is 0 or 1, no amount of evidence is going to change my degree of belief. Prejudice, deeply enough entrenched, is inviolable. It may be said that this is simply a fact of life, and that no theory of coherent belief and consistent decision can ignore it. But this answer is not very satisfying to begin with, and at any rate says nothing at all about the equivalently serious complaint that all judgements, including non-extreme ones, are infected by purely subjective and arbitrary elements that enter through the assumption of a prior probability distribution.

The answer to this objection offered by subjectivists is that in general (barring extreme prior probabilities of 1 or 0 and certain other pathological cases) prior probability distributions sooner or later get swamped by empirical evidence, as we saw in the steer example. More precisely, if my prior probability density for the random quantity X is given by dP_1, and yours by dP_2, and the corresponding conditional densities are $dP_1 \mid e$ and $dP_2 \mid e$, then under certain rather general circumstances, given a large enough body of evidence e concerning values of X, our conditional distributions will be as close together as we want them to be, in the sense that $\left|\int_a^b dP_1 \,|e - \int_a^b dP_2|\, e\right| < \varepsilon$; i.e., given P_1, P_2, and ε, we can specify the size of a body of evidence e which will insure agreement within ε.

The first such theorems were proved by De Finetti.[11] A fundamental concept that enters into many of these theorems and that often seems appropriate to experimental situations is that of *exchangeability*. Consider a sequence of random quantities, $X_0, X_1, \ldots, X_n, \ldots$. X_i may be the result of the *i*th experiment in a certain sequence of trials, or the value of a characteristic function applied to the *i*th item in a certain population, or a quantity exhibited by the *i*th item in a certain population. It will often be the case that the marginal distribution of X_i will be the same for any value of i; that is, that each X_i has the same distribution. If this is the case, we shall say that the random quantities in the sequence are 1-exchangeable. It may also be the case that every pair of random quantities, X_i and X_j, $i \neq j$, has the same *joint* (two dimensional) distribution; if that is so, we shall say that the quantities in the sequence are 2-exchangeable. And in general, we shall say that these quantities are n-exchangeable, when the joint distribution for $X_{i_0}, X_{i_1}, \ldots X_{i_{n-1}}$ is the same as the joint distribution for $X_{j_0}, X_{j_1}, \ldots X_{j_{n-1}}$, where the subscripts i_k are all distinct and the subscripts j_k are all distinct. This is just to say that

$$P(X_{i_0} \leq x_0 \; \& \; X_{i_1} \leq x_1 \; \& \ldots \& \; X_{i_{n-1}} \leq x_{n-1}) =$$

$$= P(X_{j_0} \leq x_0 \; \& \; X_{j_1} \leq x_1 \; \& \ldots \& \; X_{j_{n-1}})$$

for all n-tuples of real numbers $x_0 \ldots x_{n-1}$. Finally, we shall say that the random quantities are exchangeable just in case they are n-exchangeable for all n.

Note that exchangeability is a much weaker condition than independence; independence entails exchangeability, but the converse does not generally hold.

Let us now look at an example of one of De Finetti's convergence theorems. To be specific, suppose that we are sampling from a very large population, and that $X_i = 1$ if the ith item is an A, and $X_i = 0$ otherwise. One person assigns a probability p to $X_0 = 1$, and a probability p' to $X_0 = 1 \; \& \; X_1 = 1$. He regards the random quantities in the sequence to be 1-exchangeable and 2-exchangeable, and therefore he assigns the probability p in general to $X_i = 1$, and the probability p' in general to $X_i = 1$ & & $X_j = 1$, $i \neq j$. (Note that the condition being imposed is much weaker than general exchangeability.) Another person, with a different history, or different evidence, or different blood chemistry, or in a different mood, also regards the sequence to be 1-exchangeable and 2-exchangeable, but assigns the probability q to $X_0 = 1$ and the probability q' to $X_0 = 1$ & & $X_1 = 1$. We assume that $1 > p' > p^2 > 0$ and $1 > q' > q^2 > 0$.

We shall show that given any ε and δ, there will be numbers m and n such that both persons will assign a probability of at least $1 - \delta$ to the statement that the frequency of A's in a subsequent sample of m will differ from the observed frequency in an initial sample of n by no more than ε.

Proof: let $\bar{X}_n$ = the average value of the X_i's in the first sample, $\bar{X}_m$ the average value of the X_i's in the second sample. We compute:

$$E(\bar{X}_n - \bar{X}_m) = E(\bar{X}_n) - E(\bar{X}_m)$$
$$= \frac{1}{n}\sum_{i=0}^{n-1} E(X_i) - \frac{1}{m}\sum_{i=n}^{n+m-1} E(X_i).$$

Let E_1, D_1^2 be the mean and variance operators for the first person, and E_2, D_2^2 those for the second person. Then, since the X_i are 1-exchangeable,

$$E_1(X_i) = 1 \cdot p + 0 \cdot (1-p) = p \quad \text{for all } i$$
$$E_2(X_i) = 1 \cdot q + 0 \cdot (1-q) = q \quad \text{for all } i.$$

Therefore,

$$E_1(\bar{X}_n - \bar{X}_m) = 0 \quad \text{and} \quad E_2(\bar{X}_n - \bar{X}_m) = 0.$$

We may also compute the variance of $\bar{X}_n - \bar{X}_m$ for each person:

$$D^2(\bar{X}_n - \bar{X}_m) = E[([\bar{X}_n - \bar{X}_m] - 0)^2] = E([\bar{X}_n - \bar{X}_m]^2) =$$
$$= E(\bar{X}_n^2) - 2E(\bar{X}_n \cdot \bar{X}_m) + E(\bar{X}_m^2).$$

Since the X_i are 1-exchangeable and 2-exchangeable,

$$E(\bar{X}_n^2) = \frac{1}{n^2}\,[nE(X_i^2) + n(n-1)\,E(X_i \cdot X_i)].$$

Since $X_i = 0$ or $X_i = 1$, $X_i^2 = X_i$, and $E(X_i^2) = E(X_i)$.

Due to 2-exchangeability, $E_1(X_i \cdot X_j) = p'$, $E_2(X_i \cdot X_j) = q'$.

$$E(\bar{X}_n \cdot \bar{X}_m) = E\left(\left[\frac{1}{n}\sum_{i-0}^{n-1} X_i\right]\left[\frac{1}{m}\sum_{i=n}^{n+m} X_i\right]\right) =$$
$$= \frac{1}{n} \cdot \frac{1}{m} \cdot (m \cdot nE(X_i \cdot X_j)) =$$
$$= E(X_i \cdot X_j).$$

Thus

$$D_1^2(\bar{X}_n - \bar{X}_m) = \frac{1}{n}\,p + \frac{n-1}{n}\,p' - 2p' + \frac{1}{m}\,p + \frac{m-1}{m}\,p' =$$

$$=p\left(\frac{1}{n}+\frac{1}{m}\right)+p'\left[\frac{n-1}{n}+\frac{m-1}{m}-2\right]=$$

$$=\left(\frac{1}{n}+\frac{1}{m}\right)(p-p').$$

Similarly,

$$D_2^2(\bar{X}_n-\bar{X}_m)=\left(\frac{1}{n}+\frac{1}{m}\right)(q-q').$$

By Tchebycheff's inequality, we therefore have

$$P_1\left(|\bar{X}_n-\bar{X}_m|\geq k_1\sqrt{\left(\frac{1}{n}+\frac{1}{m}\right)(p-p')}\right)\leq\frac{1}{k_1^2}$$

$$P_2\left(|\bar{X}_n-\bar{X}_m|\geq k_2\sqrt{\left(\frac{1}{n}+\frac{1}{m}\right)(q-q')}\right)\leq\frac{1}{k_2^2}.$$

Let

$$k_1=\frac{\varepsilon}{\sqrt{\left(\frac{1}{n}+\frac{1}{m}\right)}\cdot\sqrt{p-p^1}},\qquad k_2=\frac{\varepsilon}{\sqrt{\left(\frac{1}{n}+\frac{1}{m}\right)}\cdot\sqrt{q-q^1}}.$$

Clearly, by choosing both n and m large enough, we can make both k_1 and k_2 so large that $1/k_1^2<\delta$ and $1/k_2^2<\delta$. But this is just to say both

$$P_1(\bar{X}_n-\varepsilon\leq\bar{X}_m\leq\bar{X}_n+\varepsilon)>1-\delta$$

and

$$P_2(\bar{X}_n-\varepsilon\leq\bar{X}_m\leq\bar{X}_n+\varepsilon)>1-\delta.$$

In general it is possible to show, for any sequence of exchangeable quantities X_0, $X_1,\ldots$ whatever, that for large enough n and k, the probability for the openminded (non-extreme) person that the distribution of X in the first n objects will be close to the distribution of X in the next k is as large as you please, and thus that two people will, with sufficient evidence, come to agree closely concerning the probable distribution of X over a future set of k trials.

7. Much of the literature on Bayesian inference is concerned with spelling out the conditions under which the use of arbitrary prior distributions will lead to negligible error, given a reasonable amount of data. A discussion of these problems at a relatively elementary and informal level is to be found in Edwards, Lindeman and Savage.[12]

To remark that the exact character of the prior distributions that are fed into Bayes' theorem is often unimportant, or 'swamped' by the evidence, is not to say that it is always unimportant, or that it is not needed at all. In the first place, it is clearly crucial that the prior density not be concentrated at a point; in the binomial example we just considered, it is essential that neither of the individuals be completely convinced in advance of the evidence that all the sampled items are A's, or that none of them are A's, nor may he regard the events as independent, in the sense of taking $p'=p^2$. But further it should be noted that there is always *some* prior probability distribution which will, given the evidence we in fact have, lead to a *given* posterior distribution. In the simplest case, let $P(E \mid H)$ and $P(E \mid \bar{H})$ be given, where E is the evidence we in fact have. Clearly the probability of H given E, $P(H \mid E)$, may have any value whatever, according to the value we assign to $P(H)$:

$$P(H \mid E) = \frac{P(H)\,P(E \mid H)}{P(H)\,P(E \mid H) + (1 - P(H))\,P(E \mid \bar{H})}$$

$$= \frac{1}{\dfrac{P(E \mid H) - P(E \mid \bar{H})}{P(E \mid H)} + \dfrac{1}{P(H)}\left(\dfrac{P(E \mid \bar{H})}{P(E \mid H)}\right)} =$$

$$= \frac{1}{1 - \dfrac{P(E \mid \bar{H})}{P(E \mid H)}\left(1 - \dfrac{1}{P(H)}\right)}.$$

By choosing $P(H)$ close enough to 0, we can make $P(H \mid E)$ as close as we please to 0; by choosing $P(H)$ close enough to 1, we can make $P(H \mid E)$ as close as we please to 1. In general we have, for the posterior density of x, given evidence e,

$$d(x \mid e) = \frac{v(e \mid x)\,du(x)}{\displaystyle\int v(e \mid x)\,du(x)}$$

where $du(x)$ is the prior density of x, and $v(e \mid x)$ is simply the likelihood of e, given x. Since we are in principle free to choose $du(x)$ in any way we want, we can give the posterior density almost any form we want, say $dF(x)$, by choosing a function $u_F(x)$ to represent our prior distribution, where

$$dF(x) = \frac{v(e \mid x)\, du_F(x)}{\int v(e \mid x)\, du_F(x)}.$$

Of course in real life, to work this way would lead one to suppose that one had some rather startling prior opinions. But these opinions are in no way ruled out by the logic of the subjectivistic approach. One may argue, of course, that one has the prior belief in H that one in fact has, and that one is not entitled to go around changing it after the fact to justify one's posterior probability $P(H \mid E)$. But those who adopt a subjectivistic theory cannot really argue this way in general, for it is a commonplace of this theory that sheer introspection is a pretty blunt instrument. To arrive at prior probabilities, one often has to indulge in thought experiments to see what *other* beliefs one is led to by various prior probabilities, and if those other beliefs seem more or less plausible, one will be more or less inclined to accept the original probabilities as describing his prior state of mind. There is thus nothing sacred, on this theory, about what one takes to be one's 'initial' degrees of belief; they may change as a result of introspection, as a result of calculation, as a result of blood chemistry, as well as in response to evidence.

From the point of view of many subjectivistic statisticians, it seems that this state of affairs is simply a fact of life. From a practical point of view, it isn't a serious state of affairs, because most people change their beliefs in a fairly sensible, commonsense way, so far as they can keep track of what they are doing. The Bayesian approach to statistics provides an excellent way to keep track. It also provides a lot of information about what aspects of prior probabilities have relatively little effect on posterior probabilities, and what prior probability distributions seem plausible (argued for often, be it noted, by the plausibility of the posterior distributions to which they lead!). People simply are not inclined to fiddle around with their prior probabilities retrospectively as well as introspectively, in order to avoid the consequences of what they initially suppose to be their beliefs. Besides, the fiddling would often, indeed *usually*, in ordinary statistical contexts, have to be so extreme that no one in his right mind would seriously claim to have had 'just those' prior beliefs that would allow him to continue to believe something, as we would say, 'in the face of contrary evidence'.

Philosophers, however, do not tend to regard a theory as logically adequate merely because it is popular, or because it appears to be satisfactory in a number of applications to special cases. Thus philosophers have not tended to be satisfied with the subjectivistic theory just because it is a way of arriving at decisions that business men seem willing to accept, and have sought general principles on which to found applications of the theory. Most philosophers who adopt theories closely related to this one, for example, adopt a principle of conditionality which is intended to connect prior probabilities to posterior probabilities over time. To distinguish it from a related principle of conditionality introduced by Alan Birnbaum, we shall call this the Principle of Epistemic Conditionality.

> *Principle of Epistemic Conditionality:* Let the language in which an individual expresses his beliefs be L; suppose that at t_0 there is a probability function P_0 defined over the statements of L, such that his degree of belief in any statement is the same as its probability. His beliefs are thus coherent; this is all that is demanded by the principle of coherence. Let t_1 be some later time, and let the probability function P_1, defined over the statements of L, represent his degrees of belief at t_1, so that they too are coherent. Let e be a statement of L representing the body of evidence he has acquired (say through experience or experiment) in the interval from t_0 to t_1. The principle of epistemic conditionality requires that for any statement S of L, $P_1(S)=P_0(S \mid e)$.

This principle precludes the possibility of retroactively adjusting your prior probabilities when you see where the evidence will take them. It thus constitutes an addition to the principle of coherence. One's beliefs at any time could be perfectly coherent, though totally unrelated to one's earlier beliefs. The principle of epistemic conditionality prevents this from happening.

Two problems suggest themselves. First, what shall we take to be the primordial time t_0? Clearly it can't be the 'present', because that keeps changing, and if we could start over whenever we felt like it, the force of the principle of epistemic conditionality would be nullified. On the other hand, if t_0 is supposed to be well in the past, the problem of deciding what our degrees of belief then *were*, independently of what they happen to be now (in order to examine the rationality of the beliefs we have now) seems to be a knotty one. Second, it seems artificial to suppose that between t_0 and t_1 all of our evidence can be expressed in a single sentence e, of which we are absolutely sure. Much of our data, in real life, is infected with doubt.

With regard to the second problem, a way out has been offered by Richard Jeffrey.[13] Jeffrey's principle is the following: let all of the conditions stated in the Principle of Epistemic Conditionality but the last be satisfied. In lieu of the last condition, which is that e should be a statement representing a body of acquired evidence, we say: let e be a statement of L in which a change of beliefs *originates*. That is, we need no longer regard e as accepted with certainty; it suffices that we undergo certain experiences, for example, which lead us (cause us) to assign a new probability to e. Then Jeffrey's principles requires that for any statement S of L, $P_1(S) = P_1(e)\,P_0(S \mid e) + (1 - P_1(e))\,P_0(S \mid {\sim}e)$.

Although Jeffrey's principle provides an ingenious way of handling the second problem faced by those who opt for the Universal Bayes approach, it merely aggravates the first problem. According to the principle of epistemic conditionality there are certain evidence statements that come to be accepted between t_0 and t_1. We can (in theory) spell out the kinds of statements that can qualify, and the kinds of experience that would warrant their acceptance if it occurred. On Jeffrey's theory it is not only unclear that we can spell out even the *kinds* of statements in which a probability shift may originate, but it seems impossible to characterize in detail the kinds of experience that would *warrant* a certain probability shift. Failing this, we are simply left with the bare *fact* that certain shifts occurred. The shift may not even preserve coherence, though, of course, the rational agent will then make changes in his degrees of belief in such a way as to regain coherence. How he does this seems arbitrary. Even if one accepts the principle that the origin of a probability shift (the shift in probabilities assigned to evidential propositions) cannot embody an incoherence, and if one takes this as a criterion of an *origin*, there may still be quite a number of origins one might choose. Furthermore, if the starting point is the mere fact of the probability shift, we appear to have abandoned all hope of finding a rationale for such shifts, and to be right back where we started when the only condition we imposed on beliefs was the static one of coherence.

Let us be even more explicit about this important point. Suppose that $\mathscr{L}$ is a language with a finite number of atomic sentences, and that every sentence in $\mathscr{L}$ can be expressed as a finite disjunction of state descriptions, where a state description is a conjunction containing, for every atomic sentence, either it or its negation. Let B_0 be the set of sentences assigned probability 1 at time t_0. Let B_1 be the set of sentences assigned probability 1 at t_1. $B_0 \subset B_1$, since, once a sentence is assigned probability 1, there is no way of changing that assignment by any form of conditionalization. Let P_0 be a coherent probability assignment at t_0, assigning probability 1 to all and only the consequences of B_0. Let $Cj(B_1 - B_0)$ be the conjunction of the sentences assigned probability 1 between t_0 and t_1. According to the Principle of Epistemic Conditionality, the appropriate probability function at t_1 should be

$$P_1(x) = P_0(x \mid Cj(B_1 - B_0)).$$

We show that if one prior probability is as good as another, this imposes no constraint, other than with regard to probabilities of 0 and 1, on P_1.

Let P_1' be an arbitrary coherent probability function defined on $\mathscr{L}$ such that for every statement s in B_1, $P_1'(s) = 1$. We show that for arbitrary $\varepsilon > 0$, there is an acceptable prior probability function P_0^* such that for every statement x belonging to $\mathscr{L}$, $|P_1'(x) - P_0^*(x \mid Cj(B_1 - B_0))| < \varepsilon$, where P_0^* is an *acceptable* prior probability function just in case it assigns probability 1 to exactly the same statements that P_0 assigns probability 1 to.

Since there are a finite number of atomic statements in $\mathscr{L}$, there are a finite number of state descriptions, say n. Let $\delta = \varepsilon/n$. We choose P_0^* so that for every state description s, $|P_0^*(s \mid Cj(B_1 - B_0)) - P_1'(s)| < \delta$. The possibility of satisfying the inequality for *each* state description is established on p. 115. The possibility of coherently satisfying them all simultaneously follows from the fact that both the sum of $P_1'(s)$ and the sum of $P_0^*(s \mid Cj(B_1 - B_0))$, taken over the state descriptions consistent with B_1, is 1. Now since every statement is equivalent to the disjunction of no more than n state descriptions, it follows that for no statement does its conditional P_0^* value differ from its P_1' value by more than ε. Since formally P_0^* is 'just as good' a prior probability function as P_0, it is hard to see how the Principle of Epistemic Conditionality imposes any dynamic constraint on one's belief function at all.

A similar argument may be given regarding Jeffrey's principle. Let P_1' be an arbitrary coherent probability function, agreeing with P_1 for that statement e in which the shift orginates, and also agreeing with P_1 with regard to any statements to which P_1 assigns probability 1. Then, just as before, we can find a prior probability function P_0^* such that

$$|P_1'(x) - [P_1(e) \cdot P_0^*(x \mid e) + (1 - P_1(e)) \cdot P_0^*(x \mid \sim e)]| < \varepsilon$$

for every statement x in $\mathscr{L}$.

In effect, then, if one regards one's degrees of belief as simply (and mysteriously) *given*, so that the dynamic principles have significant content, there is no way in which beliefs in conflict with the dynamic principles (or in conflict with coherence, for the matter) can be reformed; and if one regards one's degrees of belief as amenable to modification, so that they can be changed to conform (say) with the principles of coherence, then the dynamic principles are essentially empty because my 'initial' probability function can always be reformed in such a way as to do no violence either to the dynamic principles of belief change, or to my present prejudices.

8. A number of philosophers, among them Carnap,[14] Hintikka,[15] Hilpinen,[16] have responded to the first problem by proposing that it is possible to define an absolutely *a priori* logical measure function over the sentences of a language. These measures represent the degree of belief that someone who spoke the language, but who had had no experience at all, would have in the sentences. Evidence statements are then added to his body of knowledge, and, at a given time, the individual's degree of belief in any statement s ought to be $P(s \mid e)$, the conditional probability of s given e, where e is his total body of evidence.

Finding plausible principles on which to establish the original logical measure function has turned out to be a very difficult job, and indeed, one that has only been seriously attempted in detail for rather simple first order languages. Furthermore, it keeps turning out that one's most natural intuitions regarding this prior measure lead to consequences that are not at all intuitive, and thus to the search for new measures. Simply as an illustration, consider a simple first order language *L*, consisting of some one place predicates and a supply of individual constants and variables. In a finite universe (where there are a finite number of individual constants), we can construct a sentence that completely specifies the state of the universe: individual 1 has property 1, lacks property 2, lacks property 3,... etc.; individual 2 has property 1, has property 2, lacks property 3, etc.; ...; individual *i* has property 1, has property 2, lacks property 3, etc..... Such a sentence is a state description. It seems *prima facie* plausible to assign to each state description, in the absence of any knowledge at all, the same measure. This approach was adopted, for example, by Broad.[17] But a consequence of this assignment is that learning from experience is impossible: if *S* is a statement that individual $n+1$ has the first property, and *e* is a statement that completely describes the first *n* individuals, it turns out that the conditional probability of *S* given *e* is simply equal to the prior probability (the logical measure) of *S*.

This is clearly unsatisfactory. Carnap turns from this function to another which seems more satisfactory.[18] Let two state descriptions be *isomorphic* if you can obtain the one from the other by a permutation of the individual constants occuring in it. Let the set of all state descriptions isomorphic to a given state description be called a structure description. Now, instead of assigning equal measures to state descriptions, let us assign equal measures to structure descriptions. Ah! Now we can learn from experience. The degree of confirmation of *S* on *e* (with *S* and *e* as specified before) depends significantly on how many of the items mentioned in *e* turned out to have the first property.

But surely in science we need to be able to consider a language with an infinite number of individual constants. It now turns out that in any such language, the probability of any universal generalization (e.g., 'everything that has the first property also has the fifth property') is 0, regardless of the evidence we have.

A new alternative is introduced by Hintikka[19] which overcomes this problem: Hintikka's measure functions assign finite *a priori* probabilities (measures) to universal generalizations even in an infinite language. There are two free parameters in these functions, and no obvious way of deciding what values they should have is provided.

But this is far enough to pursue this program, for the moment. It is clear that no natural and universally acceptable way of assigning logical measures to the statements of even quite simple languages has yet been offered; and it is equally clear that the problem of devising such measures for languages as rich as those we need for doing what we may properly call science is going to be overwhelming.

Little has been done in statistics itself with logical probability. Gerhard Tintener presented a paper to the Royal Statistical Society in England in which he expounded Carnap's theory, and indicated its relevance for statistical inference.[20] His paper appears to have produced lively responses at the time, but has had little subsequent influence.

There is one further oddity in the attempt to establish credibilities on the basis of logical measures. It is that on the basis of rather innocuous assumptions concerning the logical measures of simple propositions, we can derive very strong commitments to degrees of belief in complex and powerful statements. For example, given simple initial assumptions concerning the *a priori* probability of getting a purple ball from an urn, and concerning the probability of getting a second purple ball from the urn, given that the first ball has been purple, we can derive a statement to the effect that the probability we *must* assign to getting less than such and such a proportion of balls in an arbitrarily long run is very close to one: *a priori* we must be practically certain that the arbitrarily long run must have a certain character. The proof of this startling result follows.

Suppose that E_0, E_1 ... is a sequence of events (or statements). Let X_i be the random quantity that has the value 1 if the ith event has the property Q (if the ith statement is true), and the value 0 otherwise. Let P be a probability measure defined over a field of events that includes all the E's. We suppose that the events are 1-exchangeable and 2-exchangeable, or, to put it more intuitively: the probability that $X_i=1$ has the same value, whatever i may be; and the probability that $X_i=1$ and $X_j=1$, for $i\neq j$, is the same whatever i and j may be. Let

$$\bar{X}_n=\frac{1}{n}\sum_{i=0}^{n-1} X_i$$

$$P(X_0=1)=m_1\neq 0$$

$$P(X_1=1 \mid X_0=1)=m_2.$$

LEMMA 1

$$P(X_i=1)=m_1.$$

Proof. Immediately by 1-exchangeability

LEMMA 2

$$i \neq j \supset P(X_i=1 \mid X_j=1)=m_2$$

Proof.

(1) $P(X_0=1 \wedge X_1=1)=P(X_i=1 \wedge X_j=1)$, by 2-exchangeability.

(2) $P(X_0=1 \wedge X_1=1)=P(X_0=1)\cdot P(X_1=1 \mid X_0=1)=m_1 \cdot m_2.$

(3) $P(X_i=1 \wedge X_j=1)=P(X_j=1)\cdot P(X_i=1 \mid X_j=1).$
$=m_1 \cdot P(X_i=1 \mid X_j=1)$ by Lemma 1.

(4) $P(X_i=1 \mid X_j=1)=m_2$ by 1, 2, 3.

LEMMA 3

$$E(\bar{X}_n)=m_1 \quad \text{(the mean of } \bar{X}_n\text{).}$$

Proof.

$$E(\bar{X}_n)=\frac{1}{n}\sum_{i=0}^{n-1} E(X_i)$$
$$E(X_i)=1\cdot m_1+(1-m_1)\cdot 0=m_1$$
$$E(\bar{X}_n)=\frac{n\cdot m_1}{n}=m_1.$$

LEMMA 4

$$D^2(\bar{X}_n)=\frac{m_1}{n}+\frac{n-1}{n}\, m_1 \cdot m_2-m_1^2 \quad \text{(the variance of } \bar{X}_n\text{).}$$

Proof:

$$D^2(\bar{X}_n)=E[(\bar{X}_n-m_1)^2] \quad \text{(definition of variance)}$$
$$=\frac{1}{n^2}E\left[\sum_{i=0}^{n-1} X_i\right]^2-2m_1E(\bar{X}_n)+m_1^2$$

$$\left(\sum_{i=0}^{n-1} X_i\right)^2 = \sum_{i=0}^{n-1} X_i^2 + \sum_{i=0}^{n-1} \sum_{\substack{j=0 \\ j \neq i}}^{n-1} X_i \cdot X_j.$$

Since $X_i = X_i^2$, $E(X_i^2) = E(X_i) = m_1$

$$E(X_i \cdot X_j) = 1 \cdot P(X_i = 1) \cdot P(X_j = 1 \mid X_i = 1) + \\ + 0 \cdot P(X_i = 0 \vee X_j = 0)$$

$$= m_1 \cdot m_2 .$$

$$D^2(\bar{X}_n) = \frac{1}{n^2} n \cdot m_1 + \frac{1}{n^2} (n)(n-1)\, m_1 m_2 - m_1^2$$

$$= \frac{m_1}{n} + \frac{n-1}{n} m_1 m_2 - m_1^2 .$$

THEOREM

$$P\left(|\bar{X}_n - m_1| \geq k \sqrt{\frac{m_1}{n} + \frac{n-1}{n} m_1 m_2 - m_1^2}\right) \leq \frac{1}{k^2}.$$

Proof. Tschebycheff's theorem says that if Y is a random quantity, $E(Y)$ its mean, and $D^2(Y)$ its variance, then

$$P(|Y - E(Y)| \geq k\sqrt{D^2(Y)}) \leq \frac{1}{k^2}$$

The theorem follows immediately by the substitution of $\bar{X}_n$ for Y, taking account of the lemmas.

COROLLARY. *If E_0, E_1,... is a sequence of exchangeable events, and the probability that E_0 has Q is* 0.01, *and the probability that E_1 has Q, given that E_0 has Q, is between* 0.01 *and* 0.02, *then, for sufficiently large n, the probability that the frequency of Q will be no more than* 0.11 *among n of the events E is at least* 0.99.

Proof:

$$D^2(\bar{X}_n) = \frac{1}{n}[m_1 - m_1 m_2] + m_1 m_2 - m_1^2$$

$$= \frac{m_1}{n}[1 - m_2] + m_1(m_2 - m_1).$$

By choosing n large enough, we can meke the first term as close to 0 as we please.

Since

$$m_1 \geq m_2 - m_1, \sqrt{m_1(m_2 - m_1)} \leq m_1$$

$$km_1 \geq k\sqrt{m_1(m_2 - m_1)}.$$

Therefore from the theorem we have

$$P(|\bar{X}_n - m_1| \geq km_1) \leq \frac{1}{k^2}.$$

Taking $k = 10$, $m_1 = 0.01$, $0.01 < m_2 \leq 0.02$, we have

$$P(|\bar{X}_n - 0.01| \geq 0.10) \leq 0.01, \quad \text{or} \quad P(X_n \leq 0.11) \geq 0.99.$$

What are we to make of this? Let us review the premises. We require that the sequence of events be 1-exchangeable and 2-exchangeable. This is a much weaker requirement than general exchangeability. Every logical theory that I know of, at least all those that derive from Carnap's work, stipulates that the measure function be symmetric, i.e., that the measure of $Q(a_i)$ is equal to the measure of $Q(a_j)$ for any individual constant a_i and a_j, and furthermore that the same thing be true in general for n-tuples of individual constants. This requirement of symmetry is therefore the same as the general requirement of exchangeability, and far stronger than we need to establish our result. We require that the prior probability that the first item have the property Q be 0.01. Surely there is no problem here; we should always be able to find such sequences, and such a property. Second, we need to suppose that the conditional probability that the second item have Q, given that the first one does, lies between 0.01 and 0.02. Again these conditions are met by any logical measure function. All measure functions that allow the possibility of learning from experience will require that the conditional probability in question be greater than 0.01. It is not true that every measure function in the Carnap or Hintikka continua will lead to the conclusion that the conditional probability will be no more than 0.02; but many of those discussed by Carnap do have this property, and it is surely intuitively acceptable that *one* instance will not more than double our prior probability. Furthermore, even if we did allow it to be somewhat more than double, we would still have a counterintuitive result: for large n the variance in any case cannot be greater than the prior probability, and we could deduce that the probability was at least 0.96 that less than half of the events in the indefinitely long run would exhibit the property Q.

For a theory which is supposed to be purely logical, and to reflect no empirical knowledge at all, this is surely a distrubing consequence, and one that we would like to avoid.

The same theorem holds, of course, for subjective probabilities that satisfy the conditions, and there are no reasons to suppose that the beliefs of actual persons may not sometimes satisfy them. The subjectivist, however, is in a much better position than the logical-measure theorist. He need merely shrug his shoulders and point out that the consequent of the theorem is just what one's prior beliefs commit one to, if one is going to be consistent. "If the consequent strikes you as implausible," he may say, "then you should change your prior beliefs: perhaps you should reject the 1- and 2-exchangeability of the sequence of events, or perhaps you should change your prior assignments regarding the first two events of the sequence." He isn't *caught* the way the logical-measure theorist is; but it is unsettling to see him recommending retroactive shifts in prior beliefs,[21] and it is also unsettling to see that such apparently innocuous prior beliefs commit one to such strong beliefs concerning such powerful statistical hypotheses as those about an arbitrarily long sequence of events.

Neither the subjectivistic nor the logical approach to statistical inference provides us with the kind of foundation for statistical inference we are looking for. There are circumstances, of course, in which the subjectivistic approach seems right and plausible: namely those in which our prior probability assignments are rationally defensible. But the defense of those prior probability assignments is not within the domain of that application of the theory. To be sure, they can be defended by reference to some yet more prior probabilities; but this defense in turn rests on *prior prior* assignments; the process cannot terminate. Furthermore, the Dutch Book Theorem on which the theory depends seems of doubtful relevance. We are never in the position of having to post odds on all our beliefs and having then to cover whatever bets an ingenious and unfriendly opponent wishes to make. Finally, what we are looking for is a dynamic theory, a theory of how evidence should force changes on our patterns of belief. The subjectivist theory of probability is in essence a static theory: it imposes conditions to be satisfied by our beliefs at a moment, but provides no connection between our beliefs at one moment and our beliefs at another moment except inertia. Inertia may be a fact of mental life, but it is a poor logical principle.

Those who adopt the logical approach have tried to grapple with a number of these problems, but without much success. This approach provides a solution to the problem of the Ur-distribution, but only for very simple worlds rather unlike ours, and only with the help of arbitrary parameters. Both the subjectivistic and logical theories, by their insistance that our degrees of belief be represented by real numbers, lead to the consequence that certain clearly empirical hypotheses can have very high probabilities relative to no evidence at all. This seems particularly damaging to the logical theory. Finally, if either the subjectivistic or the logical approach is adopted, those principles like the Minimax principle that have seemed plausible, at least in some circumstances, to writers of other persuasions, have no applicability at all except when they coincide with a Bayesian strategy. It may be that such principles have no other application; but the case for this that has so far been made from the subjectivistic point of view is far from conclusive.

NOTES

[1] F. P. Ramsey, 'Truth and Probability', in *The Foundations of Mathematics*, London 1931, pp. 156–198.

[2] A. Shimony, 'Coherence and the Axioms of Confirmation', *Journal of Symbolic Logic* **20** (1955), 1–28.

[3] R. S. Lehman, 'On Confirmation and Rational Batting', *Journal of Symbolic Logic* **20** (1955), 251–262.

[4] John G. Kemeny, 'Fair Bets and Inductive Probabilities', *Journal of Symbolic Logic* **20** (1955), 263–273.

[5] Bruno De Finetti, 'La prevision: ses lois logiques, ses sources subjectives', *Annales de l'Institut Henri Poincare* **7** (1937), 1–68. Translated as 'Foresight: Its Logical Laws, Its Subjective Sources', in *Studies in Subjective Probability* (ed. by Kyburg and Smokler), New York 1964.

[6] L. J. Savage, *Foundations of Statistics*, New York 1964.

[7] D. A. Freedman and R. A. Purves, 'Bayes Method for Bookies', *Annals of Mathematical Statistics* **40** (1969), 1177–1186.

[8] H. Jeffreys, *Theory of Probability*, Oxford 1939; and *Scientific Inference*, Cambridge 1931.

[9] See, for example, Pratt, Raiffa, and Schlaifer, *Introduction to Statistical Decision Theory*, New York 1965.

[10] The proof of this will be found in D. V. Lindley, *Probability and Statistics*, Vol. 2, Cambridge 1965, p. 3.

[11] De Finetti, 'Foresight'.
[12] Edwards, Lindeman, and Savage, 'Bayesian Statistical Inference for Psychological Research', *Psychological Review* **70** (1963), 193–242.
[13] Richard Jeffrey, *The Logic of Decision*, New York 1965.
[14] R. Carnap, *The Logical Foundations of Probability*, Chicago 1950 and 'A Basic System of Inductive Logic' in Carnap and Jeffrey (eds.), *Studies in Inductive Logic and Probability*, Berkeley 1971.
[15] J. Hintikka, 'A Two-Dimensional Continuum of Inductive Methods', in Hintikka and Suppes (eds.), *Aspects of Inductive Logic*, Amsterdam 1966.
[16] R. Hilpinen, *Rules of Acceptance and Inductive Logic*, *Acta Philosophica Fennica*, Fas. XX, Amsterdam 1968.
[17] C. D. Broad, 'The Principles of Problematic Induction', *Proceedings of Aristotelian Society* **28** (1927–28), 1–46.
[18] *The Logical Foundations of Probability*.
[19] Hintikka, *op.cit.*
[20] Gerhard Tintner, 'The Foundations of Probability and Statistical Inference', *Journal of the Royal Statistical Society, Series A* **112** (1949), 251–279.
[21] Recall the discussion of probability dynamics toward the end of Section 7.

CHAPTER SIX

COMPARISON OF APPROACHES

The variety of approaches to statistical inference and the problem of decision in the face of uncertainty is regarded by many statisticians as a Good Thing. On analogy with mathematics, statistics may be looked on as a body of theory and techniques which may be applicable to a wide variety of circumstances. Applications, however, are not the business of the theoretical statistician as such. On this view, the wider the range of available techniques, the richer and more various the body of statistical theory, the better. It is undemocratic, if not sinful, to regard one approach as wrong-headed as opposed to another.

It is hard to fault such a view as it applies to research in mathematical statistics. On the other hand, the statistician, unlike the mathematician, is called upon not only to develop his theories, but to apply them. Furthermore, the development of the theory takes place in a world in which the applicability of the theory is a pervasive constraint. The approach of the statistician to applied statistical inference has a profound effect on the conditions that he requires his theory to satisfy and on the problems he regards his theory as having to face and solve. We have observed this in the preceding chapters. It is therefore not only interesting but important to compare the applications of these theories, and to see whether there is in fact conflict among them, or whether they are perhaps merely different approaches which lead to essentially the same results. Either outcome would be of profound importance for the study of the foundations of statistics.

In a sense both outcomes are true. In the first part of the present chapter we shall examine the connections between the various approaches, and show how it is possible to construe them as leading to essentially the same results. We shall also mention the impact that this fact has had on the development of theoretical statistics. We shall then look at the ways in which the approaches are nevertheless in basic conflict, and the great need for a more general theory which will resolve the conflict. Finally, we shall examine the prospects for such a general theory.

1. The situation in which it is easiest to see the relations between a number of approaches is that in which we are faced with a decision corresponding to the choice between two simple hypotheses, such as that described on several occasions in previous chapters. We shall generalize that situation only slightly. We consider two hypotheses, H_1 and H_2, and a body of evidence e. We suppose that for every e in the sample space E, $p(e \mid H_i)$ is well determined: if E is discrete, $p(e \mid H_i)$ is just the assignment of probability to the point e by the hypothesis H_i; if E is continuous, $p(e \mid H_i)$ is the probability density attributed to the sample space by H_i at e. Furthermore, we suppose there are two actions we may consider, A_1, and A_2, and a loss function $L(A_i, H_j)$ which gives the loss incurred by action A_i when H_j is true. We also suppose that A_1 is appropriate to H_1 and A_2 to H_2 in the sense that $L(A_1, H_1) < L(A_1, H_2)$ and $L(A_2, H_2) < L(A_2, H_1)$.

A decision function, as before, is construed as a function that maps the elements of E into $\{A_1, A_2\}$; a mixed decision function maps the elements of E into $[0, 1]$, where the value of the function at e is construed as the probability of taking action A_1. The former may clearly be taken as a special case of the latter. We can represent a decision function as $d(e)$; if d is unmixed, d will simply be the characteristic function of the subset of E that is taken to lead to action A_1. Thus we may in general take $d(e)$ to be the probability of taking action A_1, given e.

For any decision function, pure or mixed, we may compute both its expected loss given the truth of H_1 and its expected loss given the truth of H_2. In particular, if d is a decision function, the expected loss under d, given H_1, $E_1(d)$, is

$$E_1(d) = L(A_1, H_1) \int_E d(e)\ p(e \mid H_1) + L(A_2, H_1) \int_E (1 - d(e))\ p(e \mid H_1).$$

Similarly:

$$E_2(d) = L(A_1, H_2) \int_E d(e)\ p(e \mid H_2) + L(A_2, H_2) \int_E (1 - d(e))\ p(e \mid H_2).$$

Clearly the maximum and minimum values of $E_1(d)$ are $L(A_2, H_1)$ and

$L(A_1, H_1)$, and similarly for $E_2(d)$ are $L(A_1, H_2)$ and $L(A_2, H_2)$. Thus we may normalize the losses by the transformation:

$$E_{1N}(d) = \frac{E_1(d) - L(A_1, H_1)}{L(A_2, H_1) - L(A_1, H_1)}; \quad 0 \le E_{1N}(d) \le 1$$

$$E_{2N}(d) = \frac{E_2(d) - L(A_2, H_2)}{L(A_1, H_2) - L(A_2, H_2)}; \quad 0 \le E_{2N}(d) \le 1.$$

The minimax solution to the decision problem is the point d^* on the boundary of the convex set of decisions for which $E_1(d^*) = E_2(d^*)$, or the point at which a 45° line through the origin intersects the boundary. In the transformed problem, it is the point d^* such that

$$E_1(d^*) = (L(A_2, H_1) - LA_1, H_1))\, E_{1N}(d) + L(A_1, H_1) =$$
$$= E_2(d^*) = ((L(A_1, H_2) - L(A_2, H_2))\, E_{2N}(d) + L(A_2, H_2);$$

that is, it is the intersection of the boundary with the line

$$E_{1N}(d) = \frac{(L(A_1, H_2) - L(A_2, H_2))}{L(A_2, H_1) - L(A_1, H_1)}\, E_{2N}(d) + \frac{L(A_2, H_2) - L(A_1, H_1)}{L(A_2, H_1) - L(A_1, H_1)}.$$

The minimax regret solution is the point d^0 on the boundary of the convex set of decisions for which $E_1(d^0) - L(A_1, H_1) = E_2(d^0) - L(A_2, H_2)$; that is the point at which the 45° line $E_1(d) = E_2(d) + (L(A_1, H_1 - - L(A_2, H_2))$ intersects the set of decision functions. In the transformed coordinates it is the point d^0 for which $E_1(d^0) - L(A_1, H_1) = (L(A_2, H_1) - - L(A_1, H_1))\, E_{1N}(d^0) = E_2(d^0) - L(A_2, H_2) = (L(A_1, H_2) - L(A_2, H_2)) \times \times E_{2N}(d^0)$. Thus it is just the intersection of the set of decision functions with the line through the origin of slope

$$\frac{L(A_1, H_2) - L(A_2, H_2)}{L(A_2, H_1) - L(A_1, H_1)}.$$

Observe that the slope of this line is the same as that of the minimax line, but that the minimax line, in the transformed coordinates, does not pass through the origin.

The lines of constant expected loss in the old coordinates had the equation

$$E_1(d)\cdot P(H_1)+E_2(d)\cdot P(H_2)=k,\quad \text{or}$$

$$E_1(d)=-\frac{P(H_2)}{P(H_1)}E_2(d)+k$$

where $P(H_1)$ is the prior probability of H_1 and $P(H_2)$ that of H_2. In the new system the equations of constant expected loss are of the form

$$E_{1N}(d)=-\frac{((L(A_1,H_2)-L(A_2,H_2))}{((L(A_2,H_1)-L(A_1,H_1))}\cdot\frac{P(H_2)}{P(H_1)}E_{2N}(d)+k.$$

Observe that if the increment of loss suffered by an erroneous act is the same under the two hypotheses, then these lines will have the same slope as the previous ones.

The Bayesian strategy picks a (any) point d^b of the set of decision functions such that no member of that set lies on a lower line of constant loss. Thus if d^b is an acceptable Bayesian decision function in the first formulation, then for every other point d in the set of decision functions

$$E_1(d)\geq-\frac{P(H_2)}{P(H_1)}E_2(d)+\left[\frac{P(H_2)}{P(H_1)}E_2(d^b)+E_1(d^b)\right].$$

The coordinates of the point d^b in the new system are $E_{1N}(d^b)$ and $E_{2N}(d^b)$. We now show that for every point d in the transformed coordinates,

$$E_{1N}(d)\geq-\frac{((L(A_1,H_2)-L(A_2,H_2))}{((L(A_2,H_1)-L(A_1,H_1))}\cdot\frac{P(H_2)}{P(H_1)}E_{2N}(d)+$$
$$+\frac{((L(A_1,H_2)-L(A_2,H_2))}{((L(A_2,H_1)-L(A_1,H_1))}\cdot\frac{P(H_2)}{P(H_1)}E_{2N}(d^b)+E_{1N}(d^b).$$

Transforming the coordinates of the initial inequality, we have

$$(L(A_2,H_1)-L(A_1,H_1))\,E_{1N}(d)+L(A_1,H_1)\geq$$
$$\geq-\frac{P(H_2)}{P(H_1)}\cdot[(L(A_1,H_2)-L(A_2,H_2))\,E_{2N}(d)+L(A_2,H_2)]+$$
$$+\frac{P(H_2)}{P(H_1)}\cdot[(L(A_1,H_2)-(L(A_2,H_2))\,E_{2N}(d^b)+L(A_2,H_2)]+$$

$$+(L(A_2, H_1)-L(A_1, H_1))\, E_{1N}(d^{\mathrm{b}})+L(A_1, H_1).$$

Substracting $L(A_1, H_1)$ from each side and dividing through by $L(A_2, H_1)-L(A_1, H_1)$, we have

$$E_{1N}(d) \geq -\frac{P(H_2)}{P(H_1)}\cdot \left\{\frac{(L(A_1, H_2)-L(A_2, H_2))}{(L(A_2, H_1)-L(A_1, H_1))} E_{2N}(d)+\frac{L(A_2, H_2)}{(L(A_2, H_1)-L(A_1, H_1))}\right\}+ +\frac{P(H_2)}{P(H_1)}\cdot\left\{\frac{(L(A_1, H_2)-L(A_2, H_2))}{(L(A_2, H_1)-L(A_1, H_1))} E_{2N}(d^{\mathrm{b}})+ +\frac{L(A_2, H_2)}{(L(A_2, H_1)-L(A_1, H_1)}\right\}+\frac{L(A_2, H_1)-L(A_1, H_1)}{L(A_2, H_1)-L(A_1, H_1)} E_{1N}(d^{\mathrm{b}}).$$

Simplifying, we have

$$E_{1N}(d) \geq -\frac{P(H_2)}{P(H_1)}\cdot\frac{L(A_1, H_2)-L(A_2, H_2)}{L(A_2, H_1)-L(A_1, H_1)} E_{2N}(d)+ +\frac{P(H_2)}{P(H_1)}\cdot\frac{L(A_1, H_2)-L(A_2, H_2)}{L(A_2, H_1)-L(A_1, H_1)} E_{2N}(d^{\mathrm{b}})+E_{1N}(d^{\mathrm{b}}).$$

which is just what we wanted to show.

Therefore every decision function which is acceptable in the first form of the problem is also acceptable in the second, as we expected.

If we take $L(A_1, H_1)=L(A_2, H_2)=0$ and $L(A_1, H_2)=L(A_2, H_1)=1$, then $E_1(d)$ is just the probability that d will lead to a false rejection of H_1 if it is true, i.e., the level of significance of d, and $1-E_2(d)$ is just the probability that d will correctly detect H_2, i.e., it is the power of d as a test of H_1 against H_2. In this case $E_{1N}(d)=E_1(d)$ and $E_{2N}(d)=E_2(d)$. To find the most powerful test of level α, we look at the intersection of the line $E_{1N}(d)=\alpha$; it intersects the boundary of the convex set of decision functions at the point d_α. The power of the test is $E_{2N}(d_\alpha)$.

All of these decision functions together with the lines of reasoning that lead to them, can be represented on the same diagram.

Figure 6 shows how, from any particular point of view, each of these points of view may be reconciled. Consider the classical hypothesis testing point of view. On this view, alpha, the level of significance of a test, is a

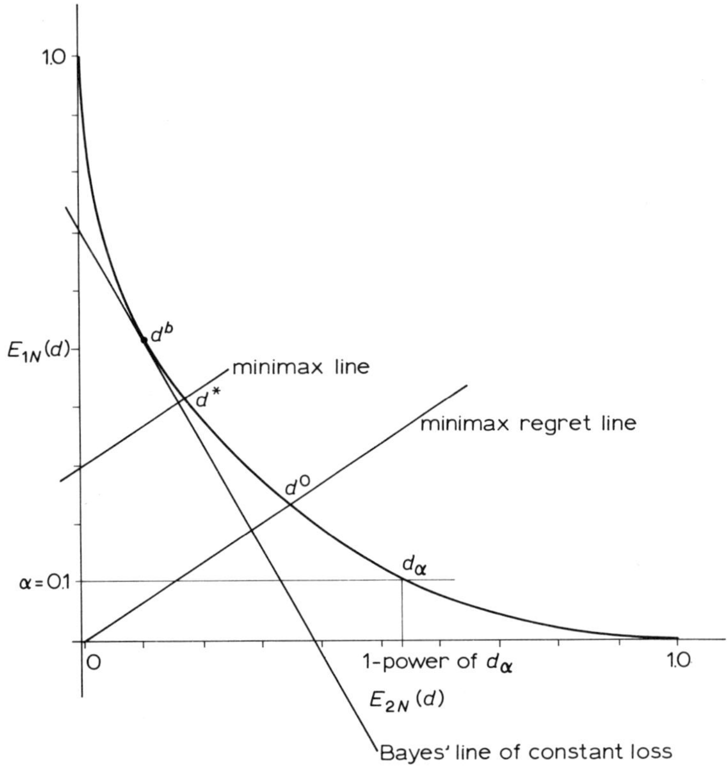

Fig. 6.

free parameter – it is what we *choose*. Its value depends on the circumstances under which we are performing the test: some circumstances will demand a very rigorous test (i.e., a very small alpha), others will allow a less rigorous test, a larger alpha. Certain circumstances will dictate that the appropriate value of alpha be $E_{1N}(d^b)$. One way of describing these circumstances may be to say that we take the various losses to be $L(A_1, H_1)$, $L(A_1, H_2)$, $L(A_2, H_2)$, and $L(A_2, H_2)$, and the prior probabilities of H_1 and H_2 to be $P(H_1)$ and $P(H_2)$. If this is the sort of information we have about the circumstances, then the Bayesian technique simply provides us with a convenient algorithm for evaluating $E_{1N}(d^b)$. In a rare case in which the Bayesian line of constant loss supporting the convex set of decision functions concides with a segment of the boundary of that set, we may say that the circumstances are not such as to determine alpha uniquely, but only to determine it to lie within a certain range.

Alternatively the hypothesis tester may simply take the Bayesian to be constructing, with his hypothetical losses and his subjective probabilities, an awkward and complicated way of expressing his *feelings* about the seriousness of making an error of the first kind.

Another way of describing the circumstances that determine an appropriate value for alpha is simply in terms of the losses $L(A_1, H_1)$, $L(A_1, H_2)$, $L(A_2, H_1)$, and $L(A_2, H_2)$, without taking into account anything corresponding to the prior probabilities of the hypotheses. Sometimes what is relevant is the absolute value of the expected loss; under these circumstances the appropriate value of alpha is $E_{1N}(d^*)$. The minimax analysis just provides a convenient method for computing $E_{1N}(d^*)$. Sometimes what is relevant is the difference between the expected loss and the minimum loss under each of the various hypotheses; if this is so, the appropriate value of alpha is $E_{1N}(d^0)$, and the minimax regret technique is just a way of finding that value of alpha.

Furthermore, *any* choice of alpha between 0 and $\frac{1}{2}$ corresponds to an appropriate statistical test, and thus every decision function on the boundary of the convex set is appropriate for some value of alpha as a test of H_1 against H_2, or of H_2 against H_1, or both. Thus every solution to the decision problem can be construed as the application of a classical statistical test.

The decision function indicated by the minimax regret procedure is the point at which the boundary of the set of decision functions is intersected by a line through the origin, with slope

$$\frac{(L(A_1, H_2) - L(A_2, H_2))}{(L(A_2, H_2) - L(A_1, H_1))}.$$

In view of the fact that this ratio can have any positive value whatsoever, every point on the boundary of the set of decision functions can be presented, by an appropriate choice of losses, as an appropriate minimax regret decision function. Thus in particular, d^b, d^*, and d_α can be indicated. The classical testing approach offers a convenient algorithm when we find it hard to evaluate the losses but when it is reasonable to regard the ratio of the difference between the losses when H_2 is true to the difference between the losses when H_1 is true as $\alpha: 1-\alpha$. If this is the ratio, d_α is the right decision function. Similarly, in lieu of supposing that the ratio is $E_{1N}(d^*)$ to $E_{N2}(d^*)$, we may find it easier to evaluate each of the four losses and to apply a minimax technique to yield the choice of d^*. Finally, it may be that in view of the fact that we have a way of evaluating the relative prior probabilities of H_1 and H_2 we feel that we can evaluate the ratio of the *expected* differences in losses; if so, this is clearly a ratio more to the point than the ratio of the absolute differences in losses. But this is just $E_{1N}(d^b)$ to $E_{2N}(d^b)$, and in this case the minimax regret procedure leads to the Bayesian decision function d^b.

A similar analysis may be undertaken from the minimax point of view. From this point of view, the minimax regret procedure applies in the special case in which $L(A_2, H_2)=L(A_1, H_1)$; in that special case the minimax and minimax regret lines coincide. We may obtain d_α either by altering the slope of the line, as we did for the minimax regret point of view, or translating the line by choosing an appropriate value of $L(A_2, H_2)-$ $-L(A_1, H_1)$. That is, to say that a certain level of significance alpha is appropriate is just to say that the situation warrants a certain loss ratio and a certain loss difference. To adopt prior probabilities which yield the decision function d^b is just to adopt a certain difference $L(A_2, H_2)-$ $-L(A_1, H_1)$.

From the Bayesian point of view, we can also account for every admissible decision function, since the ratio of $P(H_2)$ to $P(H_1)$ can have any positive value whatever, and thus the slope of the Bayesian supporting line can have any negative value whatever. To choose a level of significance alpha, is just to say that the situation is appropriate to an assignment of prior probabilities such that the Bayesian support line passes through the point d_α. To adopt a certain ratio of loss differences is to suppose that the situation is appropriate to an assignment or prior probabilities such that the Bayesian support line passes through d^0. And to adopt a certain ratio of loss differences and a certain absolute difference $L(A_2, H_2)-L(A_1, H_1)$ is just to suppose that the ratio of $P(H_2)$ to $P(H_1)$ is such that the support line passes through the point d^*.

Thus whichever point of view we adopt, the other points of view receive appropriate interpretations, and often can serve useful functions.

So far we have been considering the problem of a choice between actions. But all the same things can be said if we are considering a choice between the hypotheses H_1 and H_2. The coordinates are then the probabilities of error. In this case alpha receives its classical interpretation: it is simply the probability of committing an error of the first kind. The slope of the Bayesian support line is simply the ratio of $P(H_2)$ to $P(H_1)$. The slope of the minimax regret line is simply the ratio of the importance of committing an error given that H_2 is true to the importance of committing an error given that H_1 is true. To assign a value to this ratio is in effect to assign a value to the ratio of the prior probabilities, and conversely. Similarly, the minimax line has the same slope, and an intercept determined by $L(A_2, H_2) - L(A_1, H_1)$; to make these choices is also equivalent to choosing a ratio of prior probabilities.

These neat relationships do not, alas, hold for all decision and inference problems. This is to be expected, in view of the fact that, as we have observed earlier, there are problems in which no uniformly most powerful test of size alpha exists. But there are still a number of relationships that hold in general. One of the most important is that every admissible decision function, in any problem, is an optimal Bayes decision function for some assignment of prior probabilities. Another is that, generally, minimax procedures are Bayes solutions with respect to a least favorable *a priori* distribution.

The deep connection between the theory of testing hypotheses and the theory of (confidence) interval estimation has already been discussed in an earlier chapter. Indeed, through the device of stipulating certain loss functions for errors, the theory of hypothesis testing can be regarded not only as an instance of, but as a theory that finds its rationale in, the theory of decision under uncertainty.

There are certain circumstances in which fiducial limits and confidence limits coincide. It is easy to tell which they are: according to Fisher they are those instances in which confidence-limit theory (accidentally) yields correct results; according to Neyman, they are those instances in which the generally opaque method of fiducial inference has a reasonably clear and intelligible interpretation.

Do all of these curious relationships suggest that there is an underlying unity? Yes and no. They show that there is sometimes a formal parallelism. They do not show that there is an underlying theoretical or abstract unity. It is true that everybody can explain everything (almost everything) that everybody else says, except for the polemics. But this merely shows that in the mass of practical applications where common sense is relatively unequivocal, common sense triumphs. If all applications were of this character, it might be tempting to regard the various approaches to statistical inference as alternative formulations of the same theory. If so, it would still seem a philosophically interesting an important task to formulate that theory in neutral terms, terms that will not appear to beg important conceptual questions.

But the applications of various statistical techniques do not all yield concordant results in all circumstances. Furthermore the principles adopted in several points of view sometimes seem flatly to contradict each other. We shall now examine a relatively plausible example which brings out these incompatibilities.

2. Let us consider first a fairly straightforward hypothetical situation in which a person is faced with a problem to which statistical evidence and statistical inference seem obviously appropriate, but in which the several distinct approaches to statistical inference offer contrary advice.

The following tale is adopted from 'Probability and Decision'.[1]

Once upon a time there lived in a far away land a beautiful Princess, named Sally, the daughter of a most ingenious king. Realizing that in these days of high taxes and vengeful relatives, the traditional trial by combat or test of strength for the joy of his daughter's hand and for half his kingdom was rather out of place, and further realizing that his own treasury was seriously depleted, and yet being at heart a kind man and a good father, he caused to be published far and wide the following proclamation:

> *Attention Princes:* I, King Osbert, offer half of my lands, and the hand of my beautiful daughter Sally, to that Prince who will correctly predict the number of white balls there will be among four drawn from an urn of black and white balls. As his stake, the applicant must put up properties equivalent to two standardized castles (with moats and air conditioning). A candidate whose guess is close, within one ball, will have half of his stake returned to him.

A certain Prince, yclept Hogarth, whose scholarship was admired far and wide, decided to apply for Sally's hand. Examining the size of the

King's holdings, the fruitfulness of his vineyards, taking account of the diminishing marginal utility of dominions, and, finally unromantically throwing the fair Sally into his scale of utilities, he concluded that the utility of winning the bet was ten times as great as the utility of a single standardized castle. He went to the King to learn the details of his proposal.

The King agreed that the drawings would be conducted under circumstances insuring the strictest fairness to both parties. He further stipulated that the urn would be one of four kinds: it would either contain 20% white balls (hypothesis H_1), or half white balls (H_2), or 90% white balls (H_3), or finally, 10% white balls (H_4). Furthermore, the King, having taken rather a fancy to Hogarth, agreed that the Prince should be allowed to take a sample of two balls (with replacement) from the urn and to make his guess on the basis of that sample.

Hogarth was nobody's fool. Before agreeing to the terms of the contract, he repaired to his own country to take the matter up with the scholars resident there. At the Center for Rational Decision, he was told that he had come to the right place. The center provided him with an exhaustive analysis of his possible strategies. Five actions are open to him: to predict that 0, 1, 2, 3, or 4, of the balls will be white; let us denote these by A_0, A_1, A_2, A_3, and A_4 respectively. Given that what is of evidential import in the sample is simply thc number of white balls in the sample, a pure decision function is simply an ordered triple of these actions, the first to be taken if there are 0 white balls in the sample, the second if there is 1 white ball, the third if there are two white balls. The expectation yielded by each decision function, under each hypothesis regarding the contents of the urn can be computed. For example the decision function $\langle A_0, A_1, A_3\rangle$ has the expectation, given that H_1 is true, 3.129. This may be seen as follows:

Assuming that H_1 is true, the probabilities of obtaining 0, 1, 2, 3, or 4 white balls ar: 0.4096, 0.4096, 0.1536, 0.0256, and 0.0016, respectively. The expectation of A_0, in terms of standardized castles is thus

$$10 \times 0.4096 - 1 \times 0.4096 - 2 \times 0.1808 = 3.3248\,.$$

The expectations of A_1, A_2, A_3 and A_4 under the same hypothesis are 3.4784, 0.2784, -1.5376, and -1.9552.

If H_1 is true, we will obtain 0. 1, and 2 white balls in our sample with probabilities 0.64, 0.32, and 0.04. Thus the expectation of following the decision function $\langle A_0, A_1, A_3 \rangle$ is $0.64 \times 3.3248 + 0.32 \times 3.4784 - 0.04 \times \times 1.5376 = 3.129$. The expectation of the same decision function if H_2 is true is 0.828, if H_3 is true is 1.421, and if H_4 is true, is 4.868.

Having computed these four expectations for each of the pure decision functions, we may apply the minimax principle: for each strategy, we note the minimum of the four conditional expectations; we choose that pure strategy for which that minimal expectation is the greatest. This turns out to be: $\langle A_1, A_1, A_4 \rangle$ with expectations under H_1, H_2, H_3, and H_4 of: 3.259, 1.828, 4.127 and 2.162, respectively.

Furthermore, by employing a mixed strategy we can raise the minimum expectation well above 1.828. At this point the Prince began to look skeptical. "Since every other pure strategy has a minimum expectation under some hypothesis or other of *less* than 1.828 standardized castles, and since however many packs of cards I cut, in the one and only instance in which I make my prediction I will be in fact employing a pure strategy, how can you claim that my minimum expectation under a mixed strategy can be greater than 1.828?"

"You see, Sire," the vice-president began, "in the long run...."

The Prince shook his head, puzzled. "I only get to make one prediction, to follow one decision function. What possible relevance can the procedures I don't follow have? I understand how the long run properties of the given procedure I do follow are relevant, but I don't see how the long run properties of a mixed procedure can be relevant."

Although he failed to see the point of using a mixed strategy, the Prince was impressed by the analysis, and gratified to learn that the minimax strategy guaranteed him a minimum expected gain of 1.828 standardized castles. It would indeed be worth while to take the King up on his offer, employing the strategy $\langle A_1, A_1, A_4 \rangle$.

Still slightly perturbed by recommendation to follow a mixed strategy, however, he decided to check the analysis by consulting the Business Institute. The head of the Business Institute took him through the same analysis of each strategy. But when the table of expected gains under each hypothesis of each strategy was completed, the head of the Business Institute continued, "Now that you have decided to gamble with the King, the appropriate strategy is the minimax regret strategy. What counts is obviously not the minimum expected gain under the various hypotheses, but the difference between that and the maximum expected gain under each hypothesis. Thus if H_4 happens to be true, the maximum expected gain is 6.1648, while if H_2 is true, the maximum expected gain is 3.000. You should select that strategy which costs you as little of the maximum expected gain as possible. Since the maximum expected gain under H_1 is 3.4784, under H_2 is 3.000, under H_3 and H_4 is 6.1648, we subtract these members from the expectations, under each hypothesis, to get the expected regrets under each hypothesis. For example, the regrets under the four hypotheses, in order, for the decision function $\langle A_0, A_1, A_3\rangle$ are 0.349, 2.172, 4.744, and 1.297. (We have made all the signs positive, since we are now talking about expected regrets rather than expected gains.) We now look for that strategy for which the maximum regret is least. This turns out to be $\langle A_0, A_2, A_4\rangle$: with regrets 1.387, 2.000, 1.897, 1.897 under the four hypotheses. The corresponding expectations are 2.091, 1.000, 4.268, and 4.268. Under H_2 the expectation is less than the expectation of 1.828 offered by the previous strategy, but under H_2 the maximum expectation is only 3.000 anyway, so we aren't losing much even in that case."

Like the vice-president of the Center for Rational Decision, the Head of the Business Institute started to talk about mixed strategies. This time the Prince didn't even bother to listen; he'd been through that before. And now he really was puzzled, for the advice he had received from his two consultants was flatly inconsistent: should there be no white balls in his sample he should predict one out of four according to his first source; and should predict 0 out of four according to his second source; should there be one white ball in his sample he should predict one out of four according to his first source and two out of four according to his second source.

He therefore arranged to consult the mathematics expert of the national chapter of the Mafia. It was difficult to arrange the interview, but the advice was elegantly simple. "Look, Hogarth, you aren't going to get any information out of looking at a couple of balls. Forget that. Look at this from the King's point of view: if he can get something out of you, he's going to do it. Now he doesn't know what strategy you're going to use, does he? So he wants to protect himself, right? O.K., he gets to pick the urn, doesn't he? Now if he picks H_3 or H_4, and you happen to use $\langle A_0, A_0, A_0\rangle$ or $\langle A_4, A_4, A_4\rangle$ his expected loss is 6.1648. Who needs it? What he's got to do to cut down your gain as much as possible is to choose

H_2: the maximum expected loss he suffers then is 3.000. And you can make sure that this is his loss by playing $\langle A_2, A_2, A_2 \rangle$. No problem at all."

The Prince, still more baffled, left thoughtfully. He wasn't at all sure that the King was out to get him. Why should he have offered to make the bargain at all, if that were the case, since the Prince could be assured of having a positive expectation in any number of ways. On the other hand....

Well, there was one more group well-known for the effectiveness of its advice regarding business decisions: the Council for Coherent Credibilities. The Prince straightened his crown and knocked on their door. This interview turned out to be lengthy and difficult. The Prince had to answer batteries of questions for four hours, absurd questions such as whether he would rather have a dish of chocolate icecream if a coin landed heads three times running, or a piece of pecan pie if the King chose the urn with 50% white balls. Finally, after all the questioning, he was told to come back the next day.

Tired but game the Prince arrived early the next morning. The chief greeted him enthusiastically. "We are fortunate indeed to have such a man as you as our leader. We have subjected your choices to a careful analysis, and we find that you have an incredibly high coherence coefficient. Furthermore, your insight into your utility function is splendid; just as you claimed when you came here yesterday, you do indeed regard winning the bet with the King as worth ten times as much as a standardized castle. We are proud to be your loyal subjects."

Somewhat astonished, the Prince replied, "Thank you very much, but I really hoped you'd have some advice for me."

"Tut tut. You hardly need advice. We can do a little computation for you if you wish, but your strategy is implicit in the answers you gave to our questions yesterday."

"Well, please, then, refresh my memory?"

"From the answers you gave yesterday, it is clear that you are quite confident that the King wishes to have you for his son-in-law, and that in asking you to guess the number of *white* balls he was giving you a strong clue as to which sort of bag he would use. In fact, your present degree of belief that he would use an urn of the first sort is 0.01, and similarly your degree of belief that he would use an urn containing 90% black balls is 0.01. You think there is a real possibility that he would use an urn containing 50% white balls, since this would make your job of predicting as difficult as possible; the probability you attach to that alternative is 0.18. And you are really quite confident that he wants you to win, and that the word 'white' provides an important clue, so you have a degree of belief equal to 0.80 in the hypothesis that he will use an urn of the third sort containing 90% white balls.

"Now it is a simple matter to compute how the sample you will be allowed to draw will affect your beliefs, since you regard the draws as fair. And we can therefore compute the expected utility of any strategy in a perfectly straightforward way as follows: The utility of $\langle A_1, A_2, A_4 \rangle$ is the sum of the utility of A_1 given that 0 white balls appear in the sample, times the probability that 0 white balls will appear in the sample, the utility of A_2 given that 1 white ball appears in the sample, times the probability that 1 white ball will appear in the sample, and the utility of A_4 given that 2 white balls appear in the sample, times the probability that 2 white balls will appear in the sample. The utility of A_1 given that no white balls appear in the sample is of course just the utility of A_1 given H_1, times the probability of H_1 plus the utility of A_1 given H_2 times the probability of H_2, and so on. The probability of H_1 in turn is just the prior probability of H_1 (your present degree of belief in H_1) multiplied by the probability that 0 white balls appear in the sample, divided by the sum over all the hypotheses that exactly 0 white balls will appear in the sample. All this is just a straightforward application of Bayes theorem:

"Expected utility of A_1 under $H_1 \cdot$ probability of H_1 given 0 white balls in sample $= 3.4784 \times 0.01 \times 0.64 \times 1/(\sum P(e=0 \mid H_i)\, P(H_i)) = 0.330$.

"Expected utility of A_1, given 0 white balls in sample $= 1.33$.

"Expected utility of $\langle A_1, A_2, A_4 \rangle =$ probability of 0 white balls in sample $\times\, 1.33 +$ probability of 1 white ball $\times\, 0.448 +$ probability of 2 white balls $\times\, 5.82 = 4.23$.

"There is only one policy for you to follow that is consistent with your degrees of belief and your utility scale: it is that policy whose expectation is a maximum, namely, $\langle A_2, A_2, A_4 \rangle$. The expectation of this strategy is 4.936."

The Prince was momentarily elated, not only at getting such a clearcut answer, but at the high expectation of the recommended strategy. But then a thought occurred to him: "Suppose I had had different beliefs – I no longer feel so confident today as I did yesterday that the King will try to make things easy for me – would that make any difference?"

The Chief looked shocked. "Well, of course it would. If you have changed your prior beliefs, then we must do the analysis over again. Are you sure your beliefs have changed?"

Crestfallen, the Prince looked at the floor. "No, never mind", he said. "I was hoping to find something more objective than my own hunches."

"Aha", exclaimed the Chief. "That is just wishful thinking. Whatever policy you adopt, it will be based in one way or another on subjective assessments of one sort or another." The chief started to explain, but the Prince sadly turned on his heel and left.

As he left, the building, however, which was in the same town as a famous university, he happened to overhear some students speaking of Professor Rudolf's objective approach to statistical inference. He asked them for some more information about Rudolf, and was told that Professor Rudolf was a world-famous inductive logician, and that although his system could not yet handle complicated problems of statistical inference, it was well enough worked out to handle simple problems. "Is it objective?" the Prince asked cautiously.

There was a chorus of exclamations. "Of course, of course, as objective as deductive logic." The Prince thereupon resolved to visit Professor Rudolf.

Professor Rudolf received him kindly, and agreed wholeheartedly with the Prince's low opinion of methods of the Council for Coherent Credibility. The Prince was in luck, he announced: although his system was still in its elementary stages, yet it was quite capable of handling the Prince's problem. What was required was only an objective way of assessing the prior probabilities of the four relevant hypotheses. "Let us assume," the professor said, "that there are ten balls in each urn – that will suffice for the ratios involved. Let us consider all the possible compositions of such an urn: the first ball may be black, or may be white; the second ball may be black or may be white; and so on. We can thus describe the situation in a language L_1^{10} with one predicate, and ten individual constants." The Prince nodded, impressed. "There are thus 2^{10} possible compositions for the urn," the professor went on. "We assign equal probabilities, *a priori* to each; the proportion among the probabilities assigned to H_1, H_2, H_3 and H_4, given that all the other possibilities are ruled out, will be the same as their *a priori* probabilities, namely

0.142:0.795:0.032:0.032. These then are the logically valid prior probabilities that we use in the Bayes computation. It is then easy to see that the strategy with the highest mathematical expectation is $\langle A_1, A_2, A_2 \rangle$; its expectation is 2.398."

The Prince was rather dissappointed, since the expectation cited by the professor was so much less than that cited by the Chief of the Council for Coherent Credibilities, and somewhat confused, since this was the fifth strategy recommended to him. He decided, however, to put the whole matter before one final body: the world-reknown Count Jacob von Carp. It was his hope, of course, that Count Jacob would side with one of his previous advisors. His hope was in vain, however. No sooner had he laid his problem before the Count, than the Count got red in the face and exploded: "Professor Rudolf! That faker! He is so stupid! It is obvious that assigning equal weights to state descriptions precludes the very possibility of learning from experience." The only rational way to proceed, according to the Count, was to assign equal weights to structure descriptions – that is, to regard every hypothesis concerning the *proportion* of white balls in the urn as equally probable *a priori*. Thus, given that only four structures were possible, each should be quite simply assigned a prior probability of $\frac{1}{4}$. From this assignment, it followed that the preferred strategy was $\langle A_0, A_1, A_4 \rangle$, and that the mathematical expectation of this strategy was 3.325.

What is our poor Prince to do? He has solemnly been assured that each of six distinct strategies is the correct one for him to follow. Whatever he observes in his sample, there are at least two (and usually three) distinct predictions that he is advised to make. No one has argued with his assignment of utilities. The logical structure of the problem is what it is – no one has argued about that. The Prince has the beliefs that he has, and no others. He is in a certain situation defined by well-described utility parameters, conditional probabilities (about many of which there is no dispute), doxastic parameters (representing his beliefs),... there is no difficulty about supposing him to be in this horribly difficult practical situation. No one has argued that he should change his beliefs, or should change his utility assignment, or should reject the hypothetical fairness of the draws from the urn. What is in dispute is nothing less than the fundamental logic of statistical decision making. All of his advisors agree that he should take the King's offer; everybody agrees that his expectation is positive. But no two of his advisors agree on what his strategy should be. His first two advisors agree on the interpretation of probability (that it is empirical), but not on how to apply it. His third advisor regards probability as irrelevant. His fourth advisor offers an analysis dependent on another (the subjective) interpretation of probability. His fifth and sixth advisors offer analyses dependent on yet another interpretation of probability (the logical), but again disagree between themselves on its specific application. Here is poor Hogarth, in a definite statistical situation, almost classical in its simplicity; and each of the alternative approaches to statistical decisions leads to a different recommendation. The problem is the same, the evidence is the same, the values involved are the same, the hypotheses have only the given logical structure that they have; the opinions of Hogarth are only opinions he actually has; and yet according to the *purely abstract* interpretation of probability and abstract principles of statistical choice he adopts there will be six contrary courses of action recommended to him.

Notice that the fact that this happens to be a problem of practical decision is accidental. Given the scientific choice between four scientific hypotheses corresponding to class ratios considered, it would not be at all implausible to suppose that the epistemic importance of errors was proportional to what I have taken to be Prince Hogarth's utilities. (Thus: win 10 points if you guess the right hypothesis; lose 1 if you are a bit off; lose 2 if you are far off.) So the problem translates directly into a problem concerning the choice between scientific statistical hypotheses, if we suppose that the concept of epistemic utility makes sense, and that making an inference is deciding what to believe. Furthermore, the same difficulties arise when we attempt to deal with a problem of this nature in purely inferential terms: given the evidence, the hypothesis most probable (considered subjectively) may be H_1; the most probable according to one version of the logical interpretation of probability may be H_2; according to another logical interpretation it may be H_4; the hypothesis whose likelihood is a maximum may be H_3.

3. The various approaches to statistical inference are thus in conflict, and in many respects the conflict seems to derive from a difference in the interpretation of the underlying concept of probability. Statisticians offer two ways of accounting for this conflict.

First it is claimed that the conflict is only apparent. In one situation, one technique may be appropriate, in another situation, another technique. Thus in a *given* situation, the methods need never conflict. This pious answer cannot strike us as very plausible right now: which is the correct technique for the Prince to apply to his problem? It is not all obvious. But those who adopt this line do not maintain that the answer is always obvious. Indeed statisticians sometimes adopt a mystique analogous to that of physicians: only vast experience with a wide variety of statistical problems can lead a statistician to develop a *feel* for what technique is appropriate to what situation. And even then, it is only the man with a mysterious native talent who will become good at knowing what approach is appropriate at what juncture.

There are several things to be said about this claim. The first is that even among statisticians with good reputations and impeccable credentials, there are many instances in which there is no agreement about what the appropriate techniques are. As we have noted earlier, there are those who regard randomized decision functions as essential to good statistics (anti-Bayesians) and those who regard randomized decision functions as always irrelevant to informative inference (logical Bayesians). There are those who regard randomization in the design of experiments as absolutely crucial (Neyman-Pearson), and those who regard it as always inessential and only sometimes of psychological usefulness (subjective Bayesians). There are those who regard an experiment which is terminated at the option of the experimenter as demanding a different analysis from one of preselected size (Neyman-Pearson), and those who don't (Bayesians). None of this shows that it is false that the different techniques are appropriate to the several different kinds of statistical problems, but it does suggest that some kind of theory is called for which will characterize these different sorts of problems and explain which techniques are appropriate to which problems and why. No such theory exists, and in view of the partisan commitment of most statisticians to one view of probability or another, it is difficult to see where a unifying theory can come from. There are statisticians whose view of probability is eclectic, but this merely shows that it is possible for people, even mathematicians, to hold mutually inconsistent theses. There is little prospect of a unifying theory, capable of compelling universal assent, developing within any of the existing schools of statistical inference.

A second way of accounting for the apparent conflict of statistical advice is to draw an analogy between the science of statistics and other sciences. It has often been the case that in historically active and fruitful periods in the history of a science, a number of mutually inconsistent hypotheses have been floating around: for example, the wave and particle theories of light existed simultaneously, and were applied to different problems. According to some statisticians, this is a healthy and creative state of affairs. Sure, we have inconsistencies in the body of statistical lore now; but in time that body of lore will be honed by experience, and become simultaneously richer and more coherent. Just as in the case of other theories, statistical theories will be modified in response to experience. Techniques that work, or that work in certain sorts of situations, will be retained, or will be applied only to those sorts of situations, and other techniques will be eliminated from the corpus of statistical theory, or will be restricted in scope.

To object to such a way of accounting for the conflict of statistical theories, if not as bad as objecting to apple pie, seems to be to object to the scientific method itself, to the very techniques that have won for us so much knowledge and understanding of our world. But objection is nonetheless called for. It seems innocent enough to say that we should employ those statistical techniques that work, that it is all to the good if we have on hand a number of different (inconsistent) statistical techniques, because in due course we shall eliminate those that don't work. But it is not as innocent as it seems.

The difficulty with this idyllic picture of the progress of statistical science hinges on the ambiguity of the word 'work'. Let us suppose (what is no doubt contrary to fact) that we have a reasonably clear idea of what it is for a statistical technique to work in a given instance. Quite clearly it is not in the nature of a statistical technique to work all the time, or to work in the same degree all the time. It is conceivable, however, that we could collect a number of instances i of the application of a statistical technique, and associate with each instance a number $D(i)$ representing the degree to which that technique worked in that instance. This provides a picture of how well that technique has worked in the past. But when we talk about how well a statistical technique works, we are concerned not merely with how well it has worked in instances of its application in the past, but how well it works *in general*: that is, what concerns us is not the sample of values of the random quantity D that we have happened to observe in the past, but the general distribution of D. How do we arrive at knowledge of this distribution? Why, by statistical inference, of course!

Of course there are some notions of 'work' that render the working of certain statistical techniques tautologous. If to 'work' is to yield predictable long-run frequencies of error of the first kind, then classical testing procedures work. But this is not the only kind of 'working' that has interested people.

A number of philosophers (Braithwaite [2], Black [3]) have argued there is no vicious circularity in the inductive justification of inductive rules. These arguments, however, have not struck most philosophers as plausible. It may well be that the crystal ball tells me that consulting the crystal ball is a good way of foretelling the future; that does not increase my confidence in prophecy.

To use Wesley Salmon's example [4]: consider the inductive policy of predicting for the future that which has been rare in the past. This counter-inductive policy has rarely been successful in the past. Therefore, according to the policy, we should predict that future applications of the counter-inductive policy will be successful. This argument does not, however, make the counter-inductive policy attractive.

Furthermore, the arguments underlying most approaches to statistical inference are not of this character: it is not argued that such and such an approach has proved fruitful in the past and therefore will prove fruitful in the future. It is argued that only that approach is logically acceptable. The argument is based on a prior commitment to one or another interpretation of probability. The argument also provides an explanation of the success of alternative approaches, as we saw in the first part of this chapter. Thus although some statisticians claim that a method should be judged by its fruits, those who argue for one approach against another do not argue by results, but by fundamental principles.

This supports the intuition that what is at issue in conflicts among approaches to statistics is a profound and pervasive conception of the nature of statistical inference and the nature of probability. The fundamental principles called upon by statistical writers on the foundations of statistics are defended as perfectly clear and self-evident. We are offered what appear to be ultimate principles – principles so basic that it is difficult to see how to defend them or argue about them. We are dealing here with our fundamental intuitions concerning the principles of probabilistic argument.

That these principles lead to approaches to statistics that are in conflict may mean that they are not clearly understood, so that it is difficult to be quite sure where they lead, or it may mean that the principles themselves are in conflict. Alan Birnbaum [5] has offered convincing arguments that in certain respects the latter is the case. If that is so we must plunge yet deeper into the foundations of our topic, in order to resolve the conflict among these allegedly fundamental principles.

Birnbaum singles out three fundamental statistical principles that may be regarded as principles of evidential import underlying much of statistical inference. These are the principle of conditionality, the principle of sufficiency, and the likelihood principle. These are to be construed as principles of statistical evidence, applicable to problems of informative inference. Nevertheless, they have an obvious and direct bearing on questions of decision in the face of uncertainty, in choosing between statistical hypotheses, making tests of statistical hypotheses, and the like. He shows that these principles are in conflict with an even more broadly accepted principle to the effect that 'strong evidence against *H*' should not occur with high probability when *H* is the case.

In what follows I shall follow Birnbaum's elegant and elementary exposition. Let E be a model of an experiment, i.e., a matrix $\{p_{ij}\}\ i\in I, j\in J$, where i may have any of I values, each representing a value (possibly a vector) of a statistical parameter, and j may have any of J values, each representing a possible observation in the experiment. A model of statistical evidence is the ordered pair $\langle E, x\rangle$, where x is the observed outcome of the experiment. $\mathrm{Ev}(E, x)$ is the evidential value of the observation x in the experiment E.

The principle of conditionality states that if E is a mixture of two experiments E' and E'', in which we first perform an auxiliary experiment with outcomes 1, and 2, with associated probabilities q_1 and q_2, and then, according as the outcome is 1 or 2, perform experiment E' or E'', the outcome of the auxiliary experiment is irrelevant to the evidential import of the experiment we actually perform: $\mathrm{Ev}(E, \langle E', x\rangle)=\mathrm{Ev}(E', x)$ and $\mathrm{Ev}(E, \langle E'', y\rangle)=\mathrm{Ev}(E'', y)$. Although this principle is not universally accepted in theory, it is generally accepted in practice. At any rate, we require for our argument only the special case in which $q_1 = q_2 = \frac{1}{2}$; we might suppose that, having limited time, we simply flip a coin to decide whether to perform E' or E''. In this case, surely, the principle is fairly uncontroversial.

The principle of sufficiency is almost universally regarded as acceptable. Birnbaum's formulation is not the usual one, but it is equivalent to the usual one, and is particularly appropriate to the problem at hand. Suppose that the probability of the outcome j is proportional to the probability of the outcome j', with the same factor of proportionality, for all possible values of i, i.e., under each of the possible hypotheses. Then the information that j gives regarding i is no different from that j' gives regarding i. We might just as well then lump j and j' together as one case (whose probability will then be the sum of the probabilities of j and of j' under each hypothesis, or for each value of i). For example, if the experiment consists of tossing a bent coin twice and noting the results, and the original matrix is (assuming a binomial distribution for the outcomes):

	HH	HT	TH	TT
$q=0.2$	0.04	0.16	0.16	0.64
$q=0.3$	0.09	0.21	0.21	0.49.

An evidentially equivalent model for the experiment is:

	HH	$HT \vee TH$	TT
$q=0.2$	0.04	0.32	0.64
$q=0.3$	0.09	0.42	0.49.

Birnbaum's statement of the principle of sufficiency is:[6]

If $E = p_{ij}$ is determined from $E' = p'_{ij}$ by any function $j = j(j')$ which takes distinct values on any two values of j' which do not label proportional columns of E', then for each $\langle E', j' \rangle$ we have $\mathrm{Ev}(E', j') = \mathrm{Ev}(E, j)$.

The likelihood function has already been defined.[7] In this context it is simply the set of values kp_{ij} corresponding to a fixed value of j, k being an arbitrary constant. The likelihood principle states simply that the entire evidential import of an experiment and outcome is embodied in the likelihood function. More formally, if two models of statistical evidence $\langle E, j \rangle$ and $\langle E', j' \rangle$ determine the same likelihood function, then they have the same evidential meaning. Or, if $p_{ij} = cp'_{ij'}$ for every i, then $\mathrm{Ev}(E, j) = \mathrm{Ev}(E', j')$.

This principle has been far more controversial than the preceding two, but Birnbaum has shown that the principles of sufficiency and conditionality strictly imply the likelihood principle. The proof is as follows:

Let $\langle E', j' \rangle$ and $\langle E'', j'' \rangle$ be models of evidence determining the same likelihood function $p'_{ij'} = p''_{ij''}$. To say that these functions are the same is to say that they have the same domains, and thus that the two parameter spaces are the same: $i \in \Omega'$ if and only if $i \in \Omega''$. We can thus represent the likelihood functions by $p'_{ij'} = cp''_{ij''}$, for all $i \in \Omega'$.

Now apply the principle of conditionality to a 50-50 mixture of E' and E'', E:

$$\mathrm{Ev}(E', j') = \mathrm{Ev}(E, \langle E', j' \rangle)$$
$$\mathrm{Ev}(E'', j'') = \mathrm{Ev}(E, \langle E'', j'' \rangle).$$

The likelihood functions determined by $\langle E, \langle E', j \rangle \rangle$ and $\langle E, \langle E'', j \rangle \rangle$ are $\frac{1}{2} p'_{ij}$, and $\frac{1}{2} cp''_{ij}$ respectively, for all $i \in \Omega'$. Since these are proportional columns of E we may combine them under the new label j_0, which will have the likelihood function $p_{ij_0} = \frac{1}{2} p'_{ij'} + \frac{1}{2} cp''_{ij''}$. According to the principle of sufficiency, then, $\mathrm{Ev}(E, j_0) = \mathrm{Ev}(E, \langle E', j' \rangle)$ and $\mathrm{Ev}(E, j_0) = \mathrm{Ev}(E, \langle E'', j'' \rangle)$. But applying the principle of conditionality again, we have $\mathrm{Ev}(E'', j'') = \mathrm{Ev}(E, j')$, which is just the consequent of the statement of the likelihood principle.

The *likelihood concept* of statistical evidence carries the likelihood principle one step further: it stipulates that the whole of the evidential import of an item of statistical evidence is embodied in the likelihood function.

The even more pervasive concept of statistical evidence which is in conflict with the likelihood concept is what Birnbaum calls the confidence concept. As he realizes, this concept is not part of the formal theory of confidence region estimation; nevertheless, it is a pervasive and implicit principle which underlies the statistical practice based on the confidence region theory. This is the principle that the long-run properties of an estimator (say) are what give it its evidential status. A 99% confidence region has evidential import based on the fact that 99 times out of a hundred such a region will not fail to include the parameter in question. The interpretation of statistical evidence, in other words, must be firmly supported by known low frequencies of error. In 'The Anomalous Concept of Statistical Evidence',[8] Birnbaum expresses this as a Criterion of Unbiasedness.

Systematically misleading or inappropriate interpretations shall be impossible; that is, under no θ shall there be high probability of outcomes interpretated as strong evidence against θ.

Surely this is a plausible principle; Birnbaum himself finds it even more persuasive than the likelihood principle – he refers to it as the "one rock in a shifting scene".[9]

It is the incompatibility between the confidence concept of evidence and the likelihood concept of evidence that Birnbaum finds most unsettling. He writes "... no precise adequate concept of statistical evidence exists – and *none can exist* ...".[10]

The incompatibility between likelihood and confidence can be exhibited in a simple discrete instance, due to Birnbaum.[11] (This artificial problem is derived from a consideration of problems concerning normal distributions that have been of relatively practical concern to statisticians.) Let $\theta=(\mu, \sigma)$, where σ may have the value 0 or the value 100, representing the precision of a measurement, and μ may have any integer value between -10^{10} and 10^{10}, representing the value of a quantity to be measured. We thus have 4×10^{10} states of nature, each determining a probability distribution over the sample space of 2×10^{10} points. This function is:

$$\begin{aligned} f(x,\theta)=f(x,\mu,0)&=1, \quad \text{if} \quad x=\mu \\ f(x,\mu,0)&=0, \quad \text{if} \quad x\neq\mu \\ f(x,\mu,100)&=c[100-|x-\mu|] \quad \text{if} \quad |x-\mu|<100 \\ f(x,\mu,100)&=0 \quad \text{if} \quad |x-\mu|\geq100 \end{aligned}$$

where $c=1/10040$.

If x is an observation in the interval $(10^{-8}, 10^{8})$, the likelihood function of x is:

$$\begin{aligned} f(x,\mu,\sigma)&=1 \quad \text{for} \quad \mu=x \quad \text{and} \quad \sigma=0 \\ &=100c\doteq0.01 \quad \text{for} \quad \mu=x \text{ and } \sigma=100. \\ &=100c\doteq0.01 \quad \text{for} \quad \mu\neq x \quad \text{and} \quad \sigma=100. \end{aligned}$$

Since the likelihood ratio of the point $(x, 0)$ against any other point is at least 100:1 this seems to be strong evidence in favor of this point against any other, and in particular strong evidence in favor of $\sigma=0$ against $\sigma=100$. In other words, if the likelihood function embodies the evidential import of the experiment, its message seems clear and unambiguous.

But this hardly seems plausible: how can one measurement provide us with *any* evidence concerning the precision of the technique by which it was obtained? Furthermore, the likelihood interpretation of this experiment is flatly in contradiction to the condition of unbiasedness: if $\theta=(0, 100)$ is the true state of nature, the experiment will be misleading under the likelihood interpretation, with probability unity. That is just to say that the confidence concept of statistical evidence is flatly in contradiction to the likelihood concept of evidence.

Birnbaum also argues[12] that even the weaker likelihood *principle* is inconsistent with the confidence concept of evidence. Since the principles of sufficiency and conditionality, which seem less arguable than the likelihood principle, imply the likelihood principle, they too are incompatible with the confidence concept, and the plausibility of the confidence concept is severely undermined. The argument consists essentially in the observation that the likelihood function for an experiment in which a bent coin is tossed fifty times and forty heads are observed, and the likelihood function for an experiment in which the bent coin is tossed until forty heads are observed (and happens to have been tossed fifty times) are the same. But the evidential meaning of the two experiments, according to the confidence concept, is not at all the same.

As we showed in the chapter on hypothesis testing, however, the confidence concept itself is not without difficulties. The legitimacy of compounding utterly unrelated experiments, for example, to improve the performance characteristics of one's tests, is implied by the confidence concept, and strikes many as a *reductio ad absurdum* of that approach.

In any event, it is clear that there are basic issues of profound theoretical importance, and indeed of practical importance as well, that cannot be ignored. While it is true that in many practical applications of statistical theory these issues have no bearing on the final decision made, or the hypothesis finally accepted, it is also true that in other applications the fundamental principles turn out to be important. While it is true that within the framework of each approach the 'legitimate' applications of other approaches find justification, there are the applications regarded within one framework or another as illegitimate to be accounted for. Finally, the issues are so basic as to take on the appearance of logical, rather than of empirical questions or even vague and general methodological questions. In what follows, the attempt will be made to provide a framework in which these issues may be resolved. The framework is a logical and formal one, for it will be claimed that the issues are logical ones. It will be based on a logical interpretation of probability, for the same reason. It will be claimed that this logical interpretation of probability will provide unambigious and unique criteria for statistical evidence, and will resolve the conflicts adumbrated at length here.

NOTES

[1] H. E. Kyburg, 'Probability and Decision', *Philosophy of Science* **33** (1966), 250–261.
[2] R. B. Braitwaite, *Scientific Explanation*, Cambridge 1953.
[3] Max Black, 'Self-Supporting Inductive Arguments', *Journal of Philosophy* **55** (1958).
[4] Wesley Salmon, *The Foundations of Scientific Inference*, Pittsburgh 1967.
[5] Allan Birnbaum, 'Concepts of Statistical Evidence', in *Philosophy, Science and Method* (ed. by Morgenbesser *et al.*), New York 1969, pp. 112–143.
[6] *Ibid.*, p. 120.
[7] Chapter IV.
[8] Hectograph.
[9] 'Letters to the Editor', *Nature* **225** (1970), 1033.
[10] 'The Anomalous Concept of Statistical Evidence', p. 15.
[11] 'Concepts of Statistical Evidence', p. 128.
[12] *Ibid.*, p. 127.

CHAPTER SEVEN

THE LANGUAGE: SYNTAX

1. The concept of probability to be defined here and which will lead to a resolution of the conflicts and confusions adumbrated in the last chapter will be a logical one. Most logical interpretations of probability, Carnap's, for example, have been offered only for very simple languages. These interpretations are therefore inappropriate for languages rich enough for the expression of ordinary statistical theories and hypotheses. The claim has been, of course, that we must start somewhere, and that it is easier to start with something simple rather than with something complicated. But this is not always the case. It would be unreasonable, for example, to attempt to provide a theory of analysis exclusively in terms of integers on the grounds that integers are simpler than real numbers or sets. Somewhat analogously, I would claim that we cannot even define probability *at all* except in languages rich enough for the expression of statistical hypotheses, just as we cannot define a derivative except in a language rich enough to talk about real numbers. Statistical hypotheses come first; probabilities come after; and statistical inference is a matter both of statistical hypotheses and of probabilities.

Probability, like provability, to which it is so closely related in typography, is going to be taken as a metalinguistic concept. In particular, it will be understood as a function whose domain is a set of two-sequences, the first term of which is a body of knowledge, and the second term of which is a statement of a given language. The body of knowledge, in turn, is construed as a set of statements of the same language. Roughly speaking, we shall say that a statement has the probability p (in general p will be an interval i of reals), relative to a body of knowledge, when (a) it is known in that body of knowledge that the statement is equivalent to (has the same truth value as) a statement of the form $\ulcorner a \in b \urcorner$; (b) it is known in that body of knowledge that a belongs to c; (c) that body of knowledge contains the statistical knowledge that the proportion of objects in c that belong to b falls in the interval i; and (d) there is nothing in our body of knowledge that conflicts with this assignment of probability. This last condition will be explicated as a condition of randomness, relative to a body of knowledge; randomness in turn will be construed as a relation between a set of statements and three terms, all of the given language. (Relative to the set of statements S, a is a random member of c with respect to belonging to b.)

Since probability and randomness are going to be taken as metalinguistic concepts, and since questions of statistical knowledge, statistical inference, acceptability, evidence, and the like, are going to hinge on relations among statements and sets of statements, our first job is to establish a relatively formal metalinguistic framework.

For the metalanguage we shall adopt Zermelo-Fraenkel set theory, as axiomatized in the Appendix, together with the supplementary assertion – to which we shall hardly ever need to refer again – that there exists at least one strongly inaccessible cardinal. This assertion allows us to define models of Zermelo-Fraenkel set theory in our metalanguage.

The object languages will also employ the logical framework of the Appendix; and they will also include Zermelo-Fraenkel set theory with individuals. Punctuation, logical notation, and set theoretical notation will thus be used in three ways (but not three ways at once!). These symbols will occur in the object language, having their usual logical or set-theoretical meanings. They will be used also in the metalanguage in two ways: first as part of the logical and set-theoretical machinery of the metalanguage, and second as names of the same signs as they occur in the object language. Thus '$\sim$' is one of the primitive signs of the object language; '$\sim$' is one of the primitive signs of the metalanguage; as a sign in the object language '$\sim$' is a one-place sentential connective (expressing negation); as a sign in the metalanguage '$\sim$' is ambiguous, functioning in some contexts as a one-place sentential connective (expressing negation), and in other contexts as a term, denoting a certain expression of the object language, namely, '$\sim$'. These latter contexts will be easily identifiable by the fact that the signs in question occur in an expression enclosed in Quine's corners. Thus all of the signs in '$\ulcorner 2=3 \vee 3=3 \urcorner$' are to be construed as names in the metalanguage of the corresponding expressions ('2', '3', '$=$', and '$\vee$') in the object language. In '$\ulcorner 3 \in 4 \urcorner \in 5$', the first occurrence of '$\in$' is a term denoting the expression '$\in$' in the object language; the second occurrence of '$\in$' is as a relation expression of the metalanguage. The whole expresses the false assertion that a certain (true) expression of the object language is a member of a certain object, the number 5.

We cannot, of course, use variables in a similarly ambiguous manner; and although we could treat the constants of our object languages (other than '∈') in the same way, there is no great economy in doing so. Thus we shall suppose available in the metalanguage a denumerable list of names for variables, and a finite list of names for predicates and operation expressions of the object language.

Definitions (and theorems) in our set theory appropriate to both the object language and the metalanguage will be numbered in sequence following the enumeration of the Appendix. Definitions that are applicable only to the metalanguage will be prefixed by 'MD' rather than 'D' and will be numbered as a new sequence.

The typical object language contains (in addition to logical and set-theoretical notation) a denumerable list of variables,

$$x, y, z, \ldots$$

Their names in the metalanguage are to be:

$$v_0, v_1, v_2, \ldots$$

And the set of variables is thus the set of one-sequences

MD-7.1 $V = \{\langle x\rangle : \bigvee_{n\in\omega} x = v_n\}$

We assume that if $i \neq j$ then $v_i \neq v_j$.

We further suppose that for every $i \in \omega$, the object language contains a finite number (perhaps 0) of i-place predicates:

> $P^i_0, P^i_1, \ldots, P^i_n$ (Some or all of these may be replaced by more familiar expressions, such as '<', '∈', 'is red', etc.)

Their names in the metalanguage will be

$$Pr_{i,0}, Pr_{i,1}, \ldots, Pr_{i,n}$$

The set of i-place predicates is thus the set of one-sequences

MD-7.2 $Pr_i = \{\langle x\rangle : \bigvee_{n\in\omega} x = Pr_{i,n}\}$

We assume that if $i \neq j$, then $Pr_{k,j} \neq Pr_{k,i}$, so long as both belong to Pr_k.

If there are only n k-place predicates and both i and j are greater than n, of course $Pr_{k,i}=Pr_{k,j}=0$. The set of all predicates is thus the set of one-sequences

MD-7.3 $Pr=\bigcup_{i\in\omega} Pr_i$

Similarly we suppose that in the object language there are for every $i\in\omega$ a finite or denumerable number of i-place operation symbols:

$O_0^i, O_1^i, \ldots$ (Some or all of these may be replaced by more familiar expressions such as '+', '2', '0', etc.)

These will be named by the terms in the metalanguage,

$Op_{i,0}, Op_{i,1}, \ldots$

The set of i-place operation symbols is thus the set of one-sequences

MD-7.4 $Op_i=\{\langle x\rangle : \bigvee_{n\in\omega} x=Op_{i,n}\}$

and the set of all operation symbols will be the set of one-sequences

MD-7.5 $Op=\bigcup_{i\in\omega} Op_i$

A particular language L will be identified by and with its list of non-logical signs. To say that L is a language is just to say that it is a set of operation and predicate signs:

MD-7.6 L *is language* $\leftrightarrow \Sigma L \wedge \bigwedge_x (x\in L\rightarrow \bigvee_{i,n\in\omega} (x=Op_{i,n} \vee x=Pr_{i,n})$

The languages we shall consider will include both mathematical and empirical operation expressions and predicates. Strictly speaking, we require only two mathematical predicates, 'Σ' for 'is a set', and '$\in$' for 'belongs to', but for convenience we shall employ – as we did in the Appendix – a lot of standard mathematical terminology. The theories in which this terminology occurs may be regarded as definitional extensions of theories in which the only mathematical predicates are the two mentioned.

Each of our object languages will contain the same group of logical constants:

$\sim$ (negation)

$\wedge$ (conjunction)
$\vee$ (disjunction)
$\rightarrow$ (conditional)
$\leftrightarrow$ (biconditional)
ι (description operator)
$\bigwedge$ (universal quantification)
$\bigvee$ (existential quantification)
$=$ (identity)
( ) (parentheses)

The inference rules of the languages are the standard inference rules for first order languages with identity and the description operator. To be specific, we adopt the rules of the system of Kalish and Montague.[1]

The set of logical constants is defined by the forgoing enumeration:

MD-7.7 $LogCon = \{\text{‘}\wedge\text{’}, \text{‘}\vee\text{’}, \text{‘}\rightarrow\text{’}, \text{‘}\leftrightarrow\text{’}, \text{‘ ’}, \text{‘}\bigwedge\text{’}, \text{‘}\bigvee\text{’}, \text{‘}=\text{’}, \text{‘(’}, \text{‘)’}, \text{‘}\sim\text{’}\}$

A sign of a particular language may now be defined as one of the predicates or operation signs of that language, one of the logical constants, or one of the variable terms:

MD-7.8 x *is sign of* $L \leftrightarrow L$ *is language* $\wedge$

$$\wedge (x \in LogCon \vee x \in L \vee \bigvee_{n \in \omega} x = v_n)$$

Sequences are defined in the appendix: a sequence is a set of ordered pairs, where the first member of each pair belongs to some index set, and where, if $\langle x, y \rangle$ and $\langle z, w \rangle$ belong to the same sequence and $x = z$, then $y = w$. A sequence is thus a function whose domain is its index set, and whose range is the set of terms occurring in it. In the case of a finite sequence, the index set is usually a natural number. The concatenation of two finite sequences is defined in the appendix. The concatenation of a finite number of finite sequences can be defined as follows:

MD-7.9 $\underset{i<n}{Concat}\ x_i = \iota k[(n \in \omega \wedge \bigwedge_{i<n} \bigvee_{m \in \omega} x_i \textit{ is m-sequence} \wedge$

$$\wedge \bigvee_{y} (y \textit{ is n-sequence} \wedge y_0 = x_0 \wedge \bigwedge_{0<i<n} y_i =$$

$$= y_{i-1} \frown x_i \wedge y_{n-1} = k)) \vee (\sim (n \in \omega \wedge \bigwedge_{i<n} \bigvee_{m<\omega}$$

$$x_i \textit{ is m-sequence} \wedge \bigvee_{y} (y \textit{ is n-sequence} \wedge y_0 = x_0 \wedge$$

$$\wedge \bigwedge_{0<i<n} y_i = y_{i-1}\frown x_i)) \wedge k=0)]$$

As mentioned before we shall often use Quine's corners as a convenient informal way of expressing concatenations, as well as a signal that any logical constants of the languages occurring within the corners are to be construed as sequences of names of the corresponding constants of the object language.

An expression of L is a finite sequence of signs of L:

MD-7.10 x *is expression of* $L \leftrightarrow \bigvee_{n\in\omega} (x \textit{ is } n\textit{-sequence} \wedge \bigwedge_{i<n} x_i \textit{ is sign of } L)$

To say that the expression y occurs in the expression x, is to say that there are expressions a and b such that x is the concatenation $a\frown y\frown b$. Note that either a or b or both may be the empty expression 0.

MD-7.11 y *occurs in* $x \rightarrow \bigvee_{a,b} x = a\frown y\frown b$

The terms of L comprise the smallest set of expressions of L which includes (1) all the variables, (2) every instance of an n-place operation expression followed by n members of that set, and (3) every instance of a description operator, followed by a variable k, followed by a formula; where a formula is: (a) an identity statement, (b) an n-place predicate followed by n members of the set in question, or (c) something arrived at by constructing sentential combinations and quantifications of these.

MD-7.12 $Tm_L = \bigcap \{x: \bigwedge_{y\in x} y \textit{ is expression of } L \wedge \bigwedge_{y\in V} y\in x \wedge$

$\wedge \bigwedge_{y,z} \bigwedge_{i\in\omega} (y \textit{ is } i\textit{-sequence} \wedge z\in Op_i \wedge \mathscr{R}y \subset x \rightarrow z\frown y \in x) \wedge$

$\wedge \bigwedge_{r,w} (w \in \bigcap \{y: \bigwedge_{x\in y} x \textit{ is expression of } L \wedge$

$\wedge \bigwedge_{z,t} \bigwedge_{i\in\omega} (t \textit{ is } i\textit{-sequence} \wedge z\in Pr_i \wedge \mathscr{R}t \subset x \rightarrow z\frown t \in y) \wedge$

$\wedge \bigwedge_{s,t\in x} \ulcorner(s=t)\urcorner \in y \wedge \bigwedge_{s,t\in y} \ulcorner(s\wedge t)\urcorner, \ulcorner(s\vee t)\urcorner, \ulcorner(s\rightarrow t)\urcorner,$

$\ulcorner(s\leftrightarrow t)\urcorner, \ulcorner\sim s\urcorner \in y \wedge \bigwedge_{s\in y} \bigwedge_{t\in V} \ulcorner\bigwedge ts\urcorner, \ulcorner\bigvee ts\urcorner \in y\} \wedge$

$\wedge r\in V \rightarrow \ulcorner\iota rw\urcorner \in x)\}$

The set of basic formulas of L is easily defined. Note, however, that a basic formula may contain another formula as a part – even a non-basic one – if it is embedded in a descriptive term.

MD-7.13 $BFm_L = \{x: \bigvee_{z,w} \bigvee_{i \in \omega} (z \in Pr_i \wedge z_0 \in L \wedge w \textit{ is i-sequence} \wedge$
$\wedge \mathscr{R}w \subset Tm_L \wedge (x = \ulcorner \sim zw \urcorner \vee x = z\frown w)) \vee$
$\vee \bigvee_{z,w \in Tm_L} (x = \ulcorner (z = w) \urcorner \vee x = \ulcorner \sim (z = w) \urcorner)\}$

The set of formulas of L is then the smallest set of expressions of L that includes all of the basic formulas of L, as well as all of their disjunctions, negations, quantifications, etc.

MD-7.14 $Fm_L = \bigcap \{x: \bigwedge_{y \in x} y \textit{ is expression of } L \wedge BFm_L \subset x \wedge$
$\wedge \bigwedge_{y,z \in x} (\ulcorner \sim y \urcorner, \ulcorner (y \wedge z) \urcorner, \ulcorner (y \rightarrow z) \urcorner, \ulcorner (y \leftrightarrow z) \urcorner,$
$\ulcorner (y \vee z) \urcorner \in x) \wedge \bigwedge_{y \in x} \bigwedge_{z \in V} (\ulcorner \bigwedge zy \urcorner, \ulcorner \bigvee zy \urcorner \in x)\}$

An occurrence of a variable x in an expression z of L may be construed as an initial segment of z that ends just before that variable. A bound occurrence of a variable is an occurrence within a formula that begins with a quantifier or a description operator.

MD-7.15 $y\ Vbl_L\ x \text{ in } z \leftrightarrow z \in Fm_L \cup Tm_L \wedge \bigvee_a y\frown x\frown a = z \wedge x \in V$

MD-7.16 $y\ B_L\ x \text{ in } z \leftrightarrow y\ Vbl_L\ x \text{ in } z \wedge \bigvee_w (w \in Fm_L \cup Tm_L \wedge$
$\wedge (\ulcorner \bigwedge xw \urcorner \textit{ occurs in } z \vee \ulcorner \bigvee xw \urcorner \textit{ occurs in } z \vee \ulcorner \iota xw \urcorner$
$\textit{occurs in } z) \wedge \bigvee_{a,b} (a\ Vbl_L\ x \text{ in } w \wedge b\frown a = y)) \vee$
$\vee \bigvee_a (\ulcorner a \bigwedge \urcorner = y \vee \ulcorner a \bigvee \urcorner = y \vee \ulcorner a \iota \urcorner = y)$

Of course x is free in z just in case there is an occurrence of it in z which is not a bound occurrence, and x is bound in z just in case it has a bound occurrence in z.

MD-7.17 $x \textit{ is free in } z \leftrightarrow \bigvee_y (y\ Vbl_L\ x \text{ in } z \bigwedge \sim y\ B_L\ x \text{ in } z)$

MD-7.18 $x \textit{ is bound in } z \leftrightarrow \bigvee_y y\ B_L\ x \text{ in } z$

A statement of L is just a formula of L not containing any free variables.

MD-7.19 $St_L = \{x : x \in Fm_L \wedge \bigwedge_{z \in V} \sim z \text{ is free in } x\}$

The result of replacing all free occurrences of y in x by free occurrences of z is just what it sounds like if it can be done; if it can't be done (i.e., if some of the occurrences of y in x lie within the scope of a quantifier of z or a description operator of z), then the result is to be taken to be our original x,

MD-7.20 $$x_z^y = \imath a[\bigvee_{n \in \omega} \bigvee_s (s \text{ is } n\text{-sequence} \wedge s_0 = x \wedge \bigwedge_{0<i<n} \bigvee_{w,b}$$
$$(w\ Vbl_L\ y \text{ in } s_{i-1} \wedge \sim w\ B_L\ y \text{ in } s_{i-1} \wedge w\ Vbl_L\ z \text{ in } s_i \wedge$$
$$\wedge \sim w\ B_L\ z \text{ in } s_i \wedge \widehat{w}\widehat{y}b = s_{i-1} \wedge \widehat{w}\widehat{z}b = s_i) \wedge$$
$$\wedge \sim y \text{ is free in } s_{n-1} \wedge a = s_{n-1}) \vee$$
$$\vee (\sim \bigvee_{n \in \omega} \bigvee_s (s \text{ is } n\text{-sequence} \wedge s_0 = x \wedge$$
$$\wedge \bigwedge_{0<i<n} \bigvee_{w,b} (w\ Vbl_L\ y \text{ in } s_{i-1} \wedge \sim w\ B_L\ y \text{ in } s_{i-1} \wedge$$
$$\wedge w\ Vbl_L\ z \text{ in } s_i \wedge \sim w\ B_L\ z \text{ in } s_i \wedge \widehat{w}\widehat{y}\widehat{b} = s_{i-1} \wedge$$
$$\wedge wzb = s_i) \wedge \sim y \text{ is free in } s_{n-1}) \wedge a = x)]$$

We define the free and bound occurrences of terms of L just as we define the free and bound occurrences of variables:

MD-7.21 $y\ Tm_L\ x \text{ in } z \leftrightarrow z \in Fm_L \cup Tm_L \wedge \bigvee_a \widehat{y}\widehat{x}\widehat{a} = z \wedge x\ Tm_L$

MD-7.22 $$y\ BTm_L\ x \text{ in } z \leftrightarrow y\ Tm_L\ x \text{ in } z \wedge \bigvee_{a,b} (a\ Vbl_L\ b \text{ in } x \wedge$$
$$\wedge \sim a\ B_L\ b \text{ in } x \wedge \widehat{y}\widehat{a}\ B_L\ b \text{ in } z)$$

The result of substituting free occurrences of the term t for the variable x in z is just what it sounds like, unless something goes wrong and a variable of t would become captured, in the substituting process, by a quantifier or description operator of z; if something goes wrong, we take the result to be the original expression z.

MD-7.23 $$z_t^x = \imath k[\bigvee_{b \in V} (\sim b \in \mathscr{R}z \cup \mathscr{R}t \wedge \bigvee_{n \in \omega} \bigvee_s (s \text{ is } n\text{-sequence} \wedge$$
$$\wedge s_0 = z_b^x \wedge \bigwedge_{0<i<n} \bigvee_{w,c} (w\ Vbl_L\ b \text{ in } s_{i-1} \wedge w\ Tm_L\ t \text{ in } s_i \wedge$$

$\wedge \sim w\ BTm_L\ t$ in $s_i \wedge s_{i-1} = \widehat{w}\widehat{b}\widehat{c} \wedge s_i = \widehat{w}\widehat{t}\widehat{c}) \wedge \sim b$ is *free in* $s_{n-1} \wedge k = s_{n-1})) \vee (\sim \bigvee_{b \in V} (\sim b \in \mathscr{R}z \cup \mathscr{R}t \wedge$
$\wedge \bigvee_{n \in \omega} \bigvee_{s} (s$ *is n-sequence* $\wedge s_0 = z_b^x \wedge \bigwedge_{0<i<n} \bigvee_{w,c} (w\ Tm_L\ t$
in $s_i \wedge w\ Vbl_L\ b$ *in* $s_{i-1} \wedge \sim w\ BTm_L\ t$ in $s_i \wedge s_{i-1} = \widehat{w}\widehat{b}\widehat{c} \wedge$
$\wedge \widehat{w}\widehat{t}\widehat{c} = s_i) \wedge \sim b$ *is free in* $s_{n-1})) \wedge k = z)]$

The simultaneous substitution of $t_0, t_1, \ldots, t_{n-1}$ for $x_0, x_1, \ldots, x_{n-1}$, respectively, admits of a straightforwardly analogous definition:

MD-7.24 $z_{t_0, t_1, \ldots t_{n-1}}^{x_0, x_1, \ldots x_{n-1}} = \imath k[\bigvee_{b \in V^n} (\bigwedge_{i<n} \sim b_i \in \mathscr{R}z \cup \bigcup_{i<n} \mathscr{R}t_i \wedge$
$\wedge \bigvee_{m \in \omega} \bigvee_{s} (s$ *is m-sequence* $\wedge s_0 = z_{b_0}^{x_0} \wedge \bigwedge_{0<i<n}$
$(s_i = s_{i-1}{}_{b_i}^{x_i}) \wedge \bigwedge_{n-1<i<m} (s_i = s_{i-1}{}_{t_{i-n}}^{b_{i-n}}) \wedge \sim \bigvee_{i} (b_i$ *is free*
in $s_{m-1}) \wedge k = s_{m-1})) \vee (\sim \bigvee_{b \in V^n} (\bigwedge_{i<n} \sim b_i \in \mathscr{R}z \cup \bigcup_{i<m} \mathscr{R}t_i \wedge$
$\wedge \bigvee_{m,s} (m \in \omega \wedge s$ *is m-sequence* $\wedge s_0 = a_{b_0}^{x_0} \wedge \bigwedge_{0<i<n}$
$(s_i = s_{i-1}{}_{b_i}^{x_i}) \wedge \bigwedge_{n-1<i<m} s_i = s_{i-1}{}_{t_{i-n}}^{b_{i-n}} \wedge \sim \bigvee_{i} (b_i$ *is free*
is $s_{m-1}))) \wedge k = z)]$

Finally, we define alphabetic variance. Two formulas or terms are immediate alphabetic variants if one can be obtained from the other by rewriting the variable of an initial quantifier or description operator at each of the occurrences bound by that quantifier or operator, and conversely. Two formulas are alphabetic variants of each other when one can be obtained from the other by replacement of formulas or terms by immediate alphabetic variants of them.

MD-7.25 $IAV(x, x') \leftrightarrow \bigvee_{y, y'} \bigvee_{w, w'} [y = y'^{w'}_{w} \wedge y' = y^{w}_{w'} \wedge [(x = \ulcorner \bigwedge wy \urcorner \wedge$
$\wedge x' = \ulcorner \bigwedge w'y' \urcorner) \vee (x = \ulcorner \bigvee wy \urcorner \wedge x' = \ulcorner \bigvee w'y' \urcorner) \vee (x =$
$= \ulcorner \imath wy \urcorner \wedge x' = \ulcorner \imath w'y' \urcorner)]]$

MD-7.26 $AV(x, x') \leftrightarrow \bigvee_{n \in \omega} \bigvee_{s} (s$ *is n-sequence* $\wedge \bigwedge_{i<n} s_i \in Tm_L \cup Fm_L \wedge$
$\wedge s_0 = x \wedge s_{n-1} = x' \wedge \bigwedge_{i<n} \bigvee_{y, y'} [IAV(y, y') \wedge$
$\wedge \bigvee_{a,b} (s_i = \widehat{a}\widehat{y}\widehat{b} \wedge s_{i-1}\ \widehat{a}\widehat{y'}\widehat{b})])$

2. A theory will be construed as a language together with a set of axioms. Each of our theories will contain the same set of mathematical axioms, to be denoted by 'MA'. Among the members of MA will be included not only instances of A-1–A-8 of the appendix, but also definitions D-1 through D-52 of the appendix. (Of course 'Σ' is an informal misspelling of 'P_0^1', '$\in$' of 'P_0^2', and the formation rules have also been distorted in the direction of familiarity by writing $\ulcorner x \in y \urcorner$ for $\ulcorner Pr_{2,0}\langle x, y\rangle \urcorner$. We may, nevertheless, think of our languages as *officially* having the syntax described in this chapter.) We regard the defined expressions as primitive, and the formulas expressing the definitions are construed as axioms, and thus as members of MA. In addition, in the present section we introduce two further primitive symbols, 'D', officially written 'P_0^3' and denoted by $P_{3,0}$; and 'S', officially written 'P_0^5', and denoted by $P_{5,0}$.

Intuitively, $\ulcorner D\langle x, y, z\rangle \urcorner$ is read, 'The distribution of the random quantity y in the sequence x is given by some distribution in the set z.' This means that x is a denumerable, non-decreasing sequence of finite sets, y is a random quantity – i.e., a function whose domain includes each of the x_i and whose range is included in $\mathbb{R}^n$, for some n in ω – and z is a set of distribution functions that includes the distribution of y among the members of x_i. To say the latter is to say that there is some member f of z such that for every member s of $\mathbb{R}^n$, the limit as i increases, of the ratio of the number of members of x_i whose y value is less than or equal to s to the total number of members of x_i is f_s. Formally:

D-7.1 $$D\langle x, y, z\rangle \leftrightarrow y \text{ is fun} \wedge x \textit{ is } \omega\textit{-sequence} \wedge \bigvee_{i} x_i \neq 0 \wedge$$
$$\wedge \bigwedge_{i\in\omega} (x_i \prec \omega \wedge x_i \subset \mathscr{D}y) \wedge \bigwedge_{i,j\in\omega} (i<j \rightarrow x_i \subset x_j) \wedge$$
$$\wedge \bigvee_{n\in\omega} [\mathscr{R}y \subset \mathbb{R}^n \wedge \bigwedge_{w\in z} w \text{ is dist fun in } \mathbb{R}^n \wedge$$
$$\wedge \bigvee_{f\in z} \bigwedge_{s\in\mathbb{R}^n} \lim_{i\to\infty} \frac{[\imath m(m\in\omega \wedge m \approx \{a : a\in x_i \wedge y\langle a\rangle \leq s\})]_r}{[\imath m(m\in\omega \wedge m\approx x_i)]_r} = fs]$$

A few comments on this definition are in order. First a technical one: since division by 0_r yields 0_r and since the terms of x are finite, there is indeed exactly one natural number equipollent to x_j, and therefore one equipollent to each subset of x_j – in particular to the subset of x_j such that the value of the random quantity y for the members of that subset is less than or equal to s. Thus the definite descriptions of the right-hand side of our definition are proper. However, we have defined limits only for sequences of real numbers, and therefore we must construe the ratio whose limit we are interested in as the quotient of the *reals* corresponding to the natural numbers described. Hence the subscript 'r'.

Statements of the form $\ulcorner D\langle x, y, z\rangle\urcorner$ are going to be construed as the fundamental statements expressing our statistical knowledge. Thus the union of the x_i will often be construed as a reference set. One advantage of the foregoing definition over some treatments of limiting frequencies is that it is applicable both to finite and to denumerable reference sets. Furthermore, we can prove that if the union of the x_i is in fact finite (whether we know it or not) the particular order in which items in the union of the range of x appear in the sequence x is irrelevant to the truth of $\ulcorner D\langle x, y, z\rangle\urcorner$. More precisely, if both x and w are ω-sequences of finite sets, both are non-decreasing, and the union of the x_i is equal to the union of the w_i and is finite, then $\ulcorner D\langle x, y, z\rangle\urcorner$ will be true if and only if $\ulcorner D\langle w, y, z\rangle\urcorner$ is true.

T-7.1 $$x \textit{ is } \omega\textit{-sequence} \wedge w \textit{ is } \omega\textit{-sequence} \wedge \bigcup \mathcal{R}x = \bigcup \mathcal{R}w \wedge$$
$$\wedge \bigcup \mathcal{R}x \prec \omega \wedge \bigwedge_{i,j\in\omega} (x_i \prec \omega \wedge w_i \prec \omega \wedge (i<j \rightarrow x_i \subset x_j \wedge$$
$$\wedge\, w_i \subset w_j)) \rightarrow (D\langle x, y, z\rangle \leftrightarrow D\langle w, y, z\rangle)$$

Thus to have any reason to believe that order is relevant to a general statement of the form $\ulcorner D\langle x, y, z\rangle\urcorner$ is to have reason to believe that the underlying reference set is infinite. And if we (ever) have reason to believe that the underlying reference set is infinite, we have grounds for being more interested in some orderings of its elements than in others.

It might be thought that denumerable reference classes would not suffice for all statistical statements; in particular it might be thought that for a random quantity y to be distributed normally in a certain population, the population must have more than a denumerable number of members since the range of y is the whole set of real numbers. The following construction, however, shows that there is no problem here; a random quantity may have any distribution at all among a no more than denumerable population.

Let f be any distribution function whatever in $\mathbb{R}^n$ and y an n-dimensional random quantity. It is possible for there to be a denumerable sequence of finite sets x, such that $D(x, y, \{f\})$ is true. Let x_i have n_i members; for $i<j$, we have $x_i \subseteq x_j$ and $n_i < n_j$. For every s in $\mathbb{R}^n$ and for every i, we stipulate that x_i contain exactly m_i objects a such that $y\langle a\rangle \in s$, where m_i is the smallest integer greater than $f(s)\cdot n_i$. We therefore have:

$$\lim_{i\to\infty} \frac{[\imath m(m\in\omega \wedge m \approx \{a : a\in x_i \wedge y\langle a\rangle \leqq s],}{[\imath m(m\approx x_i)]\, r} = \lim_{i\to\infty} \frac{[m_i]\, r}{[n_i]\, r} =$$

$= \lim_{i\to\infty}$ (the real counterpart of the smallest fraction whose denominator is n_i that is greater than $f(s)$). Since f is continuous to the right, this is just $f(s)$.

For example, let f be the normal distribution with mean 1 and variance 0. To say that this describes the distribution of the random quantity y in the sequence of finite sets x, is to say that for every real number a, the limit of the relative frequency of objects whose y-values are less than $\langle a\rangle$ is exactly $f\langle a\rangle$. The domain of f is the whole real line, and for any real numbers a and b, if a is greater than b, $f\langle a\rangle$ is greater than $f\langle b\rangle$. Thus for every two real numbers a and b, the limit of the relative frequency of objects with y-values less than a is different from the relative frequency of objects with y-values less than b. Corresponding to every real number on the whole real line, then, there is a unique limiting frequency of objects with y-values less than that number. On the other hand, since the set of objects with which we are concerned, the union of the range of x, is only denumerable, there are at most a denumerable number of y-values instantiated in that set. How then can the continuum of different limiting frequency statements embodied in the assertion that f is the distribution function for y all be true? What underlies this important (and perhaps surprising) fact, of course, is simply the fact that the rational numbers are dense in the reals; thus not all of the elements of the range of the function y need be manifested in our population, and thus although the range of y will indeed comprise a non-denumerable number of 'values', only a denumerable number of these values need be instantiated in order that our statistical distribution statement be true.

There are instances in which we may be concerned with distributions in populations containing non-denumerable numbers of members. Our definition does not directly give us any way of representing such distributions. In most cases, however – perhaps in all – the required distribution statement can be reached by a limiting process. For example, suppose we want to say that the random quantity y has a uniform distribution over the points of the interval (0, 1), and that we want to be understood literally as asserting that the range of y includes the whole continuum of points in that interval. Let f be the uniform distribution in the interval (0, 1). We may interpret the claim that the distribution of y in its domain is given by f in the following way. Let d be an ω-sequence, the nth term of which is the partition of (0, 1) into $n+1$ exclusive and exhaustive subsets. Let x be a sequence of finite sets of points, x_k consisting of one point from each member of d_{k+1}. Then the distribution of y in the sequence x will be given by the distribution function f. Furthermore, this will be true of every sequence x such that x_k consists of one point from each member of d_k. The truth of the distribution statement clearly depends, as it should, on the sequence of partitions d, but not on which particular points make up the sets x_k.

We may generalize this approach to cover random quantities whose domains are sets isomorphic to a subset of $\mathbb{R}^n$; this should cover everything of relevance to natural science. Let d be an ω-sequence of partitions of the domain of y, such that d_0 is the empty set, and d_k is a partition of the domain of y into k subsets. Call the sequence x *ultimately d-dense in y* just in case (1) x_k contains exactly one element from each member of d_{k+1}, and (2) if a member of d_{k+1} contains *any* excluded points [cf. p. 000, Appendix] x_k contains *some* excluded point from that member of d_{k+1}. To say that the random quantity y is distributed in its domain, approached by the sequence d, according to the distribution function f, is then to say:

$$\bigwedge_x (x \text{ is ultimately } d\text{-dense in } y \rightarrow D\langle x, y, \{f\}\rangle)$$

The introduction of the definable mathematical predicate 'D' thus provides us with everything we need for the convenient expression of general statistical knowledge.

Probability statements, which I take to be metalinguistic, implictly involve reference to distribution statements. More directly, they involve reference to *statistical statements* employing the defined five-place predicate 'S'; $\ulcorner S\langle x, y, b, p, q\rangle \urcorner$ is to be read, 'In the limit, the the proportion of members of the reference sequence x having a y-value falling in the Borel set b, lies between p and q.' This means that, as before, x is a denumerable, non-decreasing sequence of finite sets, y is a random quantity whose domain includes each of the x_i and whose range is included in $\mathbb{R}^n$ for some value of n, b is an n-dimensional Borel set, and the limit as i increases of the ratio of the number of members of x_i having a y-value in b, to the total number of members of x_i exists and lies between p and q. Formally:

D-7.2 $$S\langle x, y, b, p, q\rangle \leftrightarrow y \text{ is fun} \wedge x \textit{ is } \omega\textit{-sequence} \wedge \bigvee_i x_i \neq 0 \wedge$$
$$\wedge \bigwedge_{i\in\omega} (x_i \prec \omega \wedge x_i \subset \mathscr{D}y) \wedge \bigwedge_{i,j\in\omega} (i<j \rightarrow x_i \subset x_j) \wedge \bigvee_{n\in\omega} (\mathscr{R}y \subset \mathbb{R}^n \wedge$$
$$\wedge\, b \in \mathbb{B}^n) \wedge p \leq \lim_{i\to\infty} \frac{[\imath m(m\in\omega \wedge m \approx \{a : a\in x_i \wedge y\langle a\rangle \in b\})]_r}{[\imath m(m\in\omega \wedge m\approx x_i)]_r} \leq q$$

The most important theorem of this section provides the connection between statements expressing our knowledge of statistical distributions, and the statistical statements which will turn out to underly our probabilities. It states that if the distribution of the random quantity y is given by one of the n-dimensional distribution functions belonging to z, then the limiting frequency in the reference sequence x of those objects whose y-value lies in the Borel set b lies between the greatest lower bound of $\{\int_b \mathrm{d}f : f\in z\}$ and the least upper bound of that set.

T-7.2 $$\mathscr{R}y \subset \mathbb{R}^n \wedge b\in \mathbb{B}^n \wedge D\langle x, y, z\rangle \rightarrow$$
$$\rightarrow S\left\langle x, y, b, glb\left\{\int_b \mathrm{d}f : f\in z\right\}, lub\left\{\int_b \mathrm{d}f : f\in z\right\}\right\rangle$$

The proof parallels the construction of the set function P_F alluded to in D-0.41.

The following theorems spell out in somewhat more detail a number of special cases embodied in Theorem 7.2. First, when we have precise knowledge of the distribution of the random quantity y in the reference sequence x,

T-7.3 $\mathscr{R}y \subset \mathbb{R}^n \wedge b \in \mathbb{B}^n \wedge D\langle x, y, \{f\}\rangle \to S\left\langle x, y, b, \int_b df, \int_b df \right\rangle$

T-7.4 $\mathscr{R}y \subset \mathbb{R}^n \wedge c \in \mathbb{R}^n \wedge D\langle x, y\{f\}\rangle \to S\langle x, y, (\{-\infty\}^n, c], fc, fc\rangle$

T-7.5 $\mathscr{R}y \subset \mathbb{R}^n \wedge (a, b] \in \mathbb{B}^n, \wedge D\langle x, y, \{f\}\rangle \to$

$$\to S\langle x, y, (a, b], \Delta f_{ab}, \Delta f_{ab}\rangle$$

T-7.6 $\mathscr{R}y \subset \mathbb{R}^n \wedge b \in \mathbb{B}^n \wedge D\langle x, y, \{f\}\rangle \wedge g$ is B-measurable $\wedge$

$$\wedge\, \mathscr{R}y \subset \mathscr{D}y \to S\left\langle x, g, b, \int_b gdf, \int_b gdf \right\rangle$$

When all we know of the distribution of the random quantity y in the reference sequence x is that it is given by one of a set z of distribution functions, we have

T-7.7 $\mathscr{R}y \subset \mathbb{R}^n \wedge t \in \mathbb{R}^n \wedge D\langle x, y, z\rangle \to$
$\to S\langle x, y, (\{-\infty\}^n, t], glb\{ft : f \in z\}, lub\{ft : f \in z\}\rangle$

T-7.8 $\mathscr{R}y \subset \mathbb{R}^n \wedge (a, b] \in \mathbb{B}^n \wedge D\langle x, y, z\rangle \to$
$\to S\langle x, y, (a, b], glb\{\Delta f_{a,b} : f \in z\}, lub\{\Delta f_{a,b} : f \in z\}\rangle$

T-7.9 $\mathscr{R}y \subset \mathbb{R}^n \wedge b \in \mathbb{B}^n \wedge D\langle x, y, z\rangle \wedge g$ is B-measurable $\wedge\, \mathscr{R}y \subset \mathscr{D}g \to$
$\to S\left\langle x, g, b, glb\left\{\int_b gdf : f \in z\right\}, lub\left\{\int_b gdf : f \in z\right\}\right\rangle$

There is a connection between statistical statements and distribution statements running the other way, but it is not as strong as one would expect or hope. What we would like to have is this:

(1) $$\bigwedge_{t \in \mathbb{R}^n} (S\langle x, y, (\{-\infty\}^n, t], f_1 t, f_2 t\rangle) \to \bigvee_f D\langle x, y, \{f\}\rangle$$

To see that (1) does not hold in general, consider the sequence x in which the random quantity y has a uniform distribution between 1 and 2. The union of the range of x is infinite, and in any arbitrary subinterval of $(1, 2)$ there is an infinite supply of objects having y values in that subinterval. We construct the sequence x' as follows: x_i contains 10^{i+1} members; $\frac{3}{4}$ of them are to have y-values in the interval $(1+\varepsilon_i, 2]$, and are to come from the first set x_i of the original sequence that has enough members answering to the description; and it is to have $\frac{1}{4}$ of its members with y-values in the interval $(1, 1+\varepsilon_i]$, where $\varepsilon_i = 10/(i+11)$. Given any real number t, the proportion of objects with y-values less than t in x' has a perfectly good limit: if t is less than or equal to 1, it is 0; if t lies between 1 and 2 it is $\frac{1}{4}+(2-t)/2$; if t is greater than or equal to 2 it is 1. But the function just described is no distribution function: it fails to be continuous to the right. Indeed, its value for 1 is 0, but for any positive epsilon, however small, its value for $1+\varepsilon$ is at least $\frac{1}{4}$. In a similar way such a function may be constructed which fails to satisfy the condition: $\lim_{t\to\{\infty\}^n} ft = 1$ and the condition $\lim_{t\to\{-\infty\}^n} ft = 0$.

When there is a function f such that (a) for every $t \in \mathbb{R}^n$ the proportion of objects with y-values less than t is ft and (b) is continuous to the right, then there is a distribution function characterizing the random quantity y in the reference sequence x:

T-7.10 $$\mathscr{R}y \subset \mathbb{R}^n \wedge \bigwedge_{t\in\mathbb{R}^n} S\langle x, y, (\{-\infty\}^n, t], ft, ft\rangle \wedge f \text{ cont. on } Rt \wedge \wedge \lim_{t\to\{-\infty\}^n} ft = 0 \quad \lim_{t\to\{\infty\}^n} ft = 1 \to D\langle x, y, \{f\}\rangle$$

Proof:

(1) $\mathscr{R}y \subset \mathbb{R}^n \wedge \bigwedge_{t\in\mathbb{R}^n} S\langle x, y, (\{-\infty\}^n, t], ft, ft\rangle \wedge f$ cont. on Rt ass.

(2) y is fun $\wedge$ x *is* ω-*sequence* $\wedge \bigvee_{i\in\omega} x_i \neq 0 \wedge \bigwedge_{i\in\omega} (x_i \prec \omega \wedge \wedge x_i \subset \mathscr{D}y) \bigwedge_{i,j\in\omega} (i<j \to x_i \subset x_j) \wedge \bigvee_{n\in\omega} (\mathscr{R}y \subset \mathbb{R}^n \wedge \wedge (\{-\infty\}^n, t] \in \mathbb{B}^n)$ 1, D 7.2

(3) $$\bigwedge_{t\in\mathbb{R}^n} \lim_{i\to\infty} \frac{[\imath m(m\in\omega \wedge m \approx \{a: a\in x_i \wedge y\langle a\rangle \leq t\})]_r}{[\imath m(m\in\omega \wedge m\approx x_i)]_r} = ft$$ 1, D 7.2

(4) $\mathscr{R}f \subset [0, 1] \wedge \mathscr{D}f = \mathbb{R}^n$ 3, 1

(5) f cont. on Rt 1

(6) $$\Delta f_{a,b} = \lim_{i\to\infty} \frac{[\imath m(m\in\omega \wedge m\approx \{d: d\in x_i \wedge a < y\langle d\rangle \leq b\})]_r}{[\imath m(m\in\omega \wedge m \approx x_i)]_r}$$ by computation

(7) $D\langle x, y, \{f\}\rangle$ 2, 3, 4, 5, 6, D-7.1

But if there are functions f_1 and f_2 such that for every $t \in \mathbb{R}^n$ $S\langle x, y, (\{-\infty\}^n, t], f_1 t, f_2 t\rangle$ and there does exist a distribution function for y in the sequence x, then for every $t \in \mathbb{R}^n$ the value of that distribution function for t will be sandwiched between $f_1 t$ and $f_2 t$.

T-7.11 $$\mathcal{R}y \subset \mathbb{R}^n \wedge \bigwedge_{t \in \mathbb{R}^n} S\langle x, y, (\{-\infty\}^n, t], f_1 t, f_2 t\rangle \wedge$$
$$\wedge \bigvee_f D\langle x, y, \{f\}\rangle \to \bigwedge_g (D\langle x, y, \{g\}\rangle \to \bigwedge_{t \in \mathbb{R}^n} f_1 t \leq gt \leq f_2 t)$$

Observe also that $D\langle x, y, \{f: \bigwedge_s (f_1 s \leq fs \leq f_2 s)\}\rangle$ does imply the existence of a distribution function f_t such that $D\langle x, y, \{f_t\}\rangle$, but also does not imply that $\Delta f_{1a,b} \leq \Delta f_{ta,b} \leq \Delta f_{2a,b}$.

3. A formula is implied by a set of formulas if there is a proof of it from those formulas as premises in our underlying first order logic; if x is a formula and S a set of formulas, a proof of x from premises S consists of a finite sequence of formulas satisfying certain familiar conditions which can be expressed syntactically. We may thus provide a definition of formal deducibility from a set of premises. What is of more interest to us in the present place, however, is the more specific notion of derivability from *empirical* (non-mathematical) premises.

We shall say that the formula x is derivable from the set of premises S, in the language L, in symbols '$S \vdash_L x$', just when x is derivable in the traditional sense from the union of S and the set MA of mathematical axioms – including, of course, mathematical definitions.

MD-7.27 $S \vdash_L x$ There is a finite sequence of formulas of L constituting a proof of x in the underlying first order logic from formulas in the union of S and MA.

The consequences in the language L of a set of formulas are just the formulas for which a proof in L from those formulas exists:

MD-7.28 $L\text{-}Cn(x) = \{y : y \in Fm_L \wedge x \vdash_L y\}$

A theory consists of a language L (represented by its list of predicates and operations) together with a set of formulas A constituting its axioms.

MD-7.29 $$T \textit{ is theory} \leftrightarrow \bigvee_{L, A} (T = \langle L, A\rangle \wedge L \textit{ is language} \wedge$$
$$\wedge \bigwedge_x (x \in A \to x \in Fm_L))$$

In addition to the mathematical axioms and definitions of our theories, there will be a certain class of statements functioning as empirical axioms. The set of empirical axioms is just the set of axioms other than the axioms (and consequences thereof) belonging to the mathematical part of the language:

MD-7.30 $EA_T = \iota x[\bigvee_{L,A} (T = \langle L, A\rangle \wedge T \text{ is theory} \wedge \wedge x = A - L\text{-}Cn(MA)) \vee (\sim T \text{ is theory} \wedge x = 0)]$

These empirical axioms will for the moment simply be taken as a set of statements in the language of the theory that are not provable on the basis of the mathematical axioms. That is to say, we may suppose them not only to include particular statements corresponding to 'Crow number 1 is black' and 'the item whose coordinates are *a*, *y*, *z*, *t* is a blue-tailed warbler', and general statements corresponding to 'all crows are black', 'the electrostatic field gradient at any point *P* in the region *R* is given by the vector *V*(*P*)', 'the weight of crows in population *Q* is roughly normally distributed with parameters and *m* and *s*'', but also 'analytic' statements corresponding to 'if something is a male crow it is not female'. Although such statements are not usually assumed to have empirical *content*, nevertheless they are empirical in the sense that they help to give the meaning of the terms that occur in them, and those are empirical terms; and further, no obvious and natural way has yet been given to distinguish between statements to be regarded as 'analytic' and those to be regarded as having 'empirical content'; and, finally, such a distinction is unnecessary for our purposes.

The set of theorems of a theory is of course just the set of *L*-consequences of its axioms:

MD-7.31 $Thm_{\langle L, A\rangle} = \{x : x \in L\text{-}Cn(A)\}$

The concept of a definitional extension of a theory (already referred to in connection with the introduction of conventional mathematical terminology) is one that will prove pervasively useful. A theory T' is a definitional extension of a theory T, when the vocabulary of T' includes that of T, and the axioms of T' are just those of T together with, for every new predicate or operation symbol, exactly one axiom in the form of a definition. More precisely:

MD-7.32 T' *is definitional extension of* $T \leftrightarrow \bigvee_{L, L', A, A'} (T$ *is theory* $\wedge$
$\wedge\ T'$ *is theory* $\wedge\ T = \langle L, A\rangle \wedge T' = \langle L', A'\rangle \wedge L \subset L' \wedge$
$\wedge\ A \subset A' \wedge \bigwedge_{x \in L' - L} \bigvee_{i \in \omega} \overset{1}{\bigvee_{y}} \overset{1}{\bigvee_{z}} (z$ *is i-sequence* $\wedge\ \mathscr{R}z \subset V \wedge$
$\wedge \bigwedge_{k, j \in i} (k \neq j \rightarrow z_j \neq z_k) \wedge \bigwedge_{w} (w$ *is free in* $y \leftrightarrow w \in \mathscr{R}z) \wedge$
$\wedge ([x \in Pr_i \wedge y \in Fm_L \wedge \ulcorner xz \leftrightarrow y \urcorner \in A'] \vee [x \in Op_i \wedge y \in Tm_L \wedge$
$\wedge \ulcorner xz = y \urcorner \in A'])) \wedge \bigwedge_{y \in A' - A} \bigvee_{x, w, z} (x \in L' - L \wedge$
$\wedge (y = \ulcorner xz = w \urcorner \vee y = \ulcorner xz \leftrightarrow w \urcorner)))$

The following theorems elucidate the definition:

MT-7.1 $\langle L', A'\rangle$ *is definitional extension of* $\langle L, A\rangle \wedge x' \in Fm_{L'} \rightarrow$
$\rightarrow \bigvee_{x} (x \in Fm_L \wedge \ulcorner x \leftrightarrow x' \urcorner \in Thm_{\langle L', A'\rangle})$

Thus a definitional extension of a theory does not essentially increase the expressive power of the theory; anything that can be said in $L' - x'$, for example – can be said also in L.

MT-7.2 T' *is definitional extension of* $T \wedge x \in Thm_{\langle L' A'\rangle} \wedge$
$\wedge\ x \in Fm_L \rightarrow x \in Thm_{\langle L, A\rangle}$

And neither does a definitional extension increase the deductive power of a theory; and therefore it cannot introduce a contradiction:

MT-7.3 $\langle L', A'\rangle$ *is definitional extension of* $\langle L, A\rangle \wedge$
$\wedge \sim \bigvee_{x} (x \in Fm_L \wedge x \in Thm_{\langle L, A\rangle} \wedge \ulcorner \sim x \urcorner \in Thm_{\langle L, A\rangle}) \rightarrow$
$\rightarrow \sim \bigvee_{x} (x \in Fm_{L'} \wedge x \in Thm_{\langle L', A'\rangle} \wedge \ulcorner \sim x \urcorner \in Thm_{\langle L', A'\rangle})$

Finally, if T_2 is a definitional extension of T_1, and T_1 of T_0, then although it may not be the case that T_2 is a definitional extension of T_0, it will be the case that *some* definitional extension of T_0 will be equivalent to T_2 in the sense that it will have the same vocabulary and the same set of theorems:

MT-7.4 $\langle L_2, A_2\rangle$ *is definitional extension of* $\langle L_1, A_1\rangle \wedge \langle L_1, A_1\rangle$ *is definitional extension of* $\langle L_0, A_0\rangle \rightarrow \bigvee_{L,A} (\langle L, A\rangle$ *is definitional extension of* $\langle L_0, A_0\rangle \wedge L = L_2 \wedge Thm_{\langle L_2, A_2\rangle} = Thm_{\langle L, A\rangle})$

4. An n-dimensional random quantity is a function from some domain to $\mathbb{R}^n$. (One might, even more plausibly, take its range to be some set of objects – 1-sequences of lengths, 2-sequences of lengths and weights, etc., isomorphic to a subset of $\mathbb{R}^n$ but not identified with it. Or one might take the range of a random quantity to be a set of 2-sequences $\langle x, y\rangle$ where x is an n-sequence of dimensions and y is a member of $\mathbb{R}^n$. For present purposes, however, the crude and classical treatment of quantities will suffice.) A random variable on the other hand will be understood as an expression (a function expression, or a functor) that, according to a theory T, denotes a random quantity, i.e., is provably a function from some domain to $\mathbb{R}^n$, for some $n \in \omega$. Thus if y is a random variable, according to the theory T, then it must not merely be true that y denotes a function, but $\ulcorner y$ is fun$\urcorner$ must be a theorem of the theory; and similarly for some numeral n it must be a theorem that the range of y is included in $\mathbb{R}^n$. (Numerals are easy to define as finite sequences of digits; we denote the set of numerals by '*Num*') '*RV* (n, x, T)' is read 'x is an n-place random variable in T'.

MD-7.33 $RV\ \langle n, x, T\rangle \leftrightarrow n \in Num \wedge T$ *is theory* $\wedge \ulcorner x$ is fun$\urcorner \in Thm_T \wedge \ulcorner \mathscr{R}x \subset \mathbb{R}^n \urcorner \in Thm_T$

We observe that what is a random variable according to one theory may not be one according to another theory. But there are two considerations that render this result not so implausible. First, it is generally a consequence of the 'meaning' postulates of the theory that a certain function has a range that is included in $\mathbb{R}^n$. Thus it would be construed as 'analytically' true of the counterpart of 'length in meters of', that its range would be included in $\mathbb{R}^1$. Second, the property of being a random variable is not generally lost by expressions: if one theory is a definitional extension of another, or if one is simply stronger than another, anything that is a random variable according to the first theory is also one according to the second theory.

MT-7.5 T *is theory* $\wedge T'$ *is theory* $\wedge T'$ *is definitional extension of* $T \wedge RV(n, x, T) \rightarrow RV(n, x, T')$

MT-7.6 $T' = \langle L', A' \rangle \wedge T$ *is theory* $\wedge T'$ *is theory* $\wedge L = L' \wedge \wedge A \subset L'\text{-}Cn(A') \wedge RV(n, x, T) \rightarrow RV(n, x, T')$

Similarly, an n-dimensional distribution functor is a term of L which necessarily denotes a distribution function in $\mathbb{R}^n$. We read '$DF(n, x, T)$' as 'x is an n-dimensional distribution functor of T'.

MD-7.34 $DF\langle n, x, T \rangle \leftrightarrow \bigvee_{L, A} (n \in Num \wedge T$ *is theory* $\wedge T = \langle L, A \rangle \wedge \wedge x \in Tm_L \wedge 0 \vdash_L \ulcorner x$ is dist fun in $\mathbb{R}^n \urcorner)$

Note that in this case we require not merely that it be provable that x is a distribution function in $\mathbb{R}^n$, but provable on logical and mathematical grounds alone. This seems a permissible strengthening, in view of the fact that typically a distribution function will be represented in purely mathematical terminology, containing no empirical terms or predicates. Whether or not the expression x is a distribution functor in a theory T is thus quite independent of the (non-mathematical) axioms of T:

MT-7.7 $\langle L, A \rangle$ *is theory* $\wedge \langle L', A' \rangle$ *is theory* $\wedge x \in Tm_{L \cap L'} \rightarrow \rightarrow (DF(n, x, \langle L, A \rangle) \leftrightarrow DF(n, x, \langle L', A' \rangle))$

The third ingredient of a distribution statement is the expression that refers to the reference class by mentioning a non-decreasing sequence of subsets of it. Actually the reference class mentioned in a distribution statement will be construed as the union of the range of the ω-sequence that is denoted by the first term of the statement. Not all sets are suitable as reference classes. For example, 'the set of prime-numbered tosses of this coin, union the set of tosses resulting in heads' is unsuitable, despite the fact that the set is potentially infinite, despite the fact that we know within close limits what the frequency of heads is in it, despite the fact that we can know perfectly well that a certain future toss belongs to it, and despite the fact that each component of the union seems to be a plausible enough reference set by itself. Informally and in general, suppose we are interested in the probability that a certain object, a, belongs to a certain set S. Given a number r between 0 and 1, there would in general be no problem in finding a set

R_1, to which we may or may not know that a belongs, which is such that the measure of S in R_1 is close to r. Furthermore, we can always find a relatively small set R_2 such that (1) we do know that a belongs to $R_1 \cup R_2$, and (2) the measure of S in $R_1 \cup R_2$ is still close to r.

There are various points at which we could relieve ourselves of the consideration of such odd reference classes. For the sake of convenience and simplicity, I have chosen to do it now, and to do it on a syntactical basis. Some of the reasons for doing this will become clear later when we discuss randomness and probability. For the time being we shall simply take our task to be that of constructing, on a syntactical basis, a canonical list of possible reference terms in a language L.

Two operations on reference sets which do seem to generate plausible new reference sets are intersection and product formation. Furthermore, an operation, uniformly applied to the objects in a reference set, should yield a set of objects that is acceptable as a reference set. Since measurement, for example, may be construed as an operation that takes 'empirical' objects into real numbers, mathematical formulas may also occur in plausible reference terms. Finally, if a certain set of n-sequences is a suitable reference set, so should the set of first terms (or of ith terms) of the n-sequences be a suitable reference set, and so should be the result of the suitable introduction of quantifiers into the defining matrix of the sequence.

As a first step toward specifying the set of canonical reference terms, we need to define the purely mathematical terms and formulas. A mathematical formula M is a formula in one of our languages L which is provably equivalent to a formula containing no operations at all, and no predicates other than '$\in$' and 'Σ'.

MD-7.35 $$\mathit{Math}\ Fm_L = \{M : \bigvee_{F,L} (F \in Fm_L \wedge \bigwedge_{x} (x \in L \wedge x \textit{ occurs in } F \rightarrow \rightarrow x = \text{`}\in\text{'} \vee x = \text{`}\Sigma\text{'}) \wedge 0 \vdash_L \ulcorner M \leftrightarrow F \urcorner)\}$$

Similarly, a mathematical term T is a term in one of our languages L which is provably equal to a term containing no operations at all and no predicates other than '$\in$' and 'Σ'.

MD-7.36 $$\mathit{Math\ Terms} = \{t : \bigvee_{x,L} (x \in Tm_L \wedge \bigwedge_{y} (y \in L \wedge y \textit{ occurs in } x \rightarrow \rightarrow y = \text{`}\in\text{'} \vee y = \text{`}\Sigma\text{'}) \wedge 0 \vdash_L \ulcorner T = x \urcorner)\}$$

A quantification of a formula F is a well-formed formula G of the form HF, in which H contains no symbols other than variables and quantifier signs: (Note that every formula is a quantification of itself.)

MD-7.37 $G \text{ is } Q_L \text{ of } F \leftrightarrow (F, G \in Fm_L \wedge \bigvee_H (G = H\widehat{\ }F \wedge \bigwedge_x (x \textit{ occurs in } H \rightarrow x \in V \vee x = `\bigwedge' \vee x = `\bigvee')))$

The final auxiliary concept we need is that of a formula-free term; the set of formula-free terms is just the smallest set containing variables and containing, for every n-place operation expression Op_n and every n-tuple of expressions in the set the concatenation of Op_n and that n-tuple of expressions.

MD-7.38 $FFT_L = \bigcap \{x: \bigwedge_y (y \in V \rightarrow y \in x) \wedge \bigwedge_{z, w, n} (z \in Op_n \wedge w \in x^n \rightarrow \rightarrow z\widehat{\ }w \in x) \wedge x \subset Tm_L\}$

Now the atomic reference terms comprise those obtained by abstraction from the basic formulas of the language, from mathematical formulas in which some variables are replaced by formula-free terms, or from quantifications of the foregoing:

MD-7.39 $ART_L = \{z: \bigvee_{s, k, x, y} [x\ BFm_L \vee \bigvee_{n \in \omega} \bigvee_{M, t} (Math\ Fm_L\ M \wedge \wedge \bigvee_{u \in V^n} x = M^{u_0, \dots u_{n-1}}_{t_0, \dots t_{n-1}} \wedge t \in FFT^n_L) \wedge z \in Tm_L \wedge \wedge y \in V^k \wedge s \text{ is } Q_L \text{ of } x \wedge \bigwedge_w (w \textit{ is free in } s \rightarrow \rightarrow \bigvee_{i \in k} w = y_i) \wedge z = \ulcorner \{y : s\} \urcorner] \vee \bigvee_t (t \in Tm_L \wedge z = \ulcorner \{t\} \urcorner)\}$

Some examples will help to make this clear. Let the non-mathematical symbols of L include:

Pr^1_1	interpreted as	is a crow
Pr^1_2	interpreted as	is ancestor of
Op^0_1	interpreted as	height in meters of
Op^0_0	interpreted as	Tom

Then the following table gives some atomic reference terms, together with their interpretations:

$\{x:P_1^1\langle x\rangle\}$	the set of crows
$\{x:\bigvee_y P_2^1\langle x, y\rangle\}$	the set of ancestors
$\{x:\bigwedge_y P_2^1\langle x, y\rangle\}$	the set of ancestors of everybody, i.e. {Adam}
$\{x:Op_1^0\langle x\rangle<2.00\}$	the set of things less than 2 m high
$\{x: \bigvee_{y, z\in\omega} y_r\cdot Op_1^0\langle x\rangle=z_r\}$	the set of things whose height is a rational number of meters
$\{x:\bigvee_y (1<Op_1^0\langle x\rangle \wedge \wedge Op_1^0\langle x\rangle<Op_1^0\langle y\rangle)\}$	the set of things more than a meter high, but shorter than something.
$\{x:Op_1^0\langle x\rangle<Op_1^0\langle Op_0^0\rangle\}$	the set of things shorter than Tom.
$\{\langle x, y\rangle:Op_1^0\langle x\rangle< <Op_1^0\langle Op_0^0\rangle \wedge Op_1^0\langle y\rangle< <Op_1^0\langle x\rangle\}$	the set of pairs of things of which the first is shorter than Tom, and the second shorter than the first.
$\{x:\sim P_1^1\langle x\rangle\}$	the set of non-crows
$\{Op_0^0\}$	the set whose only member is Tom.

What sets turn out to be reference sets depends, of course, on our language. If our language had a primitive predicate that was interpreted as applying to something if and only if that thing was a toss resulting in heads, or the seventh toss of a particular coin, the abstract formed from that predicate would be, by our definition, an atomic reference term. Furthermore it is clear that the abstract formed from (almost) any formula at all of a language L will turn out to be an atomic reference term of a definitional extension of L. But we are concerned here with a *given* language only, and within a given language not all terms will turn out to be atomic reference terms.

The problems of choosing one language over another, including the problem of choosing between a language and one of its definitional extensions, is a deep and serious problem of inductive logic. It may be conjectured that probability considerations will play a role in such choice, but the manner in which they will do so, and the nature of any other considerations that may also be involved, are at the moment open questions.

The general set of reference terms is then to be the smallest set that contains all the atomic reference terms and is closed under certain operations. As we have already noted, if x and y are appropriate reference sets, so should their intersection, $x \cap y$ be an appropriate reference set. The same is true of product formation: if x and y are appropriate reference sets, so should be the set of all two-sequences, of which the 0th item belongs to x and the 1st item belongs to y. Unions, as we have seen, are not always appropriate when their components are, and since we can construct unions from intersections and differences (if $x \subset y$ and $z \subset y$, then $x \cup z = y - [(y-x) \cap (y-z)]$), this tells us that the set of reference sets should not be closed under difference. Nevertheless there are some differences it seems unnatural to do without: if the set of crows residing in farming communities in the Northern Hemisphere is an appropriate reference set, so should the set of non-male crows residing in farming communities in the Northern Hemisphere be one. Thus we shall suppose that any reference set, less the set denoted by an atomic reference term, is also an appropriate reference set. Finally, we shall suppose that the set of reference terms is closed under marginalization: that is, if the set of ordered triples consisting of donor, gift, and recipient is an appropriate reference set, so should be the set of gifts, the set of 2-sequences of donors and recipients, etc. These conventions are all embodied in the following definition.

MD-7.40 $$RT_L = \bigcap \{x : x \subset Tm_L \wedge ART_L \subset x \wedge \bigwedge_{y,z} (y, z \in x \rightarrow \ulcorner y \cap z \urcorner \in x) \wedge$$
$$\wedge \bigwedge_{n \in \omega} \bigwedge_{y} (y \textit{ is n-sequence} \wedge \bigwedge_{i \in n} y_i \in x \rightarrow \ulcorner \prod_{i<n} y_i \urcorner \in x) \wedge$$
$$\wedge \bigwedge_{y,z} (y \in x \wedge z \in ART_L \rightarrow \ulcorner y - z \urcorner \in x) \wedge$$
$$\wedge \bigwedge_{y,w,z,t} \bigwedge_{n,m} (y \in x \wedge w \in Tm_L^n \wedge z \text{ is } Q_L \text{ of } \ulcorner w \in y \urcorner \wedge$$
$$\wedge t \in V^m \wedge \bigwedge_{i,j \in m} (i \neq j \rightarrow t_i \neq t_j) \wedge \bigwedge_{s} (s \in \mathscr{R}t \rightarrow s \textit{ is free}$$
$$\textit{in } z) \rightarrow \ulcorner \{t : z\} \urcorner \in x)\}$$

It clearly follows from our definition that if L is included in L', the reference terms of L are included in the reference terms of L':

MT-7.8 $L \subset L' \rightarrow RT_L \subset RT_{L'}$

It might be thought that the clause concerning marginalization allows an unintended case, since it is not stipulated that the quantification of $\ulcorner w \in y \urcorner$ contain only existential quantifiers. But it is not hard to see that the terms falling under the unintended case are provably empty, and so cause no problem after all. For example, consider $\{x: \bigwedge_y (\langle x, y\rangle \in t)\}$. Assume that this is not empty, i.e., that there is a z such that $z \in \{x: \bigwedge_y (\langle x, y\rangle \in t)\}$. Then $\bigwedge_y \langle z, y\rangle \in t$, and in particular $\langle z, t\rangle \in t$, contrary to the Axiom of Regularity:

$$t \in \{1, t\} \in \langle 1, t\rangle \in \langle z, t\rangle \in t.$$

Note also that this definition allows for permutations in the elements of a reference set: if $\ulcorner \{\langle x, y\rangle : z\} \urcorner$ is an acceptable reference term, so is $\ulcorner \{\langle y, x\rangle : z\} \urcorner$.

In this complicated definition I have tried to tread the narrow line between profligacy and stinginess. If we construe too narrowly the set of terms that we are going to take as plausible candidates for denoting sets, knowledge of which can provide grounds for probability statements, it will turn out that some perfectly plausible and intuitively well-grounded probability statements cannot be represented in our system. For example we may have knowledge about the proportion of female students who receive *A*'s in a certain course; given a certain state of knowledge about Sally, we would want to take this proportion as representing the probability (in this state of knowledge) that Sally will get (or did get) an *A*. In a language containing predicates corresponding to 'is female' and 'is a student', we thus want the term corresponding to '$\{x: x$ is female$\} \cap \{y: y$ is a student$\}$' to be an appropriate reference term. In general we at least want the intersections of useful reference sets to be useful reference sets.

But neither do we want to construe the set of terms in question too broadly. It is certainly not the case that any term may be taken as denoting a useful reference set, even if we have quite detailed knowledge of a distribution in it. The object is to construe this set of terms as broadly as possible, while still avoiding the pathological terms such as that corresponding to 'the set of tosses resulting in heads union the set of prime-numbered tosses.' Since it is quite probable that not all the pathological cases have been invented yet, the best we can do in establishing the adequacy of MD-7.40 is to show that it excludes at least some of the known pathological cases.

Furthermore, it is clear that, since MD-7.40 is stated relative to a given language, we could get any term we wanted – that is, any definable term – by considering a language *L′* which is a definitional extension of *L* in which a primitive expression occurs together with an axiom asserting that the object it denotes is just the same as the object denoted by the questionable term. This gambit cannot be ruled out, except by the agreement to stick to a *given* language. Thus while we may accept arbitrary extensions of our language with respect to its mathematical terminology, we do not wish to allow the same latitude with respect to the empirical part of our language.

This is not altogether implausible. But it requires that at some point we consider criteria for preferring one scientific language to another. We shall touch lightly on this problem later on. For the moment we shall simply assume that a language is given, and that the reference terms of that language are precisely those terms picked out by MD-7.40.

As we shall see later on, what counts as the 'correct' reference class in a given situation, depends on the contents of a certain set of statements, regarded as our body of knowledge, characteristic of that situation. What counts when we are dealing with probabilities is not what *is* the case, what *is* true (for then we need not be concerned with probability at all), but we have *reason to believe* or to accept. Thus if we are concerned with the probability that a certain well-balanced coin will land heads the fifth time it is tossed, the set of tosses may be an appropriate reference class, but the set of tosses that result in heads, union the unit set whose only member is the fifth toss is not, despite the fact that our statistical knowledge about that set is probably just as good as our statistical knowledge about the former set. Furthermore, it seems reasonable to regard the appropriateness of the terms featured in our statistical knowledge of the behavior of the coin as a syntactical matter rather than a semantic one; for it just might be true that the set of tosses that result in heads, union the unit set whose only member is the fifth toss was the same as the set of all tosses of that coin; but that would not make our statistical knowledge of the latter an appropriate basis for a probability statement. This whole question will be discussed in considerable detail in Chapter IX on Randomness; what has been said here is intended only as a suggestion as to why we are singling out a set of terms to be taken as reference terms on a syntactical basis.

5. There are several syntactical relations among statistical statements that underlie our whole approach; we begin by defining statistical statements themselves: a statistical statement is a statement of the form $\ulcorner S(x, y, b, p, q)\urcorner$, where $\ulcorner \bigcup \mathscr{R}x\urcorner$ is known to be equal to a reference term of the language, and b, p, and q are mathematical terms. We require y to be an n-dimensional random variable; as mentioned earlier, y may be a random variable according to one theory, and not according to another. Rather than relativize statistical statementhood to a theory, however, we relativize it to a set of statements of L; thus we stipulate that y be a random variable according to some theory whose consequences are included in the given set of statements. This set of statements also serves as a convenient container for the knowledge that the union of the range of x is an appropriate reference set. We read '$SS_{w,L}x$' as 'x is a statistical statement of L relative to the set of statements w'.

MD-7.41 $$SS_{w,L}x \leftrightarrow w \subset St_L \wedge \bigvee_A (L\text{-}Cn(A) \cap St_L \subset w \wedge \wedge \bigvee_{s,y,b,p,q} (x = \ulcorner S(s, y, b, p, q)\urcorner \wedge \bigvee_z (\ulcorner \bigcup \mathscr{R}s = z\urcorner \in w \wedge \wedge z \in RT_L) \wedge \bigvee_n RV(n, y, \langle L, A\rangle) \wedge b, p, q \in Math\ Terms_L))$$

It is an immediate consequence of our definition that if something is a statistical statement of L, relative to a set of statements, it is also a statistical statement of L relative to a larger set of statements; further that if L' is a richer language than L, the same relation holds.

MT-7.9 $$SS_{w,L}x \wedge w \subset w' \wedge L \subset L' \rightarrow SS_{w',L'}x$$

We say that two statistical statements (regardless of their subject matter) are equally strong when they mention what is *provably* the same pair of real numbers in their last two argument places:

MD-7.42 $$x\ E\text{-}STR_{w,L}\ y \leftrightarrow \bigvee_{u,z,b,p,q} \bigvee_{u',b',z',p',q'} (SS_{w,L}x \wedge SS_{w,L}y \wedge \wedge x = \ulcorner S(u, z, b, p, q)\urcorner \wedge y = \ulcorner S(u', z', b', p', q')\urcorner \wedge \wedge O \vdash_L \ulcorner p = p' \wedge q = q'\urcorner)$$

One statistical statement is stronger than another (regardless of their subject matter) just in case the real numbers mentioned by the first provably lie between the real numbers mentioned by the second.

MD-7.43 $x\, STR_{w,L}\, y \leftrightarrow \bigvee_{u,z,b,p,q}\ \bigvee_{u',b',p',q',z'} (SS_{w,L}\, x \wedge SS_{w,L}\, y \wedge$
$\wedge\, x = \ulcorner S(u, z, b, p, q)\urcorner \wedge y = \ulcorner S(u', z', b', p', q')\urcorner \wedge$
$\wedge \sim x\, E\text{-}STR_{w,L}\, y \wedge 0 \vdash_L p \geq p' \wedge q' \geq q)$

Finally, two statistical statements may differ, in the sense that they are not equally strong, and yet neither is stronger than the other.

MD-7.44 $x\, DIF_{w,L}\, y \leftrightarrow SS_{w,L}\, x \wedge SS_{w,L}\, y \wedge \sim x\, \mathbf{E\text{-}STR}_{w,L}\, y \wedge$
$\wedge \sim x\, STR_{w,L}\, y \wedge \sim y\, STR_{w,L}\, x$

Note that we do have a convenient trichotomy, in view of the way we have phrased MD-7.44, despite the fact that there are certain mathematical terms (described with the help of undecidable statements of number theory), such that it is a theorem that they belong to the interval [0, 1], but neither a theorem that the number denoted by the term is less than a half, nor a theorem that it is greater than or equal to a half (say). Such weird terms will not concern us directly. In a given model, a statistical statement using such a term will be true or false, and thus we could give a better trichotomy in semantic terms. But what really counts in the application of our statistical knowledge is what we *know*, and while we do not generally know what model (if any) of our language the universe is, we do often know that a certain statement is a theorem.

Let $\ulcorner \bigcup \mathscr{R} r_1 = t_1 \urcorner \in w$ and $\ulcorner \bigcup \mathscr{R} r_2 = t_2 \urcorner \in w$, where $t_1, t_2 \in RT_L$. Suppose that $RV(n, v_1, T)$ and $RV(n, v_2, T)$, where $T = \langle L, A\rangle$ and $L\text{-}Cn(A) \cap \cap St_L \subset w$. Let b_1 and b_2 be mathematical terms such that $0 \vdash_L \ulcorner b_1 \in \mathbb{B}^n \urcorner$ and $0 \vdash_L \ulcorner b_2 \in \mathbb{B}^n \urcorner$ and $0 \vdash_L \ulcorner m \neq n \urcorner$. Then we have:

(1) $SS_{w,L} \ulcorner S(r_1, v_1, b_1, \frac{1}{4}, \frac{1}{3})\urcorner$

(2) $SS_{w,L} \ulcorner S(r_1, v_1, b_1, \frac{1}{3}, \frac{1}{4})\urcorner$

(3) $SS_{w,L} \ulcorner S(r_1, v_1, b_1, -2, \pi)\urcorner$

(4) $SS_{w,L} \ulcorner S(r_1, v_1, b_2, \frac{1}{4}, \frac{1}{3})\urcorner$

(5) $SS_{w,L} \ulcorner S(r_1, v_1, b_1, 0_r, 1_r)\urcorner$

All of the numbered statements are statements of the metalanguage, and all of them are true. Of the *cornered* statements, however, those mentioned in (2) and (4) are logically false (cf. D-7.2), those mentioned in (3) and (5) are logically true, and that mentioned in (1) may be either true or false.

(6) $\ulcorner S(r_1, v_1, b_1, 0.3, 0.4)\urcorner\ E\text{-}STR_{w,L}\ulcorner S(r_1, v_1, b_1, \frac{6}{20}, \frac{8}{20})\urcorner$

(7) $\ulcorner S(r_1, v_1, b_1, 0.3, 0.4)\urcorner\ E\text{-}STR_{w,L}\ulcorner S(r_2, v_2, b_1, \frac{6}{20}, \frac{8}{2})\urcorner$

(8) $\ulcorner S(r_2, v_2, b_1, 0.3, 0.4)\urcorner\ E\text{-}STR_{w,L}\ulcorner S(r_1, v_1, b_2, 0.3, 0.4)\urcorner$

These statements are all true, despite the fact that some of the statements mentioned are logically false.

(9) $\ulcorner S(r_1, v_1, b_1, 0.3, 0.4)\urcorner\ STR_{w,L}\ulcorner S(r_2, v_2, b_2, 0.1, 0.9)\urcorner$

(10) $\ulcorner S(r_1, v_1, b_1, 0.3, 0.4)\urcorner\ STR_{w,L}\ulcorner S(r_2, v_2, b_2, 0.3, 0.5)\urcorner$

These are also true.

(11) $\ulcorner S(r_1, v_1, b_1, 0.3, 0.5)\urcorner\ DIF_{w,L}\ulcorner S(r_1, v_1, b_1, 0.2, 0.4)\urcorner$

Let $K = \imath w((G \rightarrow w = 0.4) \wedge (\sim G \rightarrow w = 0.5))$, where G is an unprovable Gödel sentence of L. Then

(12) $SS_{w,L}\ulcorner S(r_1, v_1, b_1, 0.2, K)\urcorner$

(13) $\ulcorner S(r_1, v_1, b_1, 0.3, 0.4)\urcorner\ STR_{w,L}\ulcorner S(r_1, v_1, b_1, 0.2, K)\urcorner$,

since it is a theorem that K is greater than or equal to 0.4. But, despite the fact that G is true, (assuming the consistency of our set theory), and therefore that $K = 0.4$, it is not *provable*, and therefore,

(14) $\sim\ulcorner S(r_1, v_1, b_1, 0.3, K)\urcorner\ STR_{w,L}\ulcorner S(r_1, v_1, b_1, 0.2, 0.4)\urcorner$

so that

(15) $\ulcorner S(r_1, v_1, b_1, 0.3, K)\ DIF_{w,L}\ulcorner S(r_1, v_1, b_1, 0.2, 0.4)\urcorner$

A number of natural theorems hold concerning these relationships:

MT-7.10 $x\ E\text{-}STR_{w,L}\ y \leftrightarrow y\ E\text{-}STR_{w,L}\ x$

$SS_{w,L}\ x \rightarrow x\ E\text{-}STR_{w,L}\ x$

$x\ E\text{-}STR_{w,L}\ y \wedge y\ E\text{-}STR_{w,L}\ z \rightarrow x\ E\text{-}STR_{w,L}\ z$

$x\ E\text{-}STR_{w,L}\ y \rightarrow \bigwedge_z ((x\ STR_{w,L}\ z \leftrightarrow y\ STR_{w,L}\ z) \wedge$

$\wedge (z\ STR_{w,L}\ x \leftrightarrow z\ STR_{w,L}\ y) \wedge (x\ DIF_{w,L}\ z \leftrightarrow y\ DIF_{w,L}\ z) \wedge$

$\wedge (x\ E\text{-}STR_{w,L}\ z \leftrightarrow y\ E\text{-}STR_{w,L}\ z))$

$w \subset w' \wedge L \subset L' \wedge x\ E\text{-}STR_{w,L}\ y \rightarrow x\ E\text{-}STR_{w',L'}\ y$

MT-7.11 $\sim x\ STR_{w,L}\ x$

$x\ STR_{w,L}\ y \wedge y\ STR_{w,L}\ z \rightarrow x\ STR_{w,L}\ z$

$x STR_{w,L}\ y \rightarrow \sim y\ STR_{w,L}\ x$

$w \subset w' \wedge L \subset L' \wedge x\ STR_{w,L}\ y \rightarrow x\ STR_{w',L'}\ y$

$\sim x\ STR_{w,L} \ulcorner S(y, z, b, p, p) \urcorner$

MT-7.12 $x\ DIF_{w,L}\ y \rightarrow y\ DIF_{w,L}\ x$

MT-7.13 $SS_{w,L}\ x \wedge SS_{w,L}\ y \rightarrow x\ E\text{-}STR_{w,L}\ y \vee x\ STR_{w,L}\ y \vee$

$\vee\ y\ STR_{w,L}\ x \vee x\ DIF_{w,L}\ y$

$x\ E\text{-}STR_{w,L}\ y \vee x\ STR_{w,L}\ y \rightarrow \sim x\ DIF_{w,L}\ y$

NOTE

[1] Kalish and Montague, *Techniques of Formal Reasoning*, Harcourt, Brace, World, New York, 1964.

CHAPTER EIGHT

RATIONAL CORPORA

1. Probability, in the sense that it is useful as a guide in life or as a guide to reasonable decisions in the face of uncertainty, must be relativized in some way to a body of knowledge. If we know that Mr. Jones is dead, we do not assign to him the canonical probability of survival for another year, as determined by mortality tables for people of his age and class. If Mr. Smith has a serious heart ailment, but lied about his health on his application for insurance, we can do nothing else than offer him insurance at the standard rates, unless we find out about his prevarication. If you know that an ordinary die was thrown and landed with an even number of spots up, the appropriate probability that the two landed uppermost to take as a basis for bets and indeed for other sorts of actions is not $\frac{1}{6}$ but $\frac{1}{3}$. It would be nice if we could always act on the basis of what is the case. As I hold a newly manufactured die in my hand, it *is* the case that it will be tossed a hundred times and land with the three up on the 100th toss, or that is *not* the case. Alas, the future is (partly) inaccessible to us. It is, on almost everyone's view, this inaccessibility of the future that renders the concept of probability important and useful.

The future is only partly inaccessible, because we do have good reasons for believing a lot of things about the future. We do know some things. What we do know clearly has a bearing on what it is reasonable to believe about the future, and therefore on how it is reasonable to act in preparation for the future. On almost everyone's view, then, the concept of a body of knowledge is central to the application of probability.

According to the classical frequentist view, the application of probability need not be regarded as an integral part of the interpretation of probability, or of its theory. It seems clear, however, that in statistics we are concerned in someway or other with the application of probability; and that if we take the classical view seriously, statistics must sit on half a foundation. The other half must be left in a limbo of common sense, customary usage, and scientific convention. I am attempting here to formalize the whole of the foundation, and for this purpose the concept of a body of knowledge is clearly central.

'Knowledge' is a word that has been so bandied by philosophers that it would be a major – perhaps impossible – undertaking to defend the thesis that we 'know' the items in what I have called a body of knowledge. I shall therefore hide my convictions under a cloak of Latin, and use the phrase 'rational corpus' in lieu of the phrase 'body of knowledge'. A rational corpus, then, will be a set of items that some entity (individual or social, real or hypothetical) is justified (in a sense to be touched on only later and briefly) in accepting. (Indeed, we shall argue later that rational corpora themselves can be explicated by reference to probability.)

The items in a rational corpus will be construed as statements of one of our languages. There is only one condition that we impose on a rational corpus: namely, that it include the sets of consequences of each of its members. The reasons underlying the choice of this rather weak condition will be adumbrated later.

2. The object x is a rational corpus, then, in symbols 'x is RC_L', just when it is a non-empty set of statements containing the statements that are consequences of each of its members:

MD-8.1 $$x \text{ is } RC_L \leftrightarrow \Sigma x \wedge x \neq 0 \wedge \bigwedge_y (y \in x \rightarrow y \in St_L \wedge \wedge St_L \cap L\text{-}Cn(\{y\}) \subset x)$$

Both mathematical (set-theoretical) and logical truths of the language L belong to every rational corpus of L:

MT-8.1 $$x \text{ is } RC_L \wedge y \in St_L \wedge 0 \vdash_L y \rightarrow y \in x$$

There is just one rational corpus that contains a contradiction, such as $\ulcorner 0 \neq 0 \urcorner$, but there are any number that are inconsistent in the limited sense that they contain both a statement and its negation.

MT-8.2 $$\overset{1}{\underset{x}{\bigvee}} (x \text{ is } RC_L \wedge \ulcorner 0 \neq 0 \urcorner \in x)$$

(Since any such rational corpus contains all statements of L, there can only be one such.)

MT-8.3 $\bigwedge_{y,x} (x \text{ is } RC_L \wedge \sim y \in x \wedge \sim \ulcorner \sim y \urcorner \in x \wedge y \in St_L \rightarrow$
$\rightarrow [x \cup (L\text{-}Cn(\{y\}) \cap St_L) \cup (L\text{-}Cn(\{\ulcorner \sim y \urcorner\}) \cap St_L)] \text{ is } RC_L)$

(By MT-8.1, neither y nor $\ulcorner \sim y \urcorner$ can be a theorem or the negation of a theorem; $\ulcorner 0 \neq 0 \urcorner$ is therefore neither an L-consequence of y nor of $\ulcorner \sim y \urcorner$. Neither can it be a consequence of any statement in x, since by hypothesis y does not belong to x, and all statements belong to the rational corpus containing $\ulcorner 0 \neq 0 \urcorner$.)

A basis for a rational corpus is a set of logically independent statements such that the union of the set of consequences of each of those statements is just the set of statements in that rational corpus. Note that we are not talking about the L-consequences of the *set*, but of each *member* of the set.

MD-8.2 y *is basis*$_L$ *for* $x \leftrightarrow (x \text{ is } RC_L \wedge y \subset x \wedge$
$\wedge \bigwedge_{z \in x} [\bigvee_{w \in y} z \in L\text{-}Cn(\{w\}) \vee (y = 0 \wedge z \in L\text{-}Cn(0))] \wedge$
$\wedge \bigwedge_{w \in y} \sim \bigvee_{z \in y - \{w\}} w \in L\text{-}Cn(\{z\}))$

While there may be a number of different bases for a given rational corpus – in general there will be – there can be only one rational corpus constructed from a given basis:

MT-8.4 y *is basis*$_L$ *for* $x \wedge y$ *is basis*$_L$ *for* $z \rightarrow x = z$.

That every rational corpus *has* a basis seems obvious, but in point of fact it is false, if, as seems plausible, there exists in the language L a denumerable sequence of logically independent empirical statements.

MT-8.5 $\bigvee_{s} (s \textit{ is } \omega\textit{-sequence} \wedge \bigwedge_{i \in \omega} s_i \in St_L \wedge \bigwedge_{i, j \in \omega} (i \neq j \rightarrow$
$\rightarrow \sim s_j \in L\text{-}Cn(\{s_i\}) \wedge \sim \ulcorner \sim s_j \urcorner \in L\text{-}Cn(\{s_i\})))\rightarrow$
$\rightarrow \bigvee_{x} (x \text{ is } RC_L \wedge \sim \bigvee_{y} (y \textit{ is basis}_L \textit{ for } x))$

Proof:

(1) Assume antecedent

(2) Let $x = \bigcup \{L\text{-}Cn(\ulcorner \{s_0 \wedge s_1 \wedge \ldots \wedge s_{n-1} \urcorner\} : n \in \omega\} \cap St_L$

(3) x is RC_L, since $y \in L\text{-}Cn(\{\ulcorner so \wedge \ldots \wedge s_{n-1}\urcorner\})$ and $z \in L\text{-}Cn(\{y\})$ imply $z \in L\text{-}Cn(\{\ulcorner so \wedge \ldots \wedge s_{n-1}\urcorner\})$

(4) sh~~ow~~ $\sim w$ *is basis*$_L$ *for* x

(5) ⌈assume w *is basis*$_L$ *for* x

(6) | sh~~ow~~: $\bigwedge_y (y \in x \rightarrow \sim y \in w)$

(7) | ⌈assume $y \in x$

(8) | | $[\bigvee_n]\ y \in L\text{-}Cn(\{\ulcorner s_0 \wedge \ldots \wedge s_{n-1}\urcorner\}) \wedge y \in St_L$ (7), (2)

(9) | | $[\bigvee_t]\ t \in w \wedge \ulcorner s_0 \wedge \ldots \wedge s_{n-1} \wedge s_n\urcorner \in L\text{-}Cn(\{t\})$ (5), (2), MD-8.2

(10) | ⌊ $\sim y \in w$ (5), (9), MD-8.2

(11) | $w = 0$ (5), (6), MD-8.2

(12) | $s_0 \in x \wedge \sim s_0 \in L\text{-}Cn(0)$ (1), (2)

(13) ⌊ $\sim w$ *is basis*$_L$ *for* x (5), (11), (12), MD-8.2

On the other hand a necessary and sufficient condition for a rational corpus to have a basis is that the union of some maximal chain of subsets in it have a basis, where a chain is just a set of sets such that of any two of its members, one is included in the other, and a maximal chain is just a chain that is not *properly* included in another chain.

MT-8.6 x is $RC_L \rightarrow [\bigvee_y\ y \text{ is basis}_L \text{ for } x \leftrightarrow \bigvee_b (\bigwedge_{p,q \in b} (p \subset q \vee q \subset p) \wedge$
$\wedge \bigwedge_{p \in b} p \subset x \wedge \bigwedge_{b^*} (\bigwedge_{p \in b^*} p \subset x \wedge \bigwedge_{p,q \in b^*} (p \subset q \vee q \subset p) \wedge b \subset b^*$
$\rightarrow b = b^*) \wedge \bigvee_y y \text{ is basis}_L \text{ for } \bigcup b)]$

The proof of this theorem is trivial, given a set-theoretical lemma; the usefulness of the theorem consists in the possibility of devising a maximal chain of subsets of x whose union has a basis constructed from the bases of the members of the chain.

LEMMA

$$\bigwedge_{p,q \in b} (p \subset q \vee q \subset p) \wedge \bigwedge_{p \in b} p \subset x \wedge \bigwedge_{b^*} (\bigwedge_{p,q \in b^*} (p \subset q \vee q \subset p) \wedge \bigwedge_{p \in b^*} p \subset x \wedge b \subset b^* \rightarrow b = b^*) \rightarrow x = \bigcup b$$

Proof:

(1)	$\bigcup b \subset x$	obvious
(2)	shøw $x \subset \bigcup b$	
(3)	$w \in x$	ass.
(4)	$\sim w \in \bigcup b$	ass.
(5)	shøw $\bigwedge_{p,q} (p, q \in b \cup \{\bigcup b \cup \{w\}\} \rightarrow p \subset q \vee q \subset p.)$	
(6)	case I: $p, q \in b$	obv.
(7)	case II: $p, q \in \{\bigcup b \cup \{w\}\}$	obv.
(8)	case III: $p \in b,\ q \in \{\bigcup b \cup \{w\}\}$	ass.
(9)	shøw $x \in p \rightarrow x \in q$	
(10)	$x \in p$	ass.
(11)	$x \in \bigcup b$	(8), (10)
(12)	$q = \bigcup b \cup \{w\}$	(8)
(13)	$x \in q$	(11), (12)
(14)	case IV: similar	
(15)	$\bigwedge_{p} (p \in b \cup \{\bigcup b \cup \{w\}\} \rightarrow p \subset x)$	obv.
(16)	shøw $\sim \bigcup b \cup \{w\} \in b$	
(17)	$\bigcup b \cup \{w\} \in b$	ass.
(18)	$\bigvee_{c} (w \in c \wedge c \in b)$	(17)
(19)	$w \in \bigcup b$	imp. by (4).
(20)	$b \neq b \cup \{\bigcup b \cup \{w\}\}$	(5), (15) imp. by hyp. of thm.

Just as there is only one rational corpus containing an inconsistent statement such as $0 \neq 0$, so there is only one rational corpus containing only logical and mathematical truths:

MT-8.7 $\overset{1}{\underset{x}{\bigvee}} (x \text{ is } RC_L \wedge \bigwedge_{w \in x} 0 \vdash_L w \wedge \sim \ulcorner 0 \neq 0 \urcorner \in x)$

0 *is basis*$_L$ *for* $\{x : x \in St_L \wedge 0 \vdash_L x\}$

3. Most philosophical writers make a stronger assumption than I have concerning bodies of knowledge, namely that they are closed under

deduction. Thus it is generally assumed that if x and y are members of a rational corpus z, and that w is a logical consequence of the set of statements consisting of x and y, then w, too, should belong to z. I have refrained from making this assumption, not out of a frivolous desire for generality, but because I think it does not appropriately hold for the kinds of bodies of knowledge we all think we have or would (realistically) like to have. It is inappropriate because it is a very strong principle indeed: it requires, for example, that a body of knowledge be strictly consistent in the sense that it has a model. But what I regard as my own body of knowledge (for many purposes) isn't consistent in that sense: among the things I am almost completely sure of, is the thesis that at least one of those things is false. Put otherwise: it seems plausible to suppose that there is a large list of statements, $s_1 \ldots s_n$, each of which I regard as practically certain, which is such that I regard it as practically certain that the denial of the conjunction of those statements holds. Should one deny access in one's body of knowledge to mere 'practical certainties'? But what of interest then could pass the test?

The principle of deductive closure is related in a number of interesting ways to a number of other principles sometimes supposed to hold for bodies of knowledge. It will help to clarify our ideas, and to make the weakness of our assumption regarding bodies of knowledge more plausible, to formulate some of these principles briefly and to explore their relations. In our symbolism, the conjunction principle may be stated thus:

The Conjunction Principle: $x \text{ is } RC_L \wedge w \in x \wedge y \in x \rightarrow \ulcorner w \wedge y \urcorner \in x$

The conjunctive closure principle states that the conjunction of any finite number of members of x also belongs to x, when x is a rational corpus.

The Conjunctive Closure Principle: $$x \text{ is } RC_L \wedge \bigvee_{n \in \omega} (w \textit{ is n-sequence} \wedge \wedge \bigwedge_{i<n} w_i \in x) \rightarrow \ulcorner w_0 \wedge w_1 \wedge \ldots \wedge \wedge w_{n-1} \urcorner \in x$$

These two principles are clearly equivalent, though one might seem intuitively stronger than the other.

The weak deduction principle is the principle embodied in our definition of a rational corpus; it simply states that if x is a rational corpus and y belongs to x, and z follows from y, then z belongs to x.

The Weak Deduction Principle: x is $RC_L \wedge y \in x \wedge z \in L\text{-}Cn(\{x\}) \cap$
$\cap St_L \rightarrow z \in x$

The Strong Deduction Principle, mentioned above, states that the consequences of any *set* of statements of a rational corpus x should also be in x.

The Strong Deduction Principle: x is $RC_L \wedge y \subset x \rightarrow L\text{-}Cn(y) \cap St_L \subset x$

Finally, let us define several degrees of consistency:

Simple Consistency Principle: x is $RC_L \rightarrow \sim \bigvee_y (y \in x \wedge \ulcorner \sim y \urcorner \in x)$

n-Consistency Principle: x is $RC_L \rightarrow \sim \bigvee_w$ (w *is n-sequence* $\wedge \bigwedge_{i<n}$
$w_i \in x \wedge \ulcorner \sim (w_0 \wedge w_1 \wedge \ldots \wedge w_{n-1}) \urcorner \in x)$

Strict Consistency Principle: x is $RC_L \rightarrow \sim \bigvee_y (y \in L\text{-}Cn(x) \wedge$
$\wedge \ulcorner \sim y \urcorner \in L\text{-}Cn(x))$

Of these principles, only the weak deduction principle holds for rational corpora as I have defined them. However, given this weak deduction principle, a number of the other principles become equivalent:

MT-8.8 x is $RC_L \rightarrow$(The Conjunction Principle holds for $x \leftrightarrow$The Conjunctive Closure Principle holds for x) $\wedge$ (The Conjunctive Closure Principle holds for $x \leftrightarrow$The Strong Deduction Principle holds for x) $\wedge$ (The Conjunction Principle holds for $x \rightarrow$(The Simple Consistency Principle holds for $x \leftrightarrow$The n-consistency Principle holds for x)) $\wedge$ (The Conjunction Principle holds for $x \rightarrow$(The Simple Consistency Principle holds for $x \leftrightarrow$The Strict Consistency Principle holds for x))

Some of these equivalences depend on the compactness theorem (e.g., the equivalence between the conjunctive closure principle and the strong deduction principle), but otherwise they are not surprising.

One consequence of the conjunctive closure principle (or the conjunction principle or the strong deduction principle) that seems not to have received much attention, despite the fact that it seems somewhat counterintuitive, is that any rational corpus satisfying it and having a basis (as any realistic body of knowledge must) must have a basis consisting of a single statement:

MT-8.9 x is $RC_L \wedge \bigwedge_{w, z \in x} \ulcorner w \wedge z \urcorner \in x \wedge y$ *is basis*$_L$ *for* $x \rightarrow y \approx 1 \vee y = 0$

Proof:

(1) assume hypothesis

(2) show $y \neq 0 \rightarrow y \approx 1$

(3) $y \neq 0$ ass.

(4) show $w_1, w_2 \in y \rightarrow w_1 = w_2$

(5) $w_1, w_2 \in y$ ass.

(6) $w_1, w_2 \in x$ (1), (5), MD-8.2

(7) $\ulcorner w_1 \wedge w_2 \urcorner \in x$ (1) (conj. princ.)

(8) $[\bigvee_{w_3}]\ w_3 \in y \wedge \ulcorner w_1 \wedge w_2 \urcorner \in L\text{-}Cn(\{w_3\})$ (7), (1), MD-8.2

(9) $w_1, w_2 \in L\text{-}Cn(\{w_3\})$ (8)

(10) $\sim w_1 \in y - \{w_3\} \wedge \sim w_2 \in y - \{w_3\}$ (7), (9), MD-8.2

(11) $w_1 = w_3 \wedge w_2 = w_3$ (5), (10)

(12) $w_1 = w_2$ (11)

(13) $y \approx 1$ (4)

Of course it may be argued that the reason a one-membered basis for a body of knowledge seems unlikely is that it would tend to be a conjunction too long to be swallowed at one gulp. Nevertheless, as most epistemologists seem to conceive of the matter, it is this one superstatement that is the primary object to which criteria for acceptance must be applied – and that, surely, is implausible.

There are thus two reasons for considering rational corpora satisfying only the weak deduction principle: Strict consistency seems an implausibly strong condition to impose, though simple consistency does not. And it does not seem plausible to suppose that every rational corpus having a basis has a one-membered basis – i.e., that everything we can claim to know follows from a single empirical statement.[1]

4. It turns out that we cannot usefully define probability relative to a body of knowledge that is not strictly consistent, so long as we accept

certain natural conditions. In particular, if we construe probability as any sort of function at all, taking sentences and bodies of belief into any sort of objects at all; if we suppose that the probability value of a contradiction is not the same as that of a tautology; if we suppose that any two statements known in the body of knowledge to have the same truth value are to have the same probability; and if the body of knowledge satisfies the weak deductive closure principle – i.e., the principle that if x belongs to the body of knowledge and y follows logically from x, then y belongs to it too – then it follows that the body of knowledge is consistent.

MT-8.10 $Prob \text{ is fun} \wedge \mathscr{D}\, Prob \subset \{w : w \text{ is } RC_L\} \times St_L \wedge w \text{ is } RC_L \wedge$
$\wedge\, Prob\langle w, \ulcorner 0=0 \urcorner\rangle \neq Prob\langle w, \ulcorner 0\neq 0 \urcorner\rangle \wedge \bigwedge_{x,y} (\ulcorner x \leftrightarrow y \urcorner \in w \rightarrow$
$\rightarrow Prob\langle w, x\rangle = Prob\langle w, y\rangle) \rightarrow \sim \bigvee_{z} (w \vdash_L z \wedge w \vdash_L \ulcorner \sim z \urcorner)$

Proof:

(1) Assume antecedent.

(2) $[\bigvee_z]\ w \vdash_L z \wedge w \vdash_L \ulcorner \sim z \urcorner$ — ass.

(3) $[\bigvee_{a_0, \ldots, a_{m-1}}]\ \{a_0, \ldots, a_{m-1}\} \vdash_L \ulcorner z \wedge \sim z \urcorner \wedge \bigwedge_{i<n} a_i \in w$ — compactness, (2)

(4) ~~show~~ $c \in St_L \rightarrow Prob\langle w, c\rangle = Prob\langle w, 0\neq 0\rangle$

(5) $c \in St_L$ — ass.

(6) $\vdash_L \ulcorner z \wedge \sim z \leftrightarrow 0\neq 0 \urcorner$ — logic

(7) $\ulcorner z \wedge \sim z \leftrightarrow 0\neq 0 \urcorner \in w$ — (6), (1), MD-8.1

(8) $Prob\langle w, \ulcorner z \vee \sim z \urcorner\rangle = Prob\langle w, \ulcorner 0\neq 0 \urcorner\rangle$ — (7), (1)

(9) $\vdash_L \ulcorner c \wedge a_0 \wedge \ldots \wedge a_{m-1} \leftrightarrow z \wedge \sim z \urcorner$ — (3), logic

(10) $\ulcorner c \wedge a_0 \wedge \ldots \wedge a_{m-1} \leftrightarrow z \wedge \sim z \urcorner \in w$ — (9), (1), MD-8.1

(11) $Prob\langle w, \ulcorner c \wedge a_0 \wedge \ldots \wedge a_{m-1} \urcorner\rangle = Prob\langle w, \ulcorner z \wedge \sim z \urcorner\rangle$ — (1), (10)

(12) $\{a_0\} \vdash_L \ulcorner c \wedge a_0 \wedge \ldots \wedge a_{m-1} \leftrightarrow c \wedge a_1 \wedge \ldots \wedge a_{m-1} \urcorner$ — logic

(13) $\ulcorner c \wedge a_0 \wedge \ldots \wedge a_{m-1} \leftrightarrow c \wedge a_1 \wedge \ldots \wedge a_{m-1} \urcorner \in w$ — (3), (12), (1)

(14) $Prob\langle w, \ulcorner c \wedge a_0 \wedge \ldots \wedge a_{m-1} \urcorner\rangle = Prob\langle w, \ulcorner c \wedge a_1 \wedge$

$$\begin{array}{ll} & \wedge \ldots \wedge a_{m-1}\urcorner\rangle \qquad (1), (13) \\ (15) & \ldots \\ (16) & Prob\langle w, \ulcorner c \wedge a_0 \wedge \ldots \wedge a_{m-1}\urcorner\rangle = Prob\langle w, c\rangle \\ (17) & Prob\langle w, c\rangle = Prob\langle w, \ulcorner 0 \neq 0\urcorner\rangle \qquad (8), \ldots, (16) \\ (18) & Prob\langle w, \ulcorner 0=0\urcorner\rangle = Prob\langle w, \ulcorner 0\neq 0\urcorner\rangle \qquad (4)\text{: imp. by (1).} \end{array}$$

Observe that there is no convenient way out of this dilemma by restricting the domain of the probability function. The rational corpus w is perfectly arbitrary, and so is the statement c. Nevertheless, we shall continue to impose only the restriction of weak deductive closure on rational corpora, and shall define probability relative to maximal consistent subsets of rational corpora. We may then stipulate that the probability of a statement, relative to a body of knowledge, has a certain value, just when it has that value relative to *every* maximal consistent subset of that body of knowledge.

5. There are still times when we wish to talk directly about the set of statements that actually comprise the rational corpus. In order to facilitate such talk – and for the sake of its intrinsic interest – I shall present some facts about the structure discoverable in a set of statements satisfying only the condition of weak deductive closure.

If a rational corpus is construed, as we have construed it, as a set of statements satisfying only the weak deduction principle, then it is clear that it may contain as members both $\ulcorner x=y\urcorner$ and $\ulcorner y=z\urcorner$, without having $\ulcorner x=z\urcorner$ as a member. Similarly it may contain $\ulcorner x\leftrightarrow y\urcorner$ and $\ulcorner y\leftrightarrow z\urcorner$ as members without having $\ulcorner x\leftrightarrow z\urcorner$ as a member. And it may contain $\ulcorner x_1\in y_1\urcorner$ and $\ulcorner x_0\in y_0\urcorner$ as members without $\ulcorner\langle x_0, x_1\rangle\in y_0\times y_1\urcorner$ being a member. In each of these three cases, however, there is a relation between the items in question that is worth taking account of, even though it is not embodied in a statement belonging to the rational corpus in question. In the first case, we shall say that x and z are related by an identity chain in the rational corpus; in the second case, we shall say that x and z are related by a biconditional chain in the rational corpus; and in the third case, we shall say that $\ulcorner\langle x_0, x_1\rangle\urcorner$ and $\ulcorner y_0\times y_1\urcorner$ are related by a membership chain in the rational corpus.

More specifically, we say that x and y are connected by a biconditional chain in a rational corpus of the language L – in symbols, $x\, B_{w,L}\, y$ –

when there exists a finite sequence s of $n-1$ statements of L belonging to w of the form $\ulcorner s_i \leftrightarrow s_{i+1} \urcorner$ where s_0 is x and s_{n-1} is y.

MD-8.3 $x\, B_{w,L}\, y \leftrightarrow w \text{ is } RC_L \wedge \bigvee_{n\in\omega} \bigvee_{s} (s \textit{ is n-sequence} \wedge 0<n \wedge$
$\wedge \bigwedge_{i<n-1} (\ulcorner s_i \leftrightarrow s_{i+1} \urcorner \in w) \wedge s_0 = x \wedge s_{n-1} = y)$

The relation of being connected in w by a biconditional chain is an equivalence relation on statements of L:

MT-8.11 $w \text{ is } RC_L \wedge x, y, z \in St_L \rightarrow x\, B_{w,L}\, x \wedge (y\, B_{w,L}\, x \rightarrow x\, B_{w,L}\, y) \wedge$
$\wedge (x\, B_{w,L}\, y \wedge y\, B_{w,L}\, z \rightarrow x\, B_{w,L}\, z)$

Proof:

(1) w is $RC_L \wedge x, y, z \in St_L$ — ass.
(2) s̸how $x\, B_{w,L}\, x$
(3) $\ulcorner x \leftrightarrow x \urcorner \in w$ — (1), MD-8.1, $\vdash \ulcorner x \leftrightarrow x \urcorner$
(4) Let $s = \{\langle 0, x\rangle, \langle 1, x\rangle\}$
(5) $x\, B_{w,L}\, x$ — (1), (4), (3), MD-8.3
(6) s̸how $x\, B_{w,L}\, y \rightarrow y\, B_{w,L}\, x$
(7) $x\, B_{w,L}\, y$ — ass.
(8) $[\bigvee_{n\in\omega} \bigvee_{s}]$ s *is n-sequence* $\wedge 0<n \wedge$
$\wedge \bigwedge_{i<n-1} (\ulcorner s_i \leftrightarrow s_{i+1} \urcorner \in w) \wedge s_0 = x \wedge s_{n-1} = y$
(7), MD-8.3
(9) Let $t = \{\langle i, r\rangle : \langle n-i-1, r\rangle \in s\}$
(10) $\bigwedge_{i<n-1} \ulcorner t_i \leftrightarrow t_{i+1} \urcorner \in w$ — (9), (8), MD-8.1
(noting that $\ulcorner s_{i+1} \leftrightarrow s_i \urcorner \in L\text{-}Cn(\{\ulcorner s_i \leftrightarrow s_{i+1} \urcorner\})$)
(11) $t_0 = s_{n-1-0} = y \wedge t_{n-1} = s_{(n-1)-(n-1)} = s_0 = x$
(12) $y\, B_{w,L}\, x$ — (11), (10), (9), (8), MD-8.3
(13) s̸how $x\, B_{w,L}\, y \wedge y\, B_{w,L}\, z \rightarrow x\, B_{w,L}\, z$
(14) $x\, B_{w,L}\, y \wedge y\, B_{w,L}\, z$ — ass.
(15) $[\bigvee_{n\in\omega} \bigvee_{s}]$ s *is n-sequence* $\wedge 0<n \wedge \bigwedge_{i<n-1} \ulcorner s_i \leftrightarrow s_{i+1} \urcorner \in w \wedge$

$\wedge s_0 = x \wedge s_{n-1} = y \wedge [\bigvee_{m \in \omega} \bigvee_t]\ t$ *is m-sequence* $\wedge\, 0 < m \wedge$

$\wedge \bigwedge_{i<m-1} \ulcorner t_i \leftrightarrow t_{i+1} \urcorner \in w \wedge t_0 = y \wedge t_{m-1} = z$

(16) $\widehat{st}$ *is* $n+m$*-sequence* $\wedge\, 0 < n+m$ (15)

(17) ~~show~~ $\bigwedge_{i<n+m-1} \ulcorner \widehat{st}_i \leftrightarrow \widehat{st}_{i+1} \urcorner \in w$

(18) $i < n+m-1$ ass.

(19) case I: $i < n-1$

(20) $\widehat{st}_i = s_i \wedge \widehat{st}_{i+1} = s_{i+1}$ (15)

(21) case II: $n \leq i \leq n+m-1$

(22) $\widehat{st}_i = t_{i-n} \wedge \widehat{st}_{i+1} = t_{i-n+1}$ (15)

(23) case III: $i = n-1$

(24) $\widehat{st}_i = s_{n-1} = y \wedge \widehat{st}_{i-1} = t_0 = y$ (15)

(25) $\widehat{st}_0 = x \wedge \widehat{st}_{n+m-1} = z$

(26) $x\ B_{w,L}\ z$ (25), (16), (17), (14), MD-8.3

Corresponding to the logical principle of the substitutivity of the biconditional, we can establish the principle that if s_1 and s_2 are connected by a biconditional chain, and t_1 is like t_2 except for containing one or more occurrences of s_1 when t_2 contains s_2, then t_1 and t_2 are connected by a biconditional chain.

MT-8.12 $t_1, t_2 \in St_L \wedge s_1\ B_{w,L}\ s_2 \wedge t_1 = \widehat{a s_1} \widehat{b} \wedge t_2 = \widehat{a s_2} \widehat{b} \rightarrow t_1\ B_{w,L}\ t_2$

Proof:

(1) $s_1\ B_{w,L}\ s_2 \wedge t_1 = \widehat{a s_1} \widehat{b} \wedge t_2 = \widehat{a s_2} \widehat{b}$ ass.

(2) $[\bigvee_n \bigvee_x]\ x$ *is n-sequence* $\wedge \bigwedge_{i<n} (x_i \in St_L) \wedge$

$\wedge \bigwedge_{i<n-1} \ulcorner x_i \leftrightarrow x_{i+1} \urcorner \in w \wedge x_0 = s_1 \wedge x_{n-1} = s_2$

(3) $\bigwedge_{i<n} \widehat{a x_i} \widehat{b} \in St_L \wedge \bigwedge_{i<n-1} \ulcorner \widehat{a x_i} \widehat{b} \leftrightarrow \widehat{a x_{i+1}} \widehat{b} \urcorner \in w$

since $\{\ulcorner x_i \leftrightarrow x_{i+1} \urcorner\} \vdash_L \ulcorner \widehat{a x_i} \widehat{b} \leftrightarrow \widehat{a x_{i+1}} \widehat{b} \urcorner$

(4) Let $y = \iota f\, (\mathscr{D} f = n \wedge \bigwedge_{i<n} f(i) = \ulcorner \widehat{a x_i} \widehat{b} \urcorner)$

(5) $t_1\ B_{w,L}\ t_2$ (3), (4), MD-8.3

The terms x and y are connected by an identity chain in a rational corpus z of the language L – in symbols, $x\ I_{w,L}\ y$ – when there is a finite sequence of n terms $s_0, \ldots s_{n-1}$ such that $\ulcorner s_i = s_{i+1} \urcorner$ is a member of z and s_0 is x and s_{n-1} is y.

MD-8.4 $x\ I_{w,L}\ y \leftrightarrow w$ is $RC_L \wedge \bigvee_{s} \bigvee_{n \in \omega} (s \textit{ is n-sequence} \wedge 0 < n \wedge$

$\wedge \bigwedge_{i<n-1} \ulcorner s_i = s_{i+1} \urcorner \in w \wedge s_0 = x \wedge s_{n-1} = y)$

Like the relation of being connected by a biconditional chain, the relation of being connected by an identity chain is an equivalence relation.

MT-8.13 w is $RC_L \wedge x, y, z \in Tm_L \rightarrow (\sim \bigvee_{s} s \textit{ is free in } x \rightarrow x\ I_{w,L}\ x) \wedge$

$\wedge (y\ I_{w,L}\ x \rightarrow x\ I_{w,L}\ y) \wedge (y\ I_{w,L}\ x \wedge x\ I_{w,L}\ z \rightarrow y\ I_{w,L}\ z)$

[proof similar to that of MT-8.11]

Corresponding to the substitutivity of identity, we have the principle that if t_1 and t_2 are connected by an identity chain and s_1, containing no free variables, is like s_2 but for containing an occurrence of t_1 where s_2 contains an occurrence of t_2, then s_1 and s_2 are connected by an identity chain.

MT-8.14 $s_1, s_2 \in Tm_L \wedge \sim \bigvee_{s_3} (s_3 \textit{ is free in } s_1 \wedge s_3 \in V) \wedge t_1\ I_{w,L}\ t_2 \wedge$

$\wedge \bigvee_{a,b} (s_1 = a\frown t_1 \frown b \wedge s_2 = a \frown t_2 \frown b) \rightarrow s_1\ I_{w,L}\ s_2$

Proof:

(1) assume hypothesis

(2) $[\bigvee_{n,x}]\ x \textit{ is n-sequence} \wedge \bigwedge_{i<n-1} (\ulcorner x_i = x_{i+1} \urcorner \in w) \wedge x_0 = t_1 \wedge$
$\wedge x_{n-1} = t_2$ — (1), MD-8.4

(3) $\{\ulcorner x_i = x_{i+1} \urcorner\} \vdash_L \ulcorner a \frown x_i \frown b = a \frown x_{i+1} \frown b \urcorner$ — logic

(4) $\bigwedge_{i<n-1} (\ulcorner a \frown x_i \frown b = a \frown x_{i+1} \frown b \urcorner \in w)$ — (3), MD-8.1

(5) Let $y = \iota f\,(\mathscr{D}f = n \wedge \bigwedge_{i<n} f_i = (a \frown x_i \frown b)$

(6) $s_1\ I_{w,L}\ s_2$ — (4), (5), MD-8.4

We also find the expected connection between identity chains and biconditional chains:

MT-8.15 $s_1, s_2 \in Tm_L \wedge \sim \bigvee_{s_3} (s_3 \textit{ is free in } s_1 \wedge s_3 \in V) \wedge t_1 \; B_{w,L} \; t_2 \wedge$

$\wedge \bigvee_{a,b} [s_1 = \widehat{a t_1} \widehat{b} \wedge s_2 = \widehat{a t_2} \widehat{b}] \rightarrow s_1 \; I_{w,L} \; s_2$

Proof:

(1) assume hypothesis

(2) $[\bigvee_{n,x}]$ x *is n-sequence* $\wedge \bigwedge_{i<n-1} \ulcorner x_i \leftrightarrow x_{i+1} \urcorner \in w) \wedge$

$\wedge x_0 = t_1 \wedge x_{n-1} = t_2$

(3) $\{\ulcorner x_i \leftrightarrow x_{i+1} \urcorner\} \vdash_L \ulcorner \widehat{a x_i} \widehat{b} = \widehat{a x_{i+1}} \widehat{b} \urcorner$ (1), logic

(4) Let $y = f(\mathscr{D} f = n \wedge \bigwedge_{i<n} (f(i) = \widehat{a x_i} \widehat{b}))$

(5) $\bigwedge_{i<n-1} (\ulcorner \widehat{a x_i} \widehat{b} = \widehat{a x_{i+1}} \widehat{b} \urcorner \in w)$ (2), (3), MD-8.1

(6) $\bigwedge_{i<n} \widehat{a x_i} \widehat{b} \in Tm_L$

(7) $s_1 \; I_{w,L} \; s_2$ (5), (6), MD-8.4

MT-8.16 $s_1, s_2 \in St_L \wedge t_1 \; I_{w,L} \; t_2 \wedge \bigvee_{a,b} (s_1 = \widehat{a t_1} \widehat{b} \wedge s_2 = \widehat{a t_2} \widehat{b}) \rightarrow s_1 \; B_{w,L} \; s_2$

(proof similar)

The notion of an inclusion chain will also come in handy. We say that the terms x and y are connected by an inclusion chain in w – in symbols, $x \; Incl_{w,L} \; y$ – when there is an n-sequence s such that s_0 is x, s_{n-1} is y, and for every i, $\ulcorner s_i \subset s_{i+1} \urcorner$ is a member of w.

MD-8.5 $x \; Incl_{w,L} \; y \leftrightarrow w$ is $RC_L \wedge \bigvee_{n \in \omega} \bigvee_{s} (s \textit{ is n-sequence} \wedge$

$\wedge \bigwedge_{i<n-1} \ulcorner s_i \subset s_{i+1} \urcorner \in w \wedge s_0 = x \wedge s_{n-1} = y)$

The relation of being connected by an inclusion chain is reflexive and transitive, but of course not symmetric.

MT-8.17 $[\sim \bigvee_{s} (s \textit{ is free in } x) \wedge x \in Tm_L \rightarrow x \; Incl_{w,L} \; x] \wedge$

$\wedge (x \; Incl_{w,L} \; y \wedge y \; Incl_{w,L} \; z \rightarrow x \; Incl_{w,L} \; z)$

Proof:

Part I $\ulcorner x \subset x \urcorner \in w$; MT-8.1

Part II

(1) $x\ Incl_{w,L}\ y \wedge y\ Incl_{w,L}\ z$ ass.

(2) $[\bigvee_{n,s}]\ s$ *is n-sequence* $\wedge \bigwedge_{i<n-1} \ulcorner s_i \subset s_{i+1} \urcorner \in w \wedge s_0 = x \wedge$
$\wedge s_{n-1} = y \wedge [\bigvee_{m,t}]\ t$ *is m-sequence* $\wedge \bigwedge_{i<m-1} \ulcorner t_i \subset t_{i+1} \urcorner \in w \wedge$
$\wedge t_0 = y \wedge t_{m-1} = z$ (1), MD-8.5

(3) $\widehat{st}$ *is* $(m+n)$*-sequence*

(4) $\bigwedge_{i<m+n} \widehat{st}(i) \in Tm_L$ if $i < n_1$, $\widehat{st}(i) = s_i$
if $n \leq i < m$, $\widehat{st}(i) = t_{i-n}$

(5) show $\bigwedge_{i<m+n-1} \ulcorner st_i \subset st_{i+1} \urcorner \in w$

(6) case I: $i < n-1$

(7) $\widehat{st}_i = s_i \wedge \widehat{st}_{i+1} = s_{i+1}$ (2)

(8) $\ulcorner \widehat{st}_i \subset \widehat{st}_{i+1} \urcorner \in w$ (7), (2)

(9) case II: $n \leq i < n+m-1$

(10) $\widehat{st}_i = t_{i-n} \wedge \widehat{st}_{i+1} = t_{i-n+1}$ (2)

(11) $\ulcorner \widehat{st}_i \subset \widehat{st}_{i+1} \urcorner \in w$ (2), (10)

(12) case III: $i = n-1$

(13) $\widehat{st}_{n-1} = y \wedge \widehat{st}_n = y$

(14) $\ulcorner \widehat{st}_i \subset \widehat{st}_{i+1} \urcorner \in w$

(15) $x\ Incl_{w,L}\ z$ (3), (4), (5), MD-8.3

Although it is not altogether obvious, the relation between inclusion chains and identity chains turns out to parallel that between inclusion and identity:

MT-8.18 $x\ Incl_{w,L}\ y \wedge y\ Incl_{w,L}\ x \leftrightarrow x\ I_{w,L}\ y$

Proof:

(1) From right to left this is obvious, since $\{s = t\} \vdash_L \ulcorner s \subset t \urcorner$.

(2) $x\ Incl_{w,L}\ y \wedge y\ Incl_{w,L}\ x$ ass.

(3) $[\bigvee_{n,s}]\ s$ *is n-sequence* $\wedge \bigwedge_{i<n-1} \ulcorner s_i \subset s_{i+1} \urcorner \in w \wedge$

$$\wedge\, s_0 = x \wedge s_{n-1} = y \wedge [\bigvee_{m,t}]\ t \text{ is } m\text{-sequence} \wedge$$

$$\wedge \bigwedge_{i<m-1} \ulcorner t_i \subset t_{i+1} \urcorner \in w \wedge t_0 = y \wedge t_{m-1} = x$$

(4) ~~show~~ $x\, I_{w,L} \ulcorner s_0 \cap s_1 \cap \ldots \cap s_{n-1} \urcorner$

(5) ~~show~~ $\bigwedge_{i<n-1} (x\, I_{w,L} \ulcorner s_0 \cap s_1 \cap \ldots \cap s_i \urcorner \rightarrow$

$\rightarrow x\, I_{w,L} \ulcorner s_0 \cap \ldots \cap s_{i+1} \urcorner)$

(6) $i < n-1 \wedge x\, I_{w,L} \ulcorner s_0 \cap \ldots \cap s_i \urcorner$ ass.

(7) $\ulcorner s_i \subset s_{i+1} \urcorner \in w$ (3), (6)

(8) $\ulcorner s_i = s_i \cap s_{i+1} \urcorner \in w$ (7), MD-8.1

(9) $\ulcorner (s_0 \cap \ldots \cap s_{i-1}) \cap s_i = (s_0 \cap \ldots \cap s_{i-1}) \cap (s_i \cap s_{i+1}) \urcorner \in w$
(8), MD-8.1

(10) $\ulcorner s_0 \cap \ldots \cap s_i = s_0 \cap \ldots \cap s_{i+1} \urcorner \in w$ (9), MD-8.1

(11) $\ulcorner s_0 \cap \ldots \cap s_i \urcorner\, I_{w,L} \ulcorner s_0 \cap \ldots \cap s_{i+1} \urcorner$ (10), MD-8.4

(12) $x\, I_{w,L} \ulcorner s_0 \cap s_1 \cap \ldots \cap s_{i+1} \urcorner$ (6), MT-8.13, (11)

(13) $x\, I_{w,L}\, s_0$ (3), MT-8.13

(14) $y\, I_{w,L} \ulcorner t_0 \cap t_1 \cap \ldots \cap t_{m-1} \urcorner$ similarly

(15) $\ulcorner y \cap x \urcorner\, I_{w,L} \ulcorner t_0 \cap t_1 \cap \ldots \cap t_{m-1} \cap x \urcorner$ (14), MT-8.14

(16) $\ulcorner x \cap y \urcorner\, I_{w,L} \ulcorner s_0 \cap s_1 \cap \ldots \cap s_{n-1} \cap y \urcorner$ (4), MT-8.14

(17) $\ulcorner t_0 \cap \ldots \cap t_{m-1} \cap x \urcorner\, I_{w,L} \ulcorner t_0 \cap \ldots \cap t_{m-1} \urcorner$ (3), MD-8.1

(18) $\ulcorner s_0 \cap \ldots \cap s_{n-1} \cap y \urcorner\, I_{w,L} \ulcorner s_0 \cap \ldots \cap s_{n-1} \urcorner$ (3), MD-8.1

(19) $\ulcorner x \cap y \urcorner\, I_{w,L} \ulcorner y \cap x \urcorner$ MD-8.1

(20) $x\, I_{w,L}\, y$ (4), (18), (16), (19), MT-8.13, (15), (17), (14)

When do we wish to take account of the membership of x in y? Surely not just when $\ulcorner x \in y \urcorner$ is a member of our rational corpus, unless, of course, we have opted for complete deductive closure. But not whenever there is a deduction of $\ulcorner x \in y \urcorner$ from statements in our rational corpus, either, for (as we remarked earlier) it seems to be begging some interesting questions to suppose that our body of knowledge is consistent in the very strong sense that no arbitrary statement is deducible from it. If we know $\ulcorner x \in y \urcorner$ and $\ulcorner y \subset z \urcorner$ it seems reasonable to feel that we should also take $\ulcorner x \in z \urcorner$ into account; and if we know $\ulcorner x \in y \urcorner$ and $\ulcorner x' \in y' \urcorner$, then surely we should take account of $\ulcorner\langle x, x'\rangle \in \{\langle z, z'\rangle : z \in y \wedge z' \in y'\}\urcorner$. And, again, if $\ulcorner x \in y\urcorner$ belongs to our body of knowledge, and it is connected therein by a biconditional chain to $\ulcorner x' \in y'\urcorner$, it seems that that, too, should be acknowledged. Putting all this together, we shall say that x and y are connected by a membership chain – in symbols $x\, M_{w,L}\, y$ – when there is an n-sequence of terms s, an n-sequence of terms t, and a term z, such that all are terms of L, and for every i $\ulcorner s_i \in t_i\urcorner$ belongs to w, $\ulcorner s_0 \in t_0 \wedge s_1 \in t_1 \wedge \ldots \wedge s_{n-1} \in t_{n-1}\urcorner\, B_{w,L}\, \ulcorner x \in z\urcorner$ and $z\, Incl_{w,L}\, y$.

MD-8.6 $x\, M_{w,L}\, y \leftrightarrow \bigvee_{n\in\omega} \bigvee_{s,t,z} (s \textit{ is } n\textit{-sequence} \wedge t \textit{ is } n\textit{-sequence} \wedge$
$\wedge \bigwedge_{i<n} \ulcorner s_i \in t_i\urcorner \in w \wedge w \text{ is } RC_L \wedge$
$\ulcorner\langle s_0, \ldots, s_{n-1}\rangle \in \prod_{i<n} t_i\urcorner\, B_{w,L} \ulcorner x \in z\urcorner \wedge z\, Incl_{w,L}\, y)$

MT-8.19a. $\ulcorner x \in y\urcorner \in w \wedge w \text{ is } RC_L \rightarrow x M_{w,L}\, y$

Proof:

Let $s = \langle x\rangle, t = \langle y\rangle; \ulcorner s_0 \in t_0\urcorner = \ulcorner x \in y\urcorner \in w;\ y\, Incl_{w,L}\, y$.

MT-8.19b. $\ulcorner x \in y\urcorner \in w \wedge \ulcorner x' \in y'\urcorner \in w \wedge w \text{ is } RC_L \rightarrow$
$\rightarrow \ulcorner\langle x, x'\rangle\urcorner\, M_{w,L} \ulcorner\{\langle z, z'\rangle : z \in y \wedge z' \in y'\}\urcorner$

Proof:

Let $s = \langle x, x'\rangle, t = \langle y, y'\rangle; \ulcorner s_0 \in t_0\urcorner = \ulcorner x \in y\urcorner, \ulcorner s_1 \in t_1\urcorner = \ulcorner x' \in y'\urcorner \cdot$
$\cdot \ulcorner x \in y \wedge x' \in y' \leftrightarrow \langle x, x'\rangle \in \{\langle z, z'\rangle : z \in y \wedge z' \in y'\}\urcorner \in w$
$\ulcorner x \in y \wedge x' \in y'\urcorner\, B_{w,L} \ulcorner\langle x, x'\rangle \in \{\langle z, z'\rangle : z \in y \wedge z' \in y'\}\urcorner \cdot$
$\cdot \ulcorner\langle x, x'\rangle \in \{\langle z, z'\rangle : z \in y \wedge z' \in y'\}\urcorner\, B_{w,L} \ulcorner\langle x, x'\rangle \in \{\langle z, z'\rangle z \in y \wedge$
$\wedge z' \in y'\}\urcorner$

MT-8.20 $x\, M_{w,L}\, y \wedge x'\, M_{w,L}\, y' \rightarrow \ulcorner\langle x, x'\rangle\urcorner\, M_{w,L} \ulcorner\{\langle z, z'\rangle : z \in y \wedge$
$\wedge z' \in y'\}\urcorner$

Proof:

(1) assume hypothesis

(2) $[\bigvee_{nstz}]\ s \textit{ is } n\textit{-sequence} \wedge t \textit{ is } n\textit{-sequence} \wedge$
$\wedge \bigwedge_{i<n} \ulcorner s_i \in t_i\urcorner \in w \wedge \ulcorner s_0 \in t_0 \wedge \ldots \wedge s_{n-1} \in t_{n-1}\urcorner$
$B_{w,L} \ulcorner x \in z\urcorner \wedge z\, Incl_{w,L}\, y \wedge [\bigvee_{n's't'z'}]\ s' \textit{ is } n'\textit{-sequence} \wedge$

$\wedge t'$ *is n'-sequence* $\wedge \bigwedge_{i<n'} \ulcorner s'_i \in t'_i \urcorner \in w \wedge \ulcorner s'_0 \in t'_0 \wedge \ldots \wedge$

$\wedge s'_{n-1} \in t'_{n-1} \urcorner B_{w,L} \ulcorner x' \in z' \urcorner \wedge z'$ $Incl_{w,L}$ y'

(3) $\widehat{ss'}$ *is n+n'-sequence* $\wedge \widehat{tt'}$ *is n+n'-sequence* $\wedge$

$\wedge \bigwedge_{i<n+m} \ulcorner \widehat{ss'_i} \in \widehat{tt'_i} \urcorner \in w$

(4) $\ulcorner \widehat{ss'_0} \in \widehat{tt'_0} \wedge \widehat{ss'_1} \in \widehat{tt'_1} \wedge \ldots \wedge \widehat{ss'}_{n+n'-1} \in \widehat{tt'}_{n+n'-1} \urcorner$

$B_{w,L} \ulcorner x \in z \wedge x' \in z' \urcorner$ (2), MT-8.12

(5) $\ulcorner x \in z \wedge x' \in z' \urcorner B_{w,L} \ulcorner \langle x, x' \rangle \in \{\langle r, r' \rangle : r \in z \wedge r' \in z'\} \urcorner$

(6) ~~show~~ z $Incl_{w,L}$ $y \wedge z'$ $Incl_{w,L}$ $y' \rightarrow \ulcorner \{\langle r, r' \rangle \text{-} r \in z \wedge$

$\wedge r' \in z'\} \urcorner$ $Incl_{w,L} \ulcorner \{\langle r, r' \rangle : r \in y \wedge r' \in y'\} \urcorner$

(7) z $Incl_{w,L}$ $y \wedge z'$ $Incl_{w,L}$ y' ass.

(8) $[\bigvee_{n\in\omega} \bigvee_{s}]$ *s is n-sequence* $\wedge \bigwedge_{i<n-1} \ulcorner s_i \subset s_{i+1} \urcorner \in w \wedge$

$\wedge s_0 = z \wedge s_{n-1} = y \wedge [\bigvee_{m\in\omega} \bigvee_{t}]$ *t is m-sequence* $\wedge$

$\wedge \bigwedge_{i<m-1} \ulcorner t_i \subset t_{i+1} \urcorner \in w \wedge t_0 = z' \wedge t_{m-1} = y'$

(9) Let $v = \imath f(\mathscr{D}f = n+m \wedge f$ is fun $\wedge \bigwedge_{i<n} f(i) =$

$= \ulcorner \{\langle r, r' \rangle : r \in s_i \wedge r' \in z'\} \urcorner \wedge \bigwedge_{n \le i < n+m} f(i) =$

$= \ulcorner \{\langle r, r' \rangle : r \in y \wedge r' \in t_{i-n}\} \urcorner)$

(10) ~~show~~ $\bigwedge_{i<n+m-1} \ulcorner v_i \subset v_{i+1} \urcorner \in w$

(11) $i < n+m-1$ ass.

(12) case I: $i < n-1$ ass.

(13) $\{\ulcorner s_i \subset s_{i+1} \urcorner\} \vdash_L \ulcorner \{\langle r, r' \rangle : r \in s_i \wedge r' \in z'\} \subset$

$\subset \{\langle r, r' \rangle : r \in s_{i+1} \wedge r' \in z'\} \urcorner$

(14) $\ulcorner r_i \subset r_{i+1} \urcorner \in w$ (13), (8), MD-8.1

(15) case II: $i = n-1$

(16) $v_i = \ulcorner \{\langle r, r' \rangle : r \in y \wedge r' \in z'\} \urcorner$; $r_{i+1} = \ulcorner \{\langle r, r' \rangle : r \in y \wedge$

$\wedge r' \in z'\} \urcorner$

(17) $0 \vdash_L \ulcorner v_i \subset v_{i+1} \urcorner$

(18) case III: $n \le i < n+m-1$

(19) $\{\ulcorner t_{i-n} \subset t_{i+1-n} \urcorner\} \vdash_L \ulcorner\{\langle r, r'\rangle : r \in y \wedge r' \in t_{i-n}\} \subset$
$\subset \{\langle r, r'\rangle : r \in y \wedge r' \in t_{i-n+1}\}\urcorner$

(20) $\ulcorner v_i \subset v_{i+1} \urcorner \in w$ (8), (19), MD-8.1

(21) $\ulcorner\{\langle r, r'\rangle : r \in z \wedge r' \in z'\}\urcorner \; Incl_{w,L} \ulcorner\{\langle r, r'\rangle : r \in y \wedge r' \in y\}\urcorner$ (9), (10)

(22) $\ulcorner\langle x, x'\rangle\urcorner \; M_{w,L} \ulcorner\{\langle r, r'\rangle : r \in y \wedge r' \in y'\}\urcorner$ (3), (4), (5), MT-8.11, (2), (6)

MT-8.21 x_0 is finite sequence $\wedge\, x_1$ is finite sequence $\wedge\, x_0 \; M_{w,L} \; y_0 \wedge$
$\wedge\, x_1 \; M_{w,L} \; y_1 \rightarrow x_0 \frown x_1 \; M_{w,L} \ulcorner y_0 \times y_1 \urcorner$

Proof: similar to that of preceding theorem.

MT-8.22 $x \; M_{w,L} \; y \wedge x \; I_{w,L} \; x' \wedge y \; I_{w,L} \; y' \rightarrow x' \; M_{w,L} \; y'$

Proof:

(1) assume antecedent

(2) $[\bigvee_{n \in \omega} \bigvee_{s,t,z}]$ (*s is n-sequence* $\wedge$ *t is n-sequence* $\wedge$
$\wedge \bigwedge_{i<n} (\ulcorner s_i \in t_i \urcorner \in w) \wedge \ulcorner s_0 \in t_0 \wedge \ldots \wedge s_{n-1} \in t_{n-1} \urcorner$
$B_{w,L} \ulcorner x \in z \urcorner \wedge z \; Incl_{w,L} \; y$

(3) $\ulcorner s_0 \in t_0 \wedge \ldots \wedge s_{n-1} \in t_{n-1} \urcorner \; B_{w,L} \ulcorner x' \in z \urcorner$ (1), (2), MT-8.16

(4) $z \; Incl_{w,L} \; y'$ (1), (2), MT-8.18, MT-8.17

(5) $x' \; M_{w,L} \; y'$ (2), (3), (4)

MT-8.23 $y \; Incl_{w,L} \; y' \wedge x \; M_{w,L} \; y \rightarrow x \; M_{w,L} \; y'$

MT-8.24 $\ulcorner x \in y \urcorner \; B_{w,L} \ulcorner x' \in y' \urcorner \wedge x \; M_{w,L} \; y \rightarrow x' \; M_{w,L} \; y'$

MT-8.25 $x \; M_{w,L} \; y \wedge x \; M_{w,L} \; y' \rightarrow x \; M_{w,L} \ulcorner y \cap y' \urcorner$

Proof:

(1) $x \; M_{w,L} \; y \wedge x \; M_{w,L} \; y'$ ass.

(2) $[\bigvee_{n \in \omega} \bigvee_{s,t,z}]$ *s is n-sequence* $\wedge$ *t is n-sequence* $\wedge$
$\wedge \bigwedge_{i \in n} \ulcorner s_i \in t_i \urcorner \in w \wedge \{\ulcorner\langle s_0, \ldots, s_{n-1}\rangle \in \prod_{i<n} t_i \urcorner\} \vdash_L \ulcorner x \in z \urcorner \wedge$
$\wedge\, z \; Incl_{w,L} \; y \wedge [\bigvee_{n' \in \omega} \bigvee_{s',t',z'}]$ *s′ is n′-sequence* $\wedge$

$\wedge t'$ *is* n'-*sequence* $\wedge \bigwedge_{i<n'} \ulcorner s'_i \in t'_i \urcorner \in w \wedge \{\ulcorner \langle s'_0, \ldots, s'_{n'-1} \rangle \in$

$\in \prod_{i<n'} t'_i \urcorner\} \vdash_L \ulcorner x \in z' \urcorner \wedge z' \; Incl_{w,L} \; y'$ (1), MD-8.6

(3) $\widehat{ss'}$ *is* $n+n'$-*sequence* $\wedge \widehat{tt'}$ *is* $n+n'$-*sequence* $\wedge$

$\wedge \bigwedge_{i<n+n'} \ulcorner \widehat{ss'_i} \in \widehat{tt'_i} \urcorner \in w$ (2), logic

(4) $\{\ulcorner \langle \widehat{ss'_0}, \ldots, \widehat{ss'}_{n+n'-1} \rangle \in \prod_{i<n+n'} \widehat{tt'_i} \urcorner\} \vdash_L \ulcorner x \in z \cap z' \urcorner$ (2), logic

(5) show $z \; Incl_{w,L} \; y \wedge z' \; Incl_{w,L} \; y' \rightarrow \ulcorner z \cap z' \urcorner \; Incl_{w,L} \ulcorner y \cap y' \urcorner$

(6) $z \; Incl_{w,L} \; y \wedge z' \; Incl_{w,L} \; y'$ ass.

(7) $[\bigvee_{m \in \omega} \bigvee_r] \; r$ *is* m-*sequence* $\wedge \bigwedge_{i<m-1} \ulcorner r_i \subset r_{i+1} \urcorner \in w \wedge$

$\wedge z = r_0 \wedge y = r_{m-1} \wedge [\bigvee_{m' \in \omega} \bigvee_{r'}] \; r'$ *is* m'-*sequence* $\wedge$

$\wedge \bigwedge_{i<m'-1} \ulcorner r'_i \subset r'_{i+1} \urcorner \in w \wedge z' = r'_0 \wedge y' = r'_{m'-1}$

(6), MD-8.5

(8) $\ulcorner z \cap z' \urcorner \; Incl_{w,L} \ulcorner y \cap z' \urcorner$ since $\{\ulcorner r_i \subset r_{i+1} \urcorner\} \vdash_L \ulcorner r_i \cap$ $\cap z' \subset r_{i+1} \cap z' \urcorner$

(9) $\ulcorner y \cap z' \urcorner \; Incl_{w,L} \ulcorner y \cap y' \urcorner$ similarly

(10) $\ulcorner z \cap z' \urcorner \; Incl_{w,L} \ulcorner y \cap y' \urcorner$ (8), (9), MT-8.17

(11) $x \; M_{w,L} \ulcorner y \cap y' \urcorner$ (3), (4), (5), MD-8.6

6. Among the statements of particular interest to us in a body of knowledge are statistical statements. Since for every positive real numbers δ and ε, if $\ulcorner S \langle x, y, z, p, q \rangle \urcorner$ belongs to a rational corpus, $\ulcorner S \langle x, y, z, p - \varepsilon, q + \delta \rangle \urcorner$ also belongs to it by weak deductive closure, we need to be able to focus on the *strongest* statement about a given subject matter. There is one small problem: a rational corpus may have as a member *no* strongest statement about a given subject matter. For example, w may be a rational corpus containing, for every n, the statement $\ulcorner S \langle x, y, z, \frac{1}{4} - 1/2^n, \frac{1}{3} + 1/2^n \rangle \urcorner$, but not containing $\ulcorner S \langle x, y, z, \frac{1}{4}, \frac{1}{3} \rangle \urcorner$. Thus we shall define a strongest statistical statement about the terms x, y, and z of a *rational* corpus to be a statement stronger than or as strong as any statement in the rational corpus and not stronger than any statement (except for equally strong ones) not in that rational corpus. We use

'$PS_{w,L}(s, x, y, z)$' to express the assertion that s is a strongest statistical statement of the rational corpus w of language L containing the subject terms x, y, and z.

MD-8.7 $PS_{w,L}(s, x, y, z) \leftrightarrow w \text{ is } RC_L \wedge \bigvee_{p,q} s = \ulcorner S\langle x, y, z, p, q\rangle \urcorner \wedge$
$\wedge \bigwedge_{p,q} ((\ulcorner S\langle x, y, z, p, q\rangle\urcorner \in w \wedge SS_{w,L} \ulcorner S\langle x, y, z, p, q\rangle\urcorner \rightarrow$
$\rightarrow s\ STR_{w,L} \ulcorner S\langle x, y, z, p, q\rangle\urcorner \vee s\ E\text{-}STR\ S\langle x, y, z, p, q\rangle\urcorner) \wedge$
$\wedge (\sim \ulcorner S\langle x, y, z, p, q\rangle\urcorner \in w \wedge SS_{w,L} \ulcorner S\langle x, y, z, p, q\rangle\urcorner \rightarrow$
$\rightarrow \sim s\ STR_{w,L} \ulcorner S\langle x, y, z, p, q\rangle\urcorner))$

After such a complicated definition it is clearly important to establish that it does what we expect it to – i.e., to show that if s and s' are strongest statements about x, y, and z of the rational corpus w in the language L, then s and s' are equally strong; and that if there is a statistical statement in w about x, y, and z, then there is (usually) a strongest one. Furthermore, we observe that if x is a reference term, y a random variable, and z a Borel term, then there is a statistical statement about x, y, and z in every rational corpus (of any language in which the hypothesis is true). From all this, it follows as the night the day, that if x is a reference term of L, y a random variable of L for the rational corpus w, and z a Borel term, then there is a statistical statement of w about x, y, and z. We always have *some* statistical knowledge.

MT-8.26 $PS_{y,L}(x, z, w, v) \wedge PS_{y,L}(x', z, w, v) \rightarrow x\ E\text{-}STR_{y,L}\ x'$

Proof:

(1) $PS_{y,L}(x, z, w, v) \wedge PS_{y,L}(x', z, w, v)$ ass.

(2) ~~show~~ $\sim x\ STR_{y,L}\ x'$

(3) $x\ STR_{y,L}\ x'$ ass.

(4) $[\bigvee_{p,q,p',q'}]\ x = \ulcorner S\langle z, w, v, p, q\rangle\urcorner \wedge$
$\wedge\ x' = \ulcorner S\langle z, w, v, p', q'\rangle\urcorner \wedge 0 \vdash_L \ulcorner p \geq p' \wedge q \leq q'\urcorner \wedge$
$\wedge \sim 0 \vdash_L \ulcorner p = p' \wedge q = q'\urcorner$ (3), (1), MD-7.41, 7.42, 7.43

(5) Let $x_0 = \ulcorner S\left\langle z, w, v, p - \frac{p-p'}{2}, q + \frac{q'-q}{2}\right\rangle\urcorner$

(6) $x\ STR_{y,L}\ x_0 \wedge x_0\ STR_{y,L}\ x'$ (4), (5), MD-7.42, 7.43

(7) case I: $x_0 \in y$

(8) $x' \; STR_{y,L} \; x_0$ — (7), (1), MD-8.7 (x')

(9) $\sim x_0 \; STR_{y,L} \; x'$ — (8), MT-7.11; imp 6.

(10) case II: $\sim x_0 \in y$

(11) $\sim x \; STR_{y,L} \; x_0$ — (10), (1), MD-8.7 (x); imp. 6

(12) $x' \; STR_{y,L} \; x$ — similarly

(13) show $\sim x \; DIF_{y,L} \; x'$

(14) $x \; DIF_{y,L} \; x'$ — ass.

(15) $[\bigvee_{p,q,p',q'}] \; x = \ulcorner S\langle z, w, v, p, q\rangle \urcorner \wedge$
$\wedge \, x' = \ulcorner S\langle z, w, v, p', q'\rangle \urcorner \wedge \sim 0 \vdash_L \ulcorner p \geq p' \wedge q \leq q' \wedge$
$\wedge \sim 0 \vdash_L \ulcorner (p' \geq p \wedge q' \leq q) \urcorner \wedge \sim 0 \vdash_L \ulcorner p = p' \wedge q = q' \urcorner$
(14), (1), MD-7.41, 7.42, 7.43, 7.44

(16) Let $x'' = \ulcorner S\left\langle z, w, v, p - \frac{|p'-p|}{2}, q + \frac{|q'-q|}{2}\right\rangle \urcorner$

(17) Let $x''' = \ulcorner S\left\langle z, w, v, p' - \frac{|p'-p|}{2}, q' + \frac{|q'-q|}{2}\right\rangle \urcorner$

(18) $x \; STR_{y,L} \; x'' \wedge x' \; STR_{y,L} \; x'''$
(15), (16), (17), MD-7.42, 7.43

(19) $x'' \in y \wedge x''' \in y$ — (18), MD-8.1, MD-7.43

(20) $(x' \; STR_{y,L} \; x'' \vee x' \; E\text{-}STR_{y,L} \; x'') \wedge (x \; STR_{y,L} \; x''' \vee$
$\vee \, x \; E\text{-}STR_{y,L} \; x''')$ — (16), (17), (19), (1), MD-8.26

(21) $0 \vdash_L \ulcorner p - \frac{|p'-p|}{2} \leq p' \wedge q' \leq q + \frac{|q'-q|}{2} \urcorner$
(16), (20), MD-7.42, 7.43

(22) $0 \vdash_L \ulcorner p' - \frac{|p'-p|}{2} \leq p \wedge q \leq q' + \frac{|q'-q|}{2} \urcorner$
(17), (20), MD-7.42, 7.43

(23) $0 \vdash_L \ulcorner p - p' \leq \frac{|p'-p|}{2} \wedge q - q' \leq \frac{|q'-q|}{2} \urcorner$ — (21), (22)

(24) $0 \vdash_L \ulcorner p = p' \wedge q = q' \urcorner$ — imp. 14, MD-7.44

(25) $x \; E\text{-}STR_{y,L} \; x'$ — (13), (12), (2), MT-7.13

There are circumstances in which a rational corpus might contain a statistical statement about z, w, and v without having a strongest one. Thus, suppose that $x = \ulcorner S\langle z, w, v, 0.4, 0.6\rangle \urcorner$ and $x' = \ulcorner S\langle z, w, v, 0.3, 0.5\rangle \urcorner$, and let $y = L\text{-}Cn(\{x\}) \cup L\text{-}Cn(\{x'\}) \cup L\text{-}Cn(\{\ulcorner \mathcal{R}w \subset \mathbb{R}^n \wedge w \text{ is function} \urcorner\})$, where $z, v, w, \varepsilon\ Tm_L$. Then y is RC_L. But there is no strongest statistical statement of y about x, v, and w: In order for it to be stronger than both x and x' it must be at least as strong as $\ulcorner S\langle z, w, v, 0.4, 0.5\rangle \urcorner$ but then there are many statements not in y that it is also stronger than – for example, $\ulcorner S\langle z, w, v, 0.35, 0.55\rangle \urcorner$.

A rational corpus in which this cannot happen we call *statistically closed.* A statistically closed rational corpus is one that contains all the logical consequences of any set of statistical statements about a given subject matter.

MD-8.8 y *is statistically closed*$_L \leftrightarrow y$ is $RC_L \wedge \bigwedge_{z,w} [L\text{-}Cn(\{x: \bigvee_{p,q,v} \ulcorner S\langle z, w, v, q, p\rangle \urcorner = x \wedge SS_{y,L}\, x \wedge x \in y\}) \subset y]$

Obviously deductive closure implies statistical closure, but not conversely:

MT-8.27 w *is deductively closed* $\wedge\, w \subset ST_L \rightarrow w$ *is statistically closed*$_L$

Statistical closure, being restricted to statements having 'the same subject matter' in a very strong sense, seems reasonable enough. Note that a statistical statement about mankind, a statistical statement about rational animals, and a statistical statement about featherless bipeds, would all have different subject matters in this sense, even if they were otherwise identical. It might well be regarded as a criterion of an inductive acceptance rule that it should lead only to statistically closed rational corpora.

To avoid the Gödelian complications alluded to in paragraph five of Chapter VII, we must restrict our attention to rational corpora that only contain statistical statements with Gödelian parameters when these are consequences of stronger, non-Gödelian statements. Such rational corpora we shall call *Normal.* This is a plausible enough restriction on bodies of knowledge; surely the only way we can come to *know* such an unusual statistical statement is to know one in the usual way that implies it. We could avoid this problem by providing a semantical criterion of strength and difference; but that seems both unnecessary (for any other purpose) and unnatural (for present purposes) at this stage of our development.

Let D_L be the set of digits of L. Thus if L' uses a binary notation $D'_L = \{\text{'0'}, \text{'1'}\}$; if L'' uses a decimal notation, $D_{L''} = \{\text{'0'}, \text{'1'}, \text{'2'}, \text{'3'}, \text{'4'}, \text{'5'}, \text{'6'}, \text{'7'}, \text{'8'}, \text{'9'}\}$. An infinite sequence of digits is then a function from ω to D_L, i.e., a member of D^{ω}.'$\bar{\bar{D}}_L$' denotes the cardinality of D_L.

MD-8.9 $$\textit{Normal}_L\ y \leftrightarrow \bigwedge_{p, q, z, w, v} (\ulcorner S\langle z, w, v, p, q\rangle\urcorner \in y \wedge$$
$$\wedge\, SS_{y,L}\ulcorner S\langle z, w, v, p, q\rangle\urcorner \rightarrow \bigvee_{p', q'} (p' \in D_L^{\omega} \wedge q' \in D_L^{\omega} \wedge$$
$$\wedge \bigwedge_{i \in \omega} \bigvee_{d_1} \bigvee_{d_2} (d_1, d_2 \in D_L \wedge 0 \vdash_L \ulcorner p'_i = d_1 \wedge q'_i = d_2\urcorner) \wedge$$
$$\wedge\, \ulcorner S\langle z, w, v, \sum_{i<\infty} p'_i \cdot \bar{\bar{D}}_L^{-(i+1)}, \sum_{i<\infty} q'_i \cdot \bar{\bar{D}}_L^{-(i+1)}\rangle\urcorner \in y \wedge$$
$$\wedge\, 0 \vdash_L \ulcorner p \leq \sum_{i<\infty} p'_i \cdot \bar{\bar{D}}_L^{-(i+1)} \wedge \sum_{i<\infty} q'_i \cdot \bar{\bar{D}}_L^{-(i+1)} \leq q\urcorner))$$

Now we can prove that if a normal and statistically closed rational corpus having a basis contains a statistical statement above z, w, and v, then it contains a strongest statement about z, w, and v.

MT-8.23 $$y \textit{ is statistically closed}_L \wedge \textit{Normal}_L\ y \wedge \bigvee_z z \textit{ is basis}_L \textit{ for } y \wedge$$
$$\wedge \bigvee_{p,q} \ulcorner S\langle z, w, v, p, q\rangle\urcorner \in y \wedge SS_{y,L}\ulcorner S\langle z, w, v, p, q\rangle\urcorner \rightarrow$$
$$\rightarrow \bigvee_x PS_{y,L}(x, z, w, v)$$

Proof:

(1) Assume the antecedent, and, for simplicity, that $D_L = D_{L''}$.

(2) Let $p = \imath k(k \in D_L^{\omega} \wedge \bigwedge_{i \in \omega} (k(i) = \imath d(d \in D_L \wedge \bigvee_q (q \in \textit{Math}$
$\textit{Terms}_L \wedge \ulcorner S\langle z, w, v, \sum_{j<i} k(j) \cdot 10^{-(j+1)} + d \cdot 10^{-(i+1)}, q\rangle\urcorner \in$
$\in y) \wedge \sim \bigvee_q (q \in \textit{Math Terms}_L \wedge \ulcorner S\langle z, w, v, \sum_{j<i} k(j) \cdot$
$\cdot 10^{-(j+1)} + d \cdot 10^{-(i+1)} + 10^{-(i+1)}, q\rangle\urcorner \in y)))$

(3) Let $q = \imath k(k \in D_L^{\omega} \wedge \bigwedge_{i \in \omega} (k(i) = \imath d(d \in D_L \wedge \sim \bigvee_q (q \in \textit{Math}$
$\textit{Terms}_L \wedge S\langle z, w, v, q, \sum_{j<i} k(j) \cdot 10^{-(j+1)} + d \cdot 10^{-(i+1)}\rangle\urcorner \in$

$\in y) \wedge \bigvee_{q} (q \in Math\ Terms_L \wedge \ulcorner S\langle z, w, v, q, \sum_{j<i} k(j) \cdot$
$\cdot 10^{-(j+1)} + d \cdot 10^{-(i+1)} + 10^{-(i+1)}\rangle\urcorner \in y)))$

(4) The descriptions of 2 and 3 are proper, i.e., $p \neq 0 \wedge q \neq 0$

(5) Let $x = \ulcorner S\langle z, w, v, \sum_{i \in \omega} p(i) \cdot 10^{-(i+1)}, \sum_{i \in \omega} q(i) \cdot 10^{-(i+1)}\rangle\urcorner$

(6) $SS_{y,L}\ x \wedge y$ is RC_L (1), (2), (3), (5), MD-7.41

(7) ~~Show~~ $\ulcorner S\langle z, w, v, p', q'\rangle\urcorner \in y \wedge SS_{y,L} \ulcorner S\langle z, w, v, p', q'\rangle\urcorner \rightarrow$
$\rightarrow x\ STR_{y,L} \ulcorner S\langle z, w, v, p', q'\rangle\urcorner \vee$
$\vee x\ E\text{-}STR_{y,L} \ulcorner S\langle z, w, v, p', q'\rangle\urcorner$

(8) $\ulcorner S\langle z, w, v, p', q'\rangle\urcorner \in y \wedge SS_{y,L} \ulcorner S\langle z, w, v, p', q'\rangle\urcorner$ ass.

(9) ~~show~~ $0 \vdash_L \ulcorner p' \leq \sum_{i \in \omega} p(i) \cdot 10^{-(i+1)} \wedge$
$\wedge \sum_{i \in \omega} q(i) \cdot 10^{-(i+1)} \leq q'\urcorner$

(10) $[\bigvee_{p'', q''}]\ p'' \in D_L^{\omega} \wedge q'' \in D_L^{\omega} \wedge \bigwedge_{i \in \omega} \bigvee_{d', d'' \in D_L} (0 \vdash_L \ulcorner p''(i) = d'$
$\wedge q''(i) = d''\urcorner) \wedge \ulcorner S\langle z, w, v, \sum_{i \in \omega} p''(i) \cdot 10^{-(i+1)},$
$\sum_{i \in \omega} q''(i) \cdot 10^{-(i+1)}\rangle\urcorner \in y \wedge 0 \vdash_L \ulcorner p' \leq \sum_{i \in \omega} p''(i) \cdot$
$\cdot 10^{-(i+1)} \wedge \sum_{i \in \omega} q''(i) \cdot 10^{-(i+1)} \leq q'\urcorner$

(1), (9), Normality

(11) ~~show~~ $0 \vdash_L \ulcorner \sum_{i \in \omega} p''(i) \cdot 10^{-(i+1)} \leq \sum_{i \in \omega} p(i) \cdot 10^{-(i+1)}\urcorner$

(12) Case I $\bigwedge_{i \in \omega} 0 \vdash_L \ulcorner p''(i) = p(i)\urcorner$

(13) Case II $\sim \bigwedge_{i \in \omega} 0 \vdash_L \ulcorner p''(i) = p(i)\urcorner$

(14) Let $n = \imath s(s \in \omega \wedge \sim 0 \vdash_L \ulcorner p''(s) = p(s)\urcorner \wedge$
$\wedge \bigwedge_{r<s} 0 \vdash_L \ulcorner p''(r) = p(r)\urcorner)$

(15) $0 \vdash_L \ulcorner p(n) < p''(n)\urcorner \vee 0 \vdash_L \ulcorner p''(n) \leq p(n)\urcorner$

since $p(n), p''(n) \in D_L$

(16) ~~show~~ $\sim 0 \vdash_L \ulcorner p(n) < p''(n)\urcorner$

(17) $0 \vdash_L \ulcorner p(n) < p''(n)\urcorner$ ass.

(18) $\ulcorner S\langle z, w, v, \sum_{i<n+1} p''(i) \cdot 10^{-(i+1)}, \sum_{i \in \omega} q''(i) \cdot$

$\cdot 10^{-(i+1)}\rangle^\urcorner \in y$ (1), (10) closure

(19) $\ulcorner S\langle z, w, v, \sum_{i<n} p(i)\cdot 10^{-(i+1)} + 10^{-(n+1)},$
$\sum_{i\in\omega} q(i)\cdot 10^{-(i+1)}\rangle^\urcorner \in y$
(1), (17), (18), closure; imp. by 2.

(20) $0 \vdash_L \ulcorner \sum_{i\in\omega} p''(i)\cdot 10^{-(i+1)} \leq \sum_{i\in\omega} p(i)\cdot 10^{-(i+1)\urcorner}$
(12), (15), (16)

(21) ~~show~~ $0 \vdash_L \ulcorner \sum_{i\in\omega} q(i)\cdot 10^{-(i+1)} \leq \sum_{i\in\omega} q''(i)\cdot 10^{-(i+1)\urcorner}$
similarly

(22) $0 \vdash_L \ulcorner p' \leq \sum_{i\in\omega} p(i)\cdot 10^{-(i+1)} \wedge \sum_{i\in\omega} q(i)\cdot 10^{-(i+1)} \leq q'^\urcorner$
(10), (11), (21)

(23) $x\ STR_{y,L}\ulcorner S\langle z, w, v, p', q'\rangle^\urcorner \vee$
$\vee\ x\ E\text{-}STR_{y,L}\ulcorner S\langle z, w, v, p', q'\rangle^\urcorner$
(22), MD-7.42, 7.43

(24) ~~Show~~ $SS_{y,L}\ulcorner S\langle z, w, v, p', q'\rangle^\urcorner \wedge$
$\wedge \sim\ulcorner S\langle z, w, v, p', q'\rangle^\urcorner \in y \rightarrow$
$\rightarrow \sim x\ STR_{y,L}\ulcorner S\langle z, w, v, p', q'\rangle^\urcorner$

(25) $SS_{y,L}\ulcorner S\langle z, w, v, p', q'\rangle^\urcorner \wedge \sim\ulcorner S\langle z, w, v, p', q'\rangle^\urcorner \in y$
ass.

(26) ~~show~~ $\sim x\ STR_{y,L}\ulcorner S\langle z, w, v, p', q'\rangle^\urcorner$

(27) $x\ STR_{y,L}\ulcorner S\langle z, w, v, p', q'\rangle^\urcorner$ ass.

(28) Let $b = \{\ulcorner S\langle z, w, v, \sum_{i<n} p(i)\cdot 10^{-(i+1)}, \sum_{i<n} q(i)\cdot$
$\cdot 10^{-(i+1)} + 10^{-(i+1)}\rangle^\urcorner : n\in\omega\}$

(29) $\bigwedge_s (s\in b \rightarrow s\in y)$ (2), (3), closure

(30) $[\bigvee_t]\ t\ is\ basis_L$ for y (1)

(31) $[\bigvee_r]\ r\in t \wedge \bigwedge_s (s\in b \rightarrow s\in L\text{-}Cn(\{r\}))$
MD-8.2, 30, 29

(32) $0 \vdash_L \ulcorner p' \leq \sum_{i\in\omega} p(i)\cdot 10^{-(i+1)} \wedge \sum_{i\in\omega} q(i)\cdot 10^{-(i+1)} \leq q'^\urcorner$
(27), MD-7.43

(33) $\ulcorner S\langle z, w, v, p', q'\rangle\urcorner \in y$ (32), (31), (30), MD-8.2 imp. by 25

(34) $\bigvee_{x} PS_{y,L}(x, z, w, v)$ (1), (24), (7), MD-8.7

If z' is a reference term of L, known in the rational corpus y to denote a finite set, w is a random variable of L according to y, and v is a mathematical term, then y contains a statistical statement about any non-decreasing sequence of subsets of z', the random quantity represented by w, and the mathematical object represented by v.

MT-8.29 $y \text{ is } RC_L \wedge z \in RT_L \wedge \bigvee_{A} \bigvee_{n} L\text{-}Cn(A) \subset y \wedge RV(n, w, \langle L, A\rangle) \wedge$
$\wedge \{\ulcorner z' \textit{ is } \omega\textit{-sequence}\urcorner, \ulcorner \bigwedge_{i,j\in\omega} (i<j \to z'_i \subset z'_j)\urcorner, \ulcorner \bigcup \mathscr{R}z' = z\urcorner,$
$\ulcorner \bigwedge_{i\in\omega} z'_i \subset \mathscr{R}w\urcorner, \ulcorner z \prec \omega\urcorner\} \subset L\text{-}Cn(A) \wedge r \in Math\ Terms_L) \to$
$\bigvee_{p,q} (\ulcorner S\langle z', w, v, p, q\rangle\urcorner \in y \wedge SS_{y,L}\ulcorner S\langle z', w, v, p, q\rangle\urcorner)$

Proof:

(1) $\ulcorner S\langle z', w, v, 0_r, 1_r\rangle\urcorner \in L\text{-}Cn(A)$

(2) $SS_{y,L}\ulcorner S\langle z', w, v, 0_r, 1_r\rangle\urcorner$

From the last two theorems it is clear that if n is any numeral of L, y any statistically closed, normal rational corpus having a basis, z any reference term, w any random variable, and v any mathematical term, there will be a strongest statement y about z' (where z' is known to denote an ω-sequence the union of whose terms is a set of n members of the set denoted by z), w, and v.

MT-8.30 y *is statistically closed*$_L \wedge$ *Normal*$_L$ $y \wedge \bigvee_{s}$ *s is basis*$_L$ *for*
$y \wedge z \in RT_L \wedge n \in Num_L \wedge v \in Math\ Terms_L \wedge$
$\wedge \bigvee_{A,m} (L\text{-}Cn(A) \subset y \wedge RV(n, w, \langle L, A\rangle) \wedge \{\ulcorner z' \textit{ is } \omega\textit{-sequence}\urcorner,$
$\ulcorner \bigwedge_{i,j\in\omega} (i<j \to z'(i) \subset z'(j))\urcorner, \ulcorner \bigcup \mathscr{R}z' \subset z \cap \{w : w \leqslant n\}\urcorner,$
$\ulcorner \bigcup \mathscr{R}z' \subset \mathscr{D}w\urcorner\} \subset y \to \bigvee_{x} PS_{y,L}(x, z', w, v)$

Proof: immediate from MT-8.28 and MT-8.29.

The antecedent of this theorem contains a lot of conjuncts, but they are not as restrictive as they appear to be. A basic desideratum of an inductive acceptance rule is that it should lead to bodies of knowledge that are statistically closed. A set of statements that was not Normal would be a very strange one indeed for someone to have reason to believe. Similarly, it is hard to imagine how one could have empirical grounds for accepting a set of statements that had no basis that would not also be grounds for accepting a slightly stronger set of statements that did have a basis. As for the rest of the clauses, they are concerned with matters that can be settled on the basis of the meanings of the terms involved.

This last theorem expresses an important practical fact, for in the ordinary course of events (and according to the statistical acceptance rule we shall outline in due course) our bodies of knowledge turn out to be Normal, and statistically closed, and to have a basis. Therefore, given any reference term, any random variable, and any mathematical term, there is a strongest statement of our body of knowledge concerning the frequency with which the value of the corresponding random quantity belongs to the set corresponding to the mathematical term, among *any* ω-sequence of finite subsets of the intersection of the set corresponding to the reference term and the set of sets containing no more than n members.

For example, we have in our body of knowledge a strongest statement concerning the frequency in any non-decreasing sequence of sets of crows whose union comprises at most the first 10^{100} crows to exist on earth, of crows whose weight falls between 50 grams and 200 grams.

The following easy metatheorems will come in handy in the next chapter.

MT-8.31 y is $RC_L \wedge SS_{y,L}\ulcorner S\langle w, x, b, p, q\rangle\urcorner \wedge \ulcorner S\langle w, x, b, p, q\rangle\urcorner \in y \wedge$
$\wedge \bigwedge_{p', q'} (\ulcorner S\langle w, x, b, p', q'\rangle\urcorner \in y \wedge SS_{y,L}\ulcorner S\langle w, x, b, p', q'\rangle\urcorner \rightarrow$
$\rightarrow 0 \vdash_L p' \leq p \wedge q \leq q') \rightarrow PS_{w,L}(\ulcorner S\langle w, x, b, p, q\rangle\urcorner, w, x, b)$

Proof: we require only to establish the last clause of MD-8.7

(1) $SS_{y,L}\ulcorner S\langle w, x, b, p', q'\rangle\urcorner \wedge \ulcorner S\langle w, x, b, p, q\rangle\urcorner$
$STR_{w,L}\ulcorner S\langle w, x, b, p', q'\rangle\urcorner$ ass.

(2) $\{\ulcorner S\langle w, x, b, p, q\rangle\urcorner\} \vdash_L \ulcorner S\langle w, x, b, p', q'\rangle\urcorner$

(3) $\ulcorner S\langle w, x, b, p', q'\rangle\urcorner \in y$

MT-8.32 y is $RC_L \wedge SS_{y,L}\ulcorner S\langle x, w, b, p, p\rangle\urcorner \wedge \ulcorner S\langle x, w, b, p, p\rangle\urcorner \in y \rightarrow$
$\rightarrow PS_{y,L}(\ulcorner S\langle x, w, b, p, p\rangle\urcorner, x, w, b)$

Proof: obvious

MT-8.33 $Normal_L\ w \wedge PS_{w,L}(t, y, z, b) \wedge PS_{w,L}(t', y', z', b') \wedge$
$\wedge t = \ulcorner S\langle y, z, b, p, q\rangle\urcorner \wedge t' = \ulcorner S\langle y', z', b', p', q'\rangle\urcorner \rightarrow$
$\rightarrow 0 \vdash_L \ulcorner p \leq q\urcorner \wedge (0 \vdash_L \ulcorner p \leq p'\urcorner \vee 0 \vdash_L \ulcorner p' \leq p\urcorner) \wedge$
$\wedge (0 \vdash_L \ulcorner q \leq q'\urcorner \vee 0 \vdash_L \ulcorner q' \leq q\urcorner)$

MT-8.34 $Normal_L\ w \wedge PS_{w,L}(t, y, z, b) \wedge PS_{w,L}(t', y', z', b') \rightarrow$
$\rightarrow t\ STR_{w,L}\ t' \vee t'\ STR_{w,L}\ t \vee t\ DIF_{w,L}\ t' \vee t\ E\text{-}STR_{w,L}\ t'$

MT-8.35 $\ulcorner b \subset \mathbb{B}^n\urcorner \in w \wedge PS_{w,L}(\ulcorner S\langle y, z, b, p, q\rangle\urcorner, y, z, b) \rightarrow$
$\rightarrow PS_{w,L}(\ulcorner S\langle y, z, \mathbb{R}^n - b, 1-q, 1-p\rangle\urcorner, y, z, \ulcorner \mathbb{R}^n - b\urcorner)$

NOTE

[1] For a more extended discussion of these matters, including a discussion of their treatment by other epistemologists, see Henry E. Kyburg, 'Conjunctivitis' in M. Swain (ed.), *Induction, Acceptance, and Rational Belief*, D. Reidel, Dordrecht-Holland, 1970.

CHAPTER NINE

RANDOMNESS

1. There is one kind of situation where the use of probability seems clear and uncontroversial: where we know a certain statistical statement; and where a certain object is 'merely' an instance of the reference class referred to in the statistical statement. For example, in the case of a well-balanced die, we know that the frequency of ones in the sequence of sets of tosses of the die is close to $\frac{1}{6}$. Furthermore, any toss of the die (other than one whose outcome we already know) is 'merely' another toss. There is nothing (ordinarily) we can know about it that will keep us from saying that the probability of its yielding a one is one sixth. This is particularly true when we refer to the toss by means of the indefinite article 'a'. Almost everybody, not in the process of defending a philosophical position regarding the interpretation of probability, will agree that it is correct to say, of a well-balanced die, tossed in a normal fashion, that the probability of getting a one on *a* toss of the die is (practically) one sixth.

Similarly, if we know the law of decomposition of a certain radioactive substance, we agree in saying that the probability that an atom, or a given atom, will decompose between t_0 and t_1 is just the number yielded by the law of decomposition. There is nothing we can know about an atom, other than that it does or does not decompose in the interval between t_0 and t_1, that will lead us to assign some other number to the probability of its decomposition in that interval of time.

Consider, on the other hand, the probability that a particular individual will survive his 35th birthday. We may have a lot of information concerning the frequency with which people in general survive their 35th birthday. But there are many considerations other than his death that would influence our willingness to take this frequency as his probability of survival. For example, we might know that he was a male, and that while in general the probability of survival was p_1, it is p_2 for males. We might then use the latter number as the probability unless, again, the individual was not 'merely a male'.

We recognize the problem here as that of choosing the correct reference class. In the case of the die and the radioactive atom, there was no problem, for there was only one candidate for the reference class. In the insurance example, there are a number of possible candidates.

The problem may be phrased in different words. In the case of a die, any throw whose outcome we don't know is a random member of the set of throws of the die, with respect to yielding a one. That is why the known frequency of ones is the best guide we have to the unknown outcome of a toss or a particular toss. We say that the probability of a one on the next toss is a sixth, because (a) we know that in general a sixth of the tosses yield a one, and (b) the 'next toss' is, relative to what we know about it, a random member of that class of tosses, i.e., it is 'merely' a member of that class. In the case of the atom, we say that the probability that it will disintegrate between t_0 and t_1 is p, because (a) we accept a theory according to which we compute p, and (b) relative to what we know, that atom is merely a member of the class to which the law applies; i.e., it is a random member of that class with respect to disintegrating between t_0 and t_1.

The idea behind the definition of probability in the next chapter is just that suggested by these examples: the probability that A will belong to B, relative to what we know, will be p, just in case (a) relative to what we know, A is a random member of some class C, with respect to belonging to B; and (b) we know that the measure of B's among the members of C is p. This measure may be a real number between 0 and 1; but it may also be an interval included in $[0, 1]$. All of this must be considerably generalized, of course, as well as made more precise. It does suggest, however, both the importance of the concept of randomness, and the fact that it is going to be a concept which is relative to our body of knowledge, which will somehow reflect what we know and what we don't know.

2. It might be thought, since the notion of randomness is often explicated in terms of probabilities, that there lurks here a potential circularity. Let us therefore explore the possibility of defining randomness in terms of probability.

We might say that an object (or event – the language is immaterial) is a random member of a certain class, if the probability that it is one member of that class is the same as the probability that it is any other member of the class. The most obvious difficulty with this approach is that it is incoherent: can the next ball to be drawn from an urn be, under any circumstances, anything but the next ball to be drawn? Can any ball, or any toss, or any application for insurance, be other than the particular thing that it is? There is an obvious sense in which the answer is 'No'. There is another sense in which we might, in certain special circumstances, be able to give another answer. For example, if we have names for the members of the class, we can ask whether or not the object denoted by a certain definite description ('the next ball to be drawn', 'the man named John Smith', 'The next toss of the coin') is the object denoted by a certain one of these names: ('ball number 15', '... *a*?...', '... *b*?...'). Only in the first case does the question seem to make good sense and to be relevant to probabilistic considerations – for example, when we know that the odd-numbered balls are black and the even-numbered ones are white. Suppose the reference class to which we are going to refer John Smith's probability of survival is *K*. What can we mean when we say that the probability of John Smith's being one member of this class is the same as the probability of his being any other member of it? Suppose the individuals in *K* are represented by policy numbers. Do we mean that John Smith is as likely to be given one policy number as another? Not likely. But even if we did mean that, how would it be relevant to his probability of survival? It would be relevant, of course, if we had a list of policy numbers that comprise the policies on which claims will be paid in the next year – as we suppose that the balls numbered so and so are black. As for the tosses of the coin, we are even worse off. Supposing – what is unrealistic in the extreme – that we had a complete enumeration of the tosses. It is surely not usually the case that we can say that the 'next' toss is as likely to be the *k*th toss as the *j*th toss – and if it were, what would that tell us that was relevant to the probability that the next toss will land heads? Unless, of course we have an enumeration of the tosses which also indicates for each toss whether it results in heads or in tails.

Furthermore, in most cases there is no interpretation of probability which makes sense of this approach to randomness. We can claim, perhaps that the next ball to be drawn is as likely to be ball number 15 as it is any other ball, in the sense that my degree of belief that it is ball 15 is the same as my degree of belief that it is any other of the N balls in the urn. But is there any way to make sense out of the claim that my degree of belief that John Smith is a certain member of the set of those insured by a company is the same as my degree of belief that he is some other member of that set? Or to make sense out of the claim that my degree of belief that the next toss is a particular toss in the sequence of tosses is the same as my degree of belief that it is some other toss?

Even in cases analogous to the balls-in-the-urn case, there are problems. We might say that an object is a random member of a class if it is chosen by a method which yields – or would yield – each member of the class with equal frequency in the long run. Or, it is chosen by a method which, before it is applied, is such that I will have the same degree of belief, with regard to each member of the class, that it will be the one chosen. In the first place it must be observed that this approach imposes a severe and unnatural restriction on the application of randomness: we can only usefully say that an object is a random member of a class when the class is finite. If it is infinite, and the method is such as to select an infinite number of members of the class in the long run, then the frequency with which any particular member is chosen must be 0, however biassed (within limits) the method. Similarly for subjective and logical interpretations of probability: if there are an infinite number of alternatives that I regard as possible at all, I must assign exactly the same probability to each of them – namely 0.

In the second (and less important) place, we are left with the problem of specifying the method, and furthermore specifying the method in such a way that we have grounds for accepting the hypothesis that the unlimited application of the method would in the long run result in the choice of each member of the population with equal frequency. Or specifying it in such a way that our degree of belief is evenly distributed among all the possibilities. I do not mean to suggest that these problems are insuperable, or even very difficult; it is merely that they must be faced and solved.

Furthermore, it is perfectly possible that we have selected a member of a population by a method which, if applied over and over, would select each member of the population with equal frequency, and that that object is *not*, in any useful sense a random one. Let T be the set of tosses of a die; these tosses can be ordered in time; let us assign to each toss a number corresponding to its ordinal position in time. Then we select a particular toss by consulting a table of random digits. If that toss turns out to be the one we just got through performing, we may well know whether or not it resulted in a one; and if we do, it is certainly not a 'random' member of the set of all tosses for the purpose of betting on whether or not it yielded a one.

But given that we have defined randomness along these lines and met the problem posed, what have we accomplished? Nothing that is of the slightest use in the application of our statistical knowledge. When I propose to bet on the fall of a die, it is *that* fall that I am going to bet on. I do not first intend to bet on some fall of a die, and then consider a method of choosing one among the many falls of the die (by an appropriate method) to bet on. I do not demand that the particular fall I am going to bet on be chosen by a method which.... Similarly, when I am going to bet on the color of a ball drawn from an urn, it is the color of the ball produced by *that* drawing that I am concerned with; I do not intend to bet on the color of some ball drawn from an urn, and then consider a method of choosing one among the many drawings of balls from that urn to bet on. That the method of drawing be such that each ball is drawn equally frequently in the long run is of course irrelevant; as Nagel pointed out long ago the (presumptive) reference set in this case must be considered to be the set of drawings from the urn, not the set of balls in the urn.

And there is no way in which we can say that Mr. Jones – the 35 year old who took out insurance with our company – is selected for survival during the coming year by a method which tends to select each individual with equal frequency. (If there is a method, which many of us doubt, it is a method which is presumably not indifferent to the virtues and vices of the individual involved.)

Furthermore, even if we do know that A is selected from C by an appropriate random method, whether this knowledge, combined with the knowledge that p of the C's are B's, would yield a probability of p that A is a B, would depend on what else we knew about A. We could know these things, and also know that in point of fact, A was a B. Then we would surely not want to say that the probability of its being a B was p. Similarly, even under the most ideal circumstances of the application of this notion of randomness, we could encounter information which would be relevant to the probability in question. Thus even supposing that Mr. Jones were somehow picked by an appropriate method to receive insurance with our company, he would cease to be a random member of the set of 35-year-old males the moment we found out that he was a wild automobile racer.

Finally, the next toss of a certain coin, although it is the toss that it is, and has been selected by me to make a point, i.e., by a method that can result only in that particular toss, is surely, as far as we are concerned now, a random member of the set of tosses of that coin; and if we know that on the whole the coin tends to land heads $\frac{2}{3}$'s of the time, we will properly attribute a probability of $\frac{2}{3}$ to its landing heads on that next toss.

What we require in order to apply our statistical knowledge is not equally distributed ignorance concerning which member of the reference class has been (or would be) chosen in the long run, but a state of knowledge in which the only relevant thing concerning the particular property of the object in question is our knowledge of the frequency of that property in the reference class.

In statistics, where the prevailing view of probability is frequentistic, randomness also plays an important role through the concept of a random sample. The classical understanding of the concept of a random sample is that it is a sample chosen by a method that will in the long run produce each possible sample equally often. Many of the classical examples of statistical argument are based on the assumption that the samples are random samples in this sense. Of course, if a method is employed that will choose each possible sample equally often in the long run, then it *must* sometimes choose samples that are patently inappropriate for statistical inference. Thus if we assign numbers to the residents of a large city, and choose a sample of 100 by means of a table of random numbers for the purpose of a political poll, it may just turn out that all the individuals in our sample belong to the same political club in a certain section of the city. To repudiate this sample (as any reasonable pollster would) by drawing another, is to undermine the whole rationale of the classical concept of a random sample. As a result theoretical statisticians (as opposed to practising pollsters) are divided as to the appropriate thing to do in a case like this.

We shall have a great deal more to say about the whole knotty problem of random sampling later on; the point here is merely that the equal probability interpretation of randomness does not yield the results that seem intuitively to be desirable. At the very least it does not appear that it is a forgone conclusion that the equal probability interpretation of randomness is the only possible or useful one, or even that it provides a sound approach to the application of statistical knowledge or to the assessment of statistical inferences.

3. We shall therefore attempt to provide an epistemological interpretation of randomness that is quite independent of the notion of probability. It is already clear that the relation of 'being a random member of' is a four-termed relation. It involves a reference to the object said to be a random member of something. It involves reference to the set of which that object is a random member. It involves an implicit reference to a body of knowledge, for whether an object is a random member of a class depends on what we know about it. For you, the applicant for insurance may be a random member of the set of 35-year-old males; for me he is not, for I happen to know that he is a test pilot. Finally, it involves an implicit reference to the property with respect to which the object is a random member of the class. The next Irish Wolfhound I see may be a random member of the set of dogs with respect to being male, but it is not a random member of the set of dogs with respect to weighing more than m pounds, where m pounds is the mean weight of dogs.

The four things we must take account of, then, are *an object*, *a class*, *a body of knowledge*, and *a property*.

This awkward collection of items may be rendered more managable by taking all of them to be, like the body of knowledge, linguistic items. We may then construe the relation of 'being a random member of' to be a metalinguistic relation holding among three terms of the object language and a set of statements of the object language.

Since most writers who have thought about the problem at all have looked as it as the problem of choosing the correct reference class on which to base an assertion of (statistical) probability, we shall begin in the same way. Some philosophers have argued that this is merely a pragmatic problem, rather than a theoretical one. The scientist is through, they say, when he has offered me a number of statistical hypotheses that he regards as acceptable. Choosing which hypothesis to use as a guide to action is not his problem, but mine, not a theoretical problem, but a practical one.

But calling a problem practical or pragmatic is no way to solve it. If I am a baffled insurance company before the christening, I shall be a baffled insurance company after it.

Some of those who regard the problem of choosing a reference class as a practical problem rather than an interesting theoretical one, nevertheless have advice to offer. Some statisticians appear to offer the advice (perhaps in self interest), 'Choose the reference class that is the most convenient mathematically', but most suggest something along the lines of Reichenbach's maxim. 'Choose the narrowest reference class about which you have adequate statistical information'. "If we are asked to find the probability holding for an individual future event, we must first incorporate the case in a suitable reference class. An individual thing or event may be incorporated in many reference classes.... We then proceed by considering *the narrowest reference class for which reliable statistics can be compiled*."[1] Reichenbach recognizes that we may have reliable statistics for a reference class A, and for a reference class C, but none for the intersection of A and C. If we are concerned about a particular individual who belongs both to A and to C, "The calculus of probability cannot help in such a case because the probabilities $P(A, B)$ and $P(C, B)$ do not determine the probability $P(A.\ C, B)$. The logician can only indicate a method by which our knowledge may be improved. This is achieved by the rule: look for a larger number of cases in the narrowest common class at your disposal."[2] This rule, aside from being obscure in application, is beside the point. If we could always follow the rule 'investigate further', we could apply it to the very individual who concerns us (wait a year and see if John Smith survives), and we would have no use for probability at all. Probability is of interest to us because we must sometimes act on incomplete information. To be sure, the correct answer to a problem may sometimes be: get more information. But from the point of view of logic this answer can be given to any empirical question. What we expect from logic is rather an indication of what epistemological stance we should adopt with respect to a given proposition, under given circumstances. Perhaps in the situation outlined by Reichenbach, our answer would be 'Suspend judgment'. That would be a proper answer. But since, as we shall see, every case can be put in the form outlined by Reichenbach, that answer, though relevant, seems wrong.

Surprisingly, Carnap does not give us an answer to this question either. The closest he comes to describing the sort of situation I have in mind in which we have a body of statistical information and we want to apply it to an individual, is when the proposition e consists of a certain structure description, i.e., a statement giving the relative frequencies with which each property is instantiated in the universe, and h is the hypothesis that an individual a has the property P. The degree of confirmation of h on e will be the relative frequency of P in the structure description e. This does not apply, however, when the individual a is mentioned in the evidence – i.e., when the evidence consists of more than the bare structure description. This is simply not an epistemologically possible state of affairs. Hilpinen, similarly, supposes that if we have encountered a at all, we know everything about it.[3] This is the principle of completely known instances. It follows from this principle that if we know that a belongs to any proper subset of our universe at all, we know already whether or not it belongs to (say) B. In real life this situation never arises. Even Reichenbach's principle is more helpful than this.

On the other hand, it is clear that if we allow the sort of approximate statistical statement discussed in the last chapter, we will always have some information – perhaps the totally uninteresting information embodied in a statistical statement of the form $\ulcorner S\langle a, b, c, 0_r, 1_r\rangle \urcorner$. Therefore no potential reference class can be ruled out merely on the grounds that we do not have 'adequate' statistics for it. In this case, then, Reichenbach's principle will direct us always to use the smallest potential reference class to which we know an object belongs, and in general, even if this doesn't turn out to be its unit set, it may turn out to be a set about which we have no interesting information at all.

To help make our ideas precise, let us look at a specific example. The principles can be made clear, if not totally persuasive, by considering just a finite class of classes, closed under intersection. To fix our ideas, let us focus on a traditional bag of counters. Let the counters be round (R) and triangular ($\bar{R}$), Blue (B) and Green ($\bar{B}$), zinc (Z) and tin ($\bar{Z}$). Let our statistical knowledge concerning counters drawn from the bag (C), be given by Table III, in which all the rational reference classes appropriate to problems concerning draws from the urn are mentioned.

TABLE III

Reference class	Subclass	Lower and upper bounds of change	
C	R	0.48	0.52
C	B	0.60	0.70
C	Z	0.45	0.55
$C \cap R$	B	0.55	0.68
$C \cap R$	Z	0.50	0.65
$C \cap R$	R	1.00	1.00
$C \cap B$	R	0.45	0.65
$C \cap B$	Z	0.30	0.50
$C \cap B$	B	1.00	1.00
$C \cap Z$	B	0.25	0.65
$C \cap Z$	R	0.35	0.70
$C \cap Z$	Z	1.00	1.00
$C \cap R \cap B$	Z	0.50	0.80
$C \cap R \cap Z$	B	0.30	0.80
$C \cap B \cap Z$	R	0.25	0.90

Now let us consider various counters. Suppose that we know that $a_1 \in C$, and, in terms of our basic categories, nothing more about it. What is the probability that $a_1 \in R$? Clearly, if we allow probabilities to be intervals, (0.48, 0.52). C is the appropriate reference class, since all we know about a_1 is that it is a member of C.

Suppose that we know that $a_2 \in C$ and also that $a_2 \in R$. (Perhaps a_2 is drawn from the bag in the dark, and we can feel that it is round.) Surely knowing that a_2 is round should have an influence on the probability we attribute to it being blue. Since we have statistical information about $C \cap R$, and since $C \cap R$ is 'narrower' than C, again there is no problem: the probability is (0.55, 0.68). Note that Carnap and Hilpinen and others who hold a logical view of probability must here abandon the attempt to base this probability on statistical measures – other things enter in. The object a_2 is neither completely known nor completely unknown.

Suppose that we know of a_3 that it belongs to both C and B. What is the probability that it belongs to R? Again we have two possible reference classes to consider: C and $C \cap B$. According to the broader reference class, the measure is (0.48, 0.52); according to the narrower one it is (0.45, 0.65). It seems reasonable, since there is no *conflict* between these measures, to regard the probability as the narrower interval, even though it is based on the broader reference class.

On the other hand, suppose we ask for the probability that $a_3 \in Z$. There the measure in the broader class is (0.45, 0.55), while the measure in the narrower class is (0.30, 0.50). Here there does seem to be a conflict, and it seems advisable to use the narrow reference class, even though the information it gives is not so precise; we would tend to say that the probability of '$a_3 \in Z$' is (0.30, 0.50).

How about an object drawn from the urn about which we can tell only that it is blue and round? Is it zinc? The probability that this counter (say, a_4) is zinc will have one value or another depending on the statistical knowledge we have. We know that between 0.45 and 0.55 of the counters are zinc, that between 0.50 and 0.65 of the round counters are zinc, that between 0.30 and 0.50 of the blue counters are zinc, and that between 0.50 and 0.80 of the round blue counters are zinc. Here again it seems natural to use the smallest possible class ($C \cap B \cap R$) as the reference class, despite the fact that we have relatively little information about it – i.e., we know the proportion of zinc counters in it only within rather wide limits.

Now take a_5 to be a counter we know to be blue and zinc. Here, as distinct from the previous example, it seems most natural to take the broadest class C as the reference class. It is true that we know something about some of the smaller classes to which we know that a_5 belongs, but what we know is very vague and indefinite, whereas our knowledge about the class C is very precise and in no way conflicts with our knowledge about the narrower classes.

Finally, let a_6 be known to belong to $C \cap R \cap Z$. What is the probability of the statement '$a_6 \in B$'? This case is a little more problematic than the preceding one. One might, by an argument analogous to the preceding one, claim that the probability was (0.60, 0.70): the smallest rational reference class to which a_6 is known to belong is one about which we have only vague information which seems not to conflict with the very strong information we have about C and B. But there is a difference in this case: a_6 is also known to belong to $C \cap R$, and our information about the measure of B in $C \cap R$ *does* conflict with our information about the measure of B in C. If one interval were included in the other, we could simply suppose that our knowledge in that case was more precise than our knowledge in the other case. But the intervals overlap: it is not merely a case of greater precision in one bit of knowledge than in the other. Thus one might want to argue (as I argued in *Probability and the Logic of Rational Belief*) that the knowledge that a_6 belongs to $C \cap R$ *prevents* a_6 from being a random member of C with respect to belonging to B – i.e., prevents C from being the appropriate reference class.

The various situations described here are summarized in Table IV.

TABLE IV

We know (altogether)	We are interested in	Possible reference class	Measure	Probability
'$a_1 \in C$'	'$a_1 \in R$'	C	(0.48, 0.52)	(0.48, 0.52)
'$a_2 \in C \cap R$'	'$a_2 \in B$'	C	(0.60, 0.70)	
		$C \cap R$	(0.55, 0.68)	(0.55, 0.68)
'$a_3 \in C \cap B$'	'$a_3 \in R$'	C	(0.48, 0.52)	(0.48, 0.52)
		$C \cap B$	(0.45, 0.65)	
	'$a_3 \in Z$'	C	(0.45, 0.55)	
		$C \cap B$	(0.30, 0.50)	(0.30, 0.50)
'$a_4 \in C \cap B \cap R$'	'$a_4 \in Z$'	C	(0.45, 0.55)	
		$C \cap B$	(0.30, 0.50)	
		$C \cap R$	(0.50, 0.65)	
		$C \cap R \cap B$	(0.50, 0.80)	(0.50, 0.80)
'$a_5 \in C \cap B \cap Z$'	'$a_5 \in R$'	C	(0.48, 0.52)	(0.48, 0.52)
		$C \cap B$	(0.45, 0.65)	
		$C \cap B$	(0.35, 0.70)	
		$C \cap B \cap Z$	(0.25, 0.90)	
'$a_6 \in C \cap R \cap Z$'	'$a_6 \in B$'	C	(0.60, 0.70)	
		$C \cap R$	(0.55, 0.68)	
		$C \cap Z$	(0.25, 0.65)	
		$C \cap R \cap Z$	(0.30, 0.80)	(0.30, 0.80)

In informal terms, the two principles that have guided the construction of the table above are:

Principle I: If a is known to belong to both X and Y, and our statistical knowledge about X and Z differs from that about Y and Z – i.e., if one interval is not included in the other – then, unless we happen to know that X is included in Y, Y prevents X from being an appropriate reference class for '$a \in Z$'. If neither is known to be included in the other, then each prevents the other from being an appropriate reference class.

Principle II: Given that we want to avoid conflict of the first sort, we want to use the most precise information we have; and therefore among all the classes not ruled out by each other in accordance with principle I, we take as our reference class that class about which we have the most precise information.

4. For the sake of simplifying future developments, we shall henceforth speak of the value of the random quantity y for the argument a falling in the Borel set b, rather than of the object a being in the set z. The two ways of talking are of course perfectly equivalent, but the former is easier to relate to our basic notion of a statistical distribution. Provided the items in corners are expressions of L, we have:

(a) $0 \vdash_L \ulcorner y\langle a\rangle \in b \leftrightarrow a \in \{x : x \in \mathscr{D}y \wedge y\langle x\rangle \in b\}\urcorner$

(b) $0 \vdash_L \ulcorner a \in z \leftrightarrow \imath f\,(f \text{ is fun} \wedge a = \mathscr{D}f \wedge \mathscr{R}f = \{0_r, 1_r\} \wedge$
$\wedge \bigwedge_{w \in \mathscr{D}f} (f\langle w\rangle = 0_r \leftrightarrow w \in z))\,\langle a\rangle \in \{0_r\}\urcorner$

A general principle regarding probability that we accept (though it has been rejected by some writers) is that two statements known to be equivalent are to have the same probability. We have in mind not merely logical (or mathematical) equivalence, but known equivalence as reflected by the existence in our rational corpus of a biconditional connecting the two statements. If two statements connected by a biconditional in our rational corpus are to have the same probability, the same is true of two statements connected by a biconditional chain.

It has already been observed that we cannot define probability relative to a body of knowledge that is inconsistent; and it has been argued that there are some reasons for wanting to take account of bodies of knowledge that are inconsistent. We shall therefore in the first instance restrict our attentions to consistent bodies of knowledge, and later extend our considerations. If x is a consistent rational corpus of the language L, so is the deductive closure of x, $L\text{-}Cn\,x \cap St_L$. We may therefore, for the sake of simplicity, formulate our initial criteria for randomness in terms of a deductively closed set of statements.

Let us suppose now that y and y' are random quantities, and that $\ulcorner y\langle a\rangle \in b\urcorner$ is connected by a biconditional to $\ulcorner y'\langle a\rangle \in b'\urcorner$ in a logically closed set of statements x. Furthermore suppose that $\ulcorner \bigcup \mathscr{R}c\urcorner$ and $\ulcorner \bigcup \mathscr{R}c'\urcorner$ are known to be identical to reference terms, and that x contains both the statements $\ulcorner a \in \bigcup \mathscr{R}c \urcorner$ and $\ulcorner a \in \bigcup \mathscr{R}c' \urcorner$. Recall that the terms c and c' appearing in statements representing our statistical knowledge denote infinite sequences of finite sets; the reference sets are the unions of the terms appearing in these sequences, i.e., the unions of the ranges of the sequences. Then we shall want to *deny* that a is a random member of c with respect to having a y-value in b, at least when a strongest statement about c and y and b differs from a strongest statement about c' and y' and b' (in the sense of 'differs' defined in Chapter VII) unless we know that $\ulcorner \bigcup \mathscr{R}c \urcorner$ is included in $\ulcorner \bigcup \mathscr{R}c' \urcorner$.

By principle II, then, we pick as the reference set a set about which we have the most precise information, and which is not prevented from being a reference set by the existence of an alternative about which we have different or more precise information.

Writing '$RAN_L(x, y, z, b, w)$' for 'x is a random member of y with respect to having a z-value in the set b, relative to the consistent rational corpus w of the language L', we have, formally,

(c)
$$\begin{aligned}
&RAN_L(x, y, z, b, w) \rightarrow \sim \ulcorner 0 \neq 0 \urcorner \in w \wedge \ulcorner x \in \bigcup \mathscr{R}y \urcorner \in L\text{-}Cn(w) \wedge \\
&\wedge \bigvee_t (\ulcorner \bigcup \mathscr{R}y = t \urcorner \in L\text{-}Cn(w) \wedge t \in RT_L) \wedge \bigwedge_{y', z', b'} \bigwedge_{s, s'} [\ulcorner z \langle x \rangle \in b \\
&\leftrightarrow z' \langle x \rangle \in b' \urcorner \in L\text{-}Cn(w) \wedge \ulcorner x \in \bigcup \mathscr{R}y' \urcorner \in L\text{-}Cn(w) \wedge \\
&\wedge PS_{w,L}(s, y, z, b) \wedge PS_{w,L}(s', y', z', b') \rightarrow s\ STR_{w,L}\ s' \vee \\
&\vee s\ E\text{-}STR_{w,L}\ s' \vee (s\ DIF_{w,L}\ s' \wedge \ulcorner \bigcup \mathscr{R}y \subset \bigcup \mathscr{R}y' \urcorner \in L\text{-}Cn(w)) \vee \\
&\vee (s'\ STR_{w,L}\ s \wedge \bigvee_{s'', y'', z'', b''} (\ulcorner z' \langle x \rangle \in b' \leftrightarrow z'' \langle x \rangle \urcorner \in L\text{-}Cn(w) \wedge \\
&\wedge \ulcorner x \in \bigcup \mathscr{R}y'' \urcorner \in L\text{-}Cn(w) \wedge PS_{w,L}(s'', y'', z'', b'') \wedge s'\ DIF_{w,L}\ s'' \wedge \\
&\wedge \sim \ulcorner \bigcup \mathscr{R}y' \subset \bigcup \mathscr{R}y'' \urcorner \in L\text{-}Cn(w))]
\end{aligned}$$

We cannot strengthen this into a biconditional and call it a definition, though. Surely we should take into account different object names, just as we took into account different random variables and different Borel terms. Thus if $\ulcorner x' \in \bigcup \mathscr{R}y' \urcorner \in w$ and $\ulcorner z \langle x \rangle \in b \leftrightarrow z' \langle x' \rangle \in b' \urcorner \in w$ and $\sim \ulcorner \bigcup \mathscr{R}y \subset \bigcup \mathscr{R}y' \urcorner \in w$, then we should also require that a strongest statement about y', z', and b' not be stronger than or different from a strongest statement about y, z, and b. But this is no longer workable. Let us suppose that we know that x belongs both to a term of y and to a term of y', that $\ulcorner z \langle x \rangle \in b \leftrightarrow z' \langle x \rangle \in b' \urcorner$, that a strongest statement about y, z, and b differs from a strongest statement about y', z', and b', and, finally that $\ulcorner\bigcup \mathscr{R}y \subset \bigcup \mathscr{R}y'\urcorner$. There seems to be no reason why we should not have $RAN_L(x, y, z, b, w)$; but there is.

Let '$\mathscr{P}_1(t)$' denote the one-membered subsets of t, 'z_s' the function whose domain is $\mathscr{P}_1(\mathscr{D}_z)$ and whose value for every object $\{x\}$ in that domain is $z\langle x\rangle$, and 'y_s' that ω-sequence whose ith term is $\mathscr{P}_1(y_i)$. We then have:

$$\text{(d)}\qquad \begin{array}{l} \ulcorner\{x\}\in\bigcup \mathscr{R}y'_s\urcorner\in w \\ PS_{w,L}(s', y', z', b')\wedge PS_{w,L}(s'', \ulcorner y'_s\urcorner, \ulcorner z'_s\urcorner, b')\rightarrow s'\ E\text{-}STR_{w,L}\ s'' \end{array}$$

and thus a strongest statement about $\ulcorner y'_s\urcorner \ulcorner z'_s\urcorner$ and b' will also differ from a strongest statement about y, z, and b. But of course we need not (and generally will not) have $\ulcorner\bigcup \mathscr{R}y \subset \bigcup \mathscr{R}y'_s\urcorner$ in our rational corpus w. Thus we would not, after all, have $RAN_L(x, y, z, b, w)$.

The way to save the situation is obvious: $\ulcorner\bigcup \mathscr{R}y \subset \bigcup \mathscr{R}y'\urcorner$ entails $\ulcorner\bigcup \mathscr{R}y_s \subset \bigcup \mathscr{R}y'_s\urcorner$, and, naturally, our statistical knowledge about $\ulcorner y_s\urcorner$, $\ulcorner z_s\urcorner$ and b agrees precisely with our statistical knowledge about y, z, and b. Thus rather than requiring that the set $\bigcup \mathscr{R}y$ itself be a subset of any set concerning which our statistical knowledge is different, we need only require that there be some subset such that our knowledge about it agrees exactly with our knowledge about y.

But even this requirement is a bit too strong. Suppose that y has the form $\ulcorner y' \times y''\urcorner$, and x is of the form $\ulcorner\langle x', x''\rangle\urcorner$. Now consider a new quintuple $\langle y''', x''', z''', b''', s'''\rangle$ such that

$$\begin{array}{l} \ulcorner x'''\in\bigcup \mathscr{R}y'''\urcorner\in w \\ PS_{w,L}(s''', y''', z''', b''') \\ PS_{w,L}(s, \ulcorner y' \times y''\urcorner, z, b) \\ \ulcorner z'''\langle x'''\rangle\in b'''\urcorner\ B_{w,L}\ulcorner z\langle x\rangle\in b\urcorner \\ s'''\ DIF_{w,L}\ s \end{array}$$

There are circumstances under which we do not want y''' to prevent the randomness of x (i.e., of $\ulcorner\langle x', x''\rangle\urcorner$) in $\ulcorner y' \times y''\urcorner$, even though there is no subset of $\ulcorner\bigcup \mathscr{R}y'''\urcorner$ which matches $\ulcorner y' \times y''\urcorner$. They may be characterized as circumstances in which there is a term y'''' (which may be y''' itself such that

$$\ulcorner\bigcup \mathscr{R}y'''' \subset \bigcup \mathscr{R}y'''\urcorner\in w$$

so that $\ulcorner \bigcup \mathscr{R} y'''' \urcorner$ is known to be a subset of $\ulcorner \bigcup \mathscr{R} y''' \urcorner$, and there are terms x'''', z'''', b'''', and a statement s'''', such that

$$\ulcorner x'''' \in \bigcup \mathscr{R} y'''' \urcorner \in w$$

$$\ulcorner z'''' \langle\langle x'''', x'' \rangle\rangle \in b'''' \leftrightarrow z \langle x \rangle \in b \urcorner \in w$$

$$PS_{w,L}(s'''', \ulcorner \imath f (f \text{ is } \omega\text{-sequence} \wedge \bigwedge_{i \in \omega} f \langle i \rangle = y''' \langle i \rangle \times \times y'' \langle i \rangle) \urcorner, z'''', b'''')$$

$$s'''' \; E\text{-}STR_{w,L} \; s$$

An example will help to make this clear. Suppose that y is a sequence of trials of a two-stage experiment: one of two urns, U_0 and U_1, is selected, and then a ball is drawn from that urn. We may know that in the long run, urn U_0 is selected $\frac{1}{4}$ of the time. If we know the frequency with which black balls are drawn from each of the urns, we may also know that the frequency with which urn U_0 is selected on those occasions when one ball has been drawn and has been black is $\frac{1}{3}$. Clearly the appropriate reference class for the probability that urn U_0 has been chosen when we have drawn one black ball is the cross product of the set of selections of an urn, and the set of drawings of a single ball which turns out to be black. We must be sure that the set of selections of an urn, even though the frequency with which urn U_0 is selected in this set differs from $\frac{1}{3}$, does not prevent the pair ⟨selection of an urn, draw of a black ball⟩ from being a random member of the cross product. The following definition ensures just this.

We shall say that (the object denoted by) x is a random member of (the union of the range of the ω-sequence denoted by) y, with respect to having a z-value in the set denoted by b, relative to the consistent body of knowledge w, just in case

(1) $\ulcorner x \in \bigcup \mathscr{R} y \urcorner \in w$

(2) $\bigvee_{s} PS_{w,L}(s, y, z, b)$

(3) $\sim \ulcorner 0 \neq 0 \urcorner \in L\text{-}Cn(w)$

(4) $\bigwedge_{s, s'} \bigwedge_{x', y', z', b'} [\ulcorner z \langle x \rangle \in b \leftrightarrow z' \langle x' \rangle \in b' \urcorner \in w \wedge \ulcorner x' \in \bigcup \mathscr{R} y' \urcorner \in w \wedge$

$\wedge PS_{w,L}(s, y, z, b) \wedge PS_{w,L}(s', y', z', b') \rightarrow$

(i) $s \; STR_{w,L} \; s' \vee$

(ii) $s \; E\text{-}STR_{w,L} \; s' \vee$

(iii) $(s \; DIF_{w,L} \; s' \wedge$

(iii-a) $\bigvee_{s'', y'', z'', b''} (\ulcorner z''\langle x'\rangle \in b'' \leftrightarrow z\langle x\rangle \in b\urcorner \in w \wedge$
$\wedge \ulcorner x' \in \bigcup \mathscr{R} y''\urcorner \in w \wedge \ulcorner \bigcup \mathscr{R} y'' \subset \bigcup \mathscr{R} y'\urcorner \in w \wedge \ulcorner \bigcup \mathscr{R} y'' \neq$
$\neq \bigcup \mathscr{R} y'\urcorner \in w \wedge PS_{w,L}(s'', y'', z'', b'') \wedge s\ E\text{-}STR_{w,L}\ s'') \vee$
(iii-b) $\bigvee_{x'', x''', y'', y''', y'''', z'', b''} (\ulcorner y = \imath f(f \text{ is } \omega\text{-sequence} \wedge$
$\wedge \bigwedge_{i \in \omega} f\langle i\rangle = y''\langle i\rangle \times y'''\langle i\rangle)\urcorner \in w \wedge \ulcorner \bigcup \mathscr{R} y'''' \subset \bigcup \mathscr{R} y'\urcorner \in w \wedge$
$\wedge \ulcorner x \in \bigcup \mathscr{R} y''''\urcorner \in w \wedge \ulcorner x = \langle x'', x'''\rangle\urcorner \in w \wedge$
$\wedge \ulcorner z''\langle\langle x', x'''\rangle\rangle \in b'' \leftrightarrow z\langle x\rangle \in b\urcorner \in w \wedge PS_{w,L}(s'', \ulcorner \imath f(f \text{ is}$
$\omega\text{-sequence} \wedge \bigwedge_{i \in \omega} f\langle i\rangle = y''''\langle i\rangle \times y'''\langle i\rangle)\urcorner, z'', b'') \wedge$
$\wedge s''\ E\text{-}STR_{w,L}\ s))) \vee$
(iv) $(s'\ STR_{w,L}\ s \wedge \bigvee_{s'', x'', y'', z'', b''} (\ulcorner z''\langle x''\rangle \in b'' \leftrightarrow z\langle x\rangle \in b\urcorner \in w \wedge$
$\wedge \ulcorner x'' \in \bigcup \mathscr{R} y''\urcorner \in w \wedge PS_{w,L}(s'', y'', z'', b'') \wedge s''\ DIF_{w,L}\ s' \wedge$
$\wedge \sim$(iii-a) $\wedge \sim$(iii-b)))]

Rather than packing all this into one definition, we may break it down into two parts: one part says that no term y' prevents $RAN_L(x, y, z, b, w)$ by being a source of *different* information; the other part says that no term y' prevents $RAN_L(x, y, z, b, w)$ by being a source of *more precise* information. Let us write '$y' D_L(x, y, z, b, w)$' for the first relation, and '$y' S_L(x, y, z, b, w)$' for the second. Since we wish to take account of bodies of knowledge that are not deductively closed, we replace talk of biconditionals by talk of biconditional chains, talk of membership by talk of membership chains, talk of inclusion by talk of inclusion chains, and talk of identity by talk of identity chains. We further simplify the definition of the relation D_L (which is, roughly, the denial of part (iii) above) by defining separately the two cases in which, though our knowledge about y' differs from our knowledge about y, yet we do not say that y' prevents x from being a random member of y with respect to having a *z-value* in b, relative to what we know.

The first case is that in which there is a proper subset of y' matching our knowledge about y, z, and b. We shall write '$(x', y', z', b')\ SND_L(x, y, z, b)$' to mean that y' fails to prevent $RAN_L(x, y, z, b, w)$ by virtue of our knowledge of the existence of a proper subset of y' matching y. In words, this will be the case when there are terms y'', z'', b'' such that we know that the z''-value of x' is in b'', just in case the z-value of x is in b; that x' is a member of some term of the reference sequence y''; that either the union of the range of the sequence denoted by y'' is the unit set of x' or it is properly included in the union of the range of the sequence denoted by y'; and that the strongest statistical statement we know about y, z, and b is exactly as strong as the strongest statistical statement we know about y'', z'', and b''.

MD-9.1 $$(x', y', z', b')\, SND_L(x, y, z, b, w) \leftrightarrow \bigvee_{s, s''} \bigvee_{y'', z'', b''} (\ulcorner z''\langle x'\rangle \in b''\urcorner\, B_{w,L} \ulcorner z\langle x\rangle \in b\urcorner \wedge x'\, M_{w,L} \ulcorner \bigcup \mathscr{R} y''\urcorner \wedge (\ulcorner \bigcup \mathscr{R} y''\urcorner\, I_{w,L} \ulcorner\{x'\}\urcorner \vee$$
$$\vee (\ulcorner \bigcup \mathscr{R} y''\urcorner\, Incl_{w,L} \ulcorner \bigcup \mathscr{R} y'\urcorner \wedge \sim\ulcorner \bigcup \mathscr{R} y''\urcorner\, I_{w,L} \ulcorner \bigcup \mathscr{R} y'\urcorner)) \wedge$$
$$\wedge\, PS_{w,L}(s'', y'', z'', b'') \wedge PS_{w,L}(s, y, z, b) \wedge s\, E\text{-}STR_{w,L}\, s'')$$

Similarly, we consider the case in which there is a subset of y' which may replace a factor in y (where y is a cross product) to yield something that matches y. Since we wish not to be bound to any order in reference sets that are composed of n-sequences, we introduce, into the object language, a relation of permutation: One set, y is a permutation of another set y', just in case, for some n, both y and y' are sets of n-sequences, and there is a 1-1 function f, with range and domain n, such that every member of y' is a permutation by f of some member of y:

D-9.1 $$y'\ Perm\ y \leftrightarrow \bigvee_{f, n} (n \in \omega \wedge \bigwedge_{x} (x \in y \cup y' \rightarrow x \text{ is } n\text{-sequence}) \wedge$$
$$\wedge f \text{ is } 1-1 \wedge \mathscr{D} f = n \wedge \bigwedge_{x \in y} \bigvee_{x' \in y'} \bigwedge_{i \in n} x\langle i\rangle = x'\langle f\langle i\rangle\rangle)$$

We write '$(x', y', z', b')\, PND_L(x, y, z, b, w)$' to mean that y' fails to prevent $RAN_L(x, y, z, b, w)$ by virtue of our knowledge of a subset of y' that can replace an initial factor in a permutation of the union of the range of y to yield something matching y. In words, for given terms x', y', z', and b', this will be the case when there are terms x''', y''', y'', z'', b'', y'''', y^*, and $y^{*\prime}$ such that the union of the range of y^* is a permutation of the union of the range of our original y; the union of the range of y^* is a subset of the cross product of the union of the range of y'' and the union of the range of y'''; the union of the range of $y^{*\prime}$ is a subset of the cross product of the union of the range of a subset y'''' of y' and the union of the range of y'''; the pair $\langle x', x'''\rangle$ belongs to the union of the range of $y^{*\prime}$; $\ulcorner z''\langle\langle x', x'''\rangle\rangle \in b''\urcorner$ and $\ulcorner z\langle x\rangle \in b\urcorner$ are connected by a biconditional chain; the cardinality of x' is strictly less than that of x; and the strongest statistical statement we know about $y^{*\prime}$, z'', and b'' is just as strong as the strongest statement we know about y, z, and b.

MD-9.2 $$(x', y', z', b')\ PND_L(x, y, z, b, w) \leftrightarrow \bigvee_{s, s''} \bigvee_{x''', y''', y'', z'', b'', y'''', y^*, y^{*\prime}} (\ulcorner \bigcup \mathscr{R} y^*\ Perm \bigcup \mathscr{R} y\urcorner \in w \wedge \ulcorner \bigcup \mathscr{R} y^*\urcorner\ Incl_{w,L} \ulcorner \bigcup \mathscr{R} y'' \times \times \bigcup \mathscr{R} y'''\urcorner \wedge \ulcorner \bigcup \mathscr{R} y^{*\prime}\urcorner\ Incl_{w,L} \ulcorner \bigcup \mathscr{R} y'''' \times \bigcup \mathscr{R} y'''\urcorner \wedge \wedge \ulcorner \bigcup \mathscr{R} y''''\urcorner\ Incl_{w,L} \ulcorner \bigcup \mathscr{R} y'\urcorner \wedge \ulcorner \langle x', x'''\rangle \urcorner\ M_{w,L} \ulcorner \bigcup \mathscr{R} y^*\urcorner \wedge \wedge \ulcorner z''\langle\langle x', x'''\rangle\rangle \in b''\urcorner\ B_{w,L} \ulcorner z\langle x\rangle \in b\urcorner \wedge \ulcorner x' \prec x\urcorner \in w \wedge \wedge PS_{w,L}(s, y, z, b) \wedge PS_{w,L}(s'', y^{*\prime}, z'', b'') \wedge s\ E\text{-}STR_{w,L}\ s'')$$

Now we can say that y' prevents the randomness of x in y with respect to having a z-value in b, relative to w, due to difference – in symbols, $y'\ D_L(x, y, z, b, w)$ – just in case the randomness is saved neither by the existence of a matching subset, nor by the existence of a product set that matches y.

MD-9.3 $$y'\ D_L(x, y, z, b, w) \leftrightarrow \bigvee_{s, s', x', z', b'} (\ulcorner z'\langle x'\rangle \in b'\urcorner\ B_{w,L} \ulcorner z\langle x\rangle \in b\urcorner \wedge x'\ M_{w,L} \ulcorner \bigcup \mathscr{R} y'\urcorner \wedge PS_{w,L}(s, y, z, b) \wedge PS_{w,L}(s', y', z', b') \wedge \wedge s\ DIF_{w,L}\ s' \wedge \sim (x', y', z', b')\ SND_L(x, y, z, b, w) \wedge \wedge \sim (x', y', z', b')\ PND_L(x, y, z, b, w))$$

Similarly, we say that y' prevents the randomness of x in y with respect to having a z-value in b, relative to w, due to greater precision – in symbols $y'\ S_L(x, y, z, b, w)$ – just in case there are terms x', z', and b' such that we know that x' is a member of the reference sequence y'; that x' has a z'-value in b' if and only if x has a z-value in b; that a strongest statement about y', z', and b' is stronger than a strongest statement about y, z, and b; and there is no reference term y'' that prevents x' from being a random member of y' with respect to having a z'-value in b', by virtue of different statistics.

MD-9.4 $$y'\ S_L(x, y, z, b, w) \leftrightarrow \bigvee_{s, s', x', z', b'} (x'\ M_{w,L} \ulcorner \bigcup \mathscr{R} y'\urcorner \wedge \wedge PS_{w,L}(s, y, z, b) \wedge \ulcorner z\langle x\rangle \in b\urcorner\ B_{w,L} \ulcorner z'\langle x'\rangle \in b'\urcorner \wedge$$

$$\wedge PS_{w,L}(s', y', z', b') \wedge s' \, STR_{w,L} \, s \wedge$$
$$\wedge \sim \bigvee_{y''} y'' \, D_L(x', y', z', b', w))$$

The definition of randomness is now obvious: x is a random member of y with respect to having a z-value in b, relative to the body of knowledge w, just in case w is strictly consistent, x is known to be a member of w, there is a strongest statement about y, z, and b in w – which entails that the union of the range of y is known to be equal to a standard reference term – and nothing prevents that randomness either by difference or by extra strength, i.e., precision.

MD-9.5 $$RAN_L(x, y, z, b, w) \leftrightarrow \sim \ulcorner 0 \neq 0 \urcorner \in L\text{-}Cn(w) \wedge x \, M_{w,L} \ulcorner \bigcup \mathscr{R} y \urcorner \wedge$$
$$\wedge \bigvee_{s} PS_{w,L}(s, y, z, b) \wedge \sim \bigvee_{y'} (y' \, D_L(x, y, z, b, w) \vee$$
$$\vee \, y' \, S_L(x, y, z, b, w))$$

A few obvious consequences of these definitions will help to make clear their intuitive content:

MT-9.1 $$RAN_L(x, y, z, b, w) \rightarrow x \, M_{w,L} \ulcorner \bigcup \mathscr{R} y \urcorner$$
$$\sim \ulcorner 0 \neq 0 \urcorner \in L\text{-}Cn(w)$$
$$x, y, z \in Tm_L$$
$$b \in Math \; Terms_L$$
$$w \text{ is } RC_L$$
$$\bigvee_{n, A} (A \subset w \wedge RV(n, z, \langle L, A \urcorner))$$
$$\bigvee_{y'} (y' \in RT_L \wedge \ulcorner y' = \bigcup \mathscr{R} y \urcorner \in w)$$
$$\sim y' \, D_L(x, y, z, b, w)$$
$$\sim y' \, S_L(x, y, z, b, w)$$

A very important consequence of our definitions is that if $\ulcorner z \langle x \rangle \in b \urcorner$ and $\ulcorner z' \langle x' \rangle \in b' \urcorner$ are connected by a biconditional chain, $RAN_L(x, y, z, b, w)$, and $RAN_L(x', y', z', w)$, then our statistical knowledge about y, z, and b will usually be just the same as our statistical knowledge about y', z', and b'.

There is one peculiar and unnatural circumstance under which this will not be true. Suppose that there is an infinite sequence of terms t

and an infinite sequence of statements s such that (1) $PS_{w,L}(s_i, t_i, y, b)$ and (2) if $i-j$ is divisible by 2, s_i $E\text{-}STR_{w,L}$ s_j, and if not, then s_i $DIF_{w,L}$ s_j. Further, suppose that there is an infinite sequence of reference terms z such that $\ulcorner \bigcup \mathscr{R}t_i = z_i \urcorner$ belongs to w, and such that for every i, $\ulcorner \bigcup \mathscr{R}t_{i+1} \subset \subset \bigcup \mathscr{R}t_i \urcorner$ belongs to w. We may then have both $RAN_L(x, t_i, y, b, w)$ and $RAN_L(x, t_{i+1}, y, b, w)$, despite the fact that (for every i) our knowledge about t_i conflicts with our knowledge about t_{i+1}. Speaking loosely, for every i, t_{i+1} $D_L(x, t_i, y, b, w)$ is falsified by the existence of an appropriate term (namely t_{i+2}) known to match t_i statistically, and known also to denote a set included in t_{i+1}.

There is also a circumstance that does not seem quite so unnatural and unusual. Let y be a reference term and let t_1 and t_2 be terms denoting sequences of finite subsets of the set denoted by y, such that we know $\ulcorner \bigcup \mathscr{R}t_1 = y \urcorner$ and $\ulcorner \bigcup \mathscr{R}t_2 = y \urcorner$ and let s_1 and s_2 be such that $PS_{w,L}(s_1, t_1, z, b)$, $PS_{w,L}(s_2, t_2, z, b)$, and s_1 $DIF_{w,L}$ s_2. To show that all this is possible (for some, not unusual, language) let $y=$ 'the set of tosses of coin c', $t_1=$ 'that function whose domain is ω, and whose value for every natural number i is the set of tosses consisting of the first i tosses resulting in heads, together with the first $2 \cdot i$ tosses resulting in tails', and $t_2=$ 'that function whose domain is ω, and whose value for every i is the set of tosses consisting of the first $2 \cdot i$ tosses resulting in heads together with the first i tosses resulting in tails.' Let z denote the characteristic function of the set of heads, and $b = \ulcorner\{1\}\urcorner$. It seems plausible to suppose that we have $\ulcorner \bigcup \mathscr{R}t_1 = y \urcorner$ and $\ulcorner \bigcup \mathscr{R}t_2 = y \urcorner$ in our body of knowledge, so that we have both $t_1 D_L(x, t_2, z, b, w)$ and $t_2 D_L(x, t_1, z, b, w)$. But now consider the sequence of reference sequences: $t_{10} = t_1$; $t_{21} = t_2 - t_1(1)$; $t_{12} = t_1 - \{t_1(1), t_1(2)\}$; etc. At each step we disregard one more coin-toss, and thus ensure that $\ulcorner t_{1i} \neq t_{2(i+1)} \urcorner$ is in our body of knowledge. Now we have neither $t_1 D_L(x, t_2, z, b, w)$ nor $t_2 D_L(x, t_1, z, b, w)$, nor indeed $t_{1i} D_L(x, t_{2(i+1)}, z, b, w)$. Furthermore, isn't it the case that by construction we have $\ulcorner S(t_1, z, b, \frac{1}{3}, \frac{1}{3}) \urcorner$ and $\ulcorner S(t_2, z, b, \frac{2}{3}, \frac{2}{3}) \urcorner$ in our body of knowledge? All this we have only under one circumstance: that we *know* that the set denoted by y is infinite.

In order that these difficulties should not arise, it suffices to make two stipulations: First, that the rational corpus in question be statistically terminating, and second that it be statistically consistent.

We say that a set w is statistically terminating in symbols – w is

$S\text{-}Term_L$ – when, given any object, any sequence whose terms comprise a reference set, any random variable and any borel term, and any strongest statistical statement about these objects, there is a sequence comprising a *minimal* equally useful subset of the reference set – by which we mean that any set known to be included in it, about which we have the same statistical information, is known to be identical to it. Putting all this in uniformly metalinguistic terminology, we have:

MD-9.6 $$w \text{ is } S\text{-}Term_L \leftrightarrow \bigwedge_{x,t,y,b,s} (x\, M_{w,L} \ulcorner \bigcup \mathscr{R}t \urcorner \wedge PS_{w,L}(s,t,y,b) \rightarrow \\ \rightarrow \bigvee_{s',t',y',b'} (x\, M_{w,L} \ulcorner \bigcup \mathscr{R}t' \urcorner \wedge PS_{w,L}(s',t',y',b') \wedge \\ \wedge \ulcorner y\langle x\rangle \in b \urcorner\, B_{w,L} \ulcorner y'\langle x\rangle \in b' \urcorner \wedge s\, E\text{-}STR_{w,L}\, s' \wedge \\ \wedge \ulcorner \bigcup \mathscr{R}t' \urcorner\, Incl_{w,L} \ulcorner \bigcup \mathscr{R}t \urcorner \wedge \bigwedge_{s'',t'',y'',b''} (x\, M_{w,L} \ulcorner \bigcup \mathscr{R}t'' \urcorner \wedge \\ \wedge \ulcorner y''\langle x\rangle \in b'' \urcorner\, B_{w,L} \ulcorner y'\langle x\rangle \in b' \urcorner \wedge PS_{w,L}(s'',t'',y'',b'') \wedge \\ \wedge s''\, E\text{-}STR_{w,L}\, s' \wedge \ulcorner \bigcup \mathscr{R}t'' \urcorner\, Incl_{w,L} \ulcorner \bigcup \mathscr{R}t' \urcorner \rightarrow \\ \rightarrow \ulcorner \bigcup \mathscr{R}t'' \urcorner\, I_{w,L} \ulcorner \bigcup \mathscr{R}t' \urcorner)))$$

It is convenient to have a short way of saying that the reference sequence y is minimal with respect to a statement s:

MD-9.7 $$y\, MIN_{w,L}\, s \leftrightarrow \bigvee_{z,x,b,t} (s\, B_{w,L} \ulcorner t\langle x\rangle \in b \urcorner \wedge PS_{w,L}(z,y,t,b) \wedge \\ \wedge x\, M_{w,L} \ulcorner \bigcup \mathscr{R}y \urcorner \wedge \bigwedge_{z',y',b',t'} (x\, M_{w,L} \ulcorner \bigcup \mathscr{R}y' \urcorner \wedge \\ \wedge \ulcorner t'\langle x\rangle \in b' \urcorner\, B_{w,L} \ulcorner t\langle x\rangle \in b \urcorner \wedge PS_{w,L}(z',y',t',b') \wedge \\ \wedge \ulcorner \bigcup \mathscr{R}y' \urcorner\, Incl_{w,L} \ulcorner \bigcup \mathscr{R}y \urcorner \wedge z'\, E\text{-}STR_{w,L}\, z \rightarrow \\ \rightarrow \ulcorner \bigcup \mathscr{R}y' \urcorner\, I_{w,L} \ulcorner \bigcup \mathscr{R}y \urcorner))$$

It follows almost immediately from definitions 9.6 and 9.7 that if w is statistically terminating in L, then if x is known to belong to the union of the range of t, and s is a strongest statistical statement about the sequence t, the random quantity y, and the Borel set b, then there is a minimal sequence t' with respect to the statement $\ulcorner y\langle x\rangle \in b \urcorner$.

MT-9.2 $$w \text{ is } S\text{-}Term_L \rightarrow \bigwedge_{x,y,t,b,s} (x\, M_{w,L} \ulcorner \bigcup \mathscr{R}t \urcorner \wedge PS_{w,L}(s,t,y,b) \rightarrow \\ \rightarrow \bigvee_{t'} t'\, MIN_{w,L} \ulcorner y\langle x\rangle \in b \urcorner)$$

Of course it is clear that if y is minimal with respect to s, it will also be minimal with respect to the denial of s:

MT-9.3 $y\ MIN_{w,L}\ s \rightarrow y\ MIN_{w,L} \ulcorner \sim s \urcorner$

It follows from definition MD-9.4 that except in extreme cases, no reference class can be *known*, in a statistically terminating rational corpus, to be infinite. The arguments following the proof of the theorem attempt to show that this is not an implausible result.

MT-9.4 $w \text{ is } S\text{-}Term_L \wedge RAN_L(x, t, y, b, w) \wedge PS_{w,L}(s, t, y, b) \wedge$
$\wedge \{s\} \vdash \ulcorner \sim S(t, y, b, 0_r, 0_r) \wedge \sim S(t, y, b, 1_r, 1_r) \urcorner \rightarrow$
$\rightarrow \sim \bigvee_f \ulcorner f \text{ is } 1-1 \wedge \mathscr{D}f = \bigcup \mathscr{R}t \wedge \omega = \mathscr{R}f \wedge s \urcorner \in w$

Proof:

(1) Antecedent of 9.4 — ass.

(2) $[\bigvee_f] \ulcorner f \text{ is } 1-1 \wedge \mathscr{D}f = \bigcup \mathscr{R}t \wedge \omega = \mathscr{R}f \wedge s \urcorner \in w$ — ass.

(3) Let $g = \ulcorner \imath g(\mathscr{D}g = \mathscr{D}f \cap \{a : y\langle a \rangle \in b\} \wedge$
$\wedge \bigwedge_{a, b \in \mathscr{D}g} (g\langle a \rangle < g\langle b \rangle \leftrightarrow f\langle a \rangle < f\langle b \rangle)) \urcorner$

(4) Let $h = \ulcorner \imath h(\mathscr{D}h = \mathscr{D}f \cap \{a : y\langle a \rangle \notin b\} \wedge$
$\wedge \bigwedge_{a, b \in \mathscr{D}h} (h\langle a \rangle < h\langle b \rangle \leftrightarrow f\langle a \rangle < f\langle b \rangle) \wedge \mathscr{R}h = \omega) \urcorner$

(5) $\ulcorner f \text{ is } 1-1 \wedge \mathscr{D}f = \bigcup \mathscr{R}t \wedge \omega = \mathscr{R}f \wedge s \urcorner \vdash \ulcorner \mathscr{R}g = \omega \wedge$
$\wedge \mathscr{R}h = \omega \wedge \mathscr{D}g \cup \mathscr{D}h = \mathscr{D}f \wedge g \text{ is } 1-1 \wedge h \text{ is } 1-1 \urcorner$
(1), (3), (4)

(6) Let $t_p = \ulcorner \imath r(\mathscr{D}r = \omega \wedge \bigwedge_{i \in \omega} (r_i = \{g^{-1}(j) : j < i+1\} \cup$
$\cup \{h^{-1}(j) : j < [\sqrt{2}i - i] + 1\})) \urcorner$
(We use '$[p]$' to denote the largest integer less than p.)

(7) $\bigvee_r \ulcorner \bigcup \mathscr{R}t_p = r \urcorner \in w \wedge r \in RT_L$ — (2), (5), (6)

(8) Let $s_p = \ulcorner S(t_p, y, b, 1/\sqrt{2}, 1/\sqrt{2}) \urcorner$

(9) $PS_{w,L}(s_p, t_p, y, b)$ — (5), (7), (8), MT-8.32

(10) $x\ M_{w,L} \ulcorner \bigcup \mathscr{R}t_p \urcorner$ — (1), (7), MT-8.22

(11) $[\bigvee_{s', t', y', b'}]\ x\ M_{w,L} \ulcorner \bigcup \mathscr{R}t' \urcorner \wedge PS_{w,L}(s', t', y', b') \wedge$

$\wedge \ulcorner y\langle x\rangle \in b \urcorner\ B_{w,L} \ulcorner y'\langle x\rangle \in b' \urcorner \wedge s_p\ E\text{-}STR\ s' \wedge$
$\wedge \ulcorner \bigcup \mathscr{R}t' \urcorner\ Incl_{w,L} \ulcorner \bigcup \mathscr{R}t_p \urcorner \wedge \bigwedge_{s'', t'', y'', b''} (x\ M_{w,L} \ulcorner \bigcup \mathscr{R}t'' \urcorner \wedge$
$\wedge \ulcorner y''\langle x\rangle \in b'' \urcorner\ B_{w,L} \ulcorner y'\langle x\rangle \in b' \urcorner \wedge$
$\wedge\ PS_{w,L}(s'', t'', y'', b'') \wedge s''\ E\text{-}STR\ s' \wedge$
$\wedge \ulcorner \bigcup \mathscr{R}t'' \urcorner\ Incl_{w,L} \ulcorner \bigcup \mathscr{R}t' \urcorner \rightarrow \ulcorner \bigcup \mathscr{R}t' \urcorner\ I_{w,L} \ulcorner \bigcup \mathscr{R}t'' \urcorner)$
(1), (9), (10), MD-9.4

(12) Let $b = \iota a(a \in \bigcup \mathscr{R}t' \wedge \sim a = x \wedge \bigwedge_c (c \in \bigcup \mathscr{R}t' \wedge$
$\wedge \sim c = x \rightarrow f\langle a\rangle \leq f\langle c\rangle))$

(13) $\ulcorner \{b\} \urcorner \in ART_L$ MD-7.39

(14) $\ulcorner r - \{b\} \urcorner \in RT_L$ (13), MD-7.40

(15) Let $t'' = \ulcorner \iota r(\mathscr{D}r = \omega \wedge \bigwedge_{i \in \omega} r_i = t'_i - \{b\}) \urcorner$

(16) $\ulcorner \bigcup \mathscr{R}t' = r \urcorner \vdash \ulcorner \bigcup \mathscr{R}t'' = r - \{b\} \urcorner$ (15)

(17) $\ulcorner \bigcup \mathscr{R}t'' = r - \{b\} \urcorner \in w$ (12), (16)

(18) Let $s'' = \ulcorner S(t'', y', b', 1/\sqrt{2}, 1/\sqrt{2}) \urcorner$

(19) $x\ M_{w,L} \ulcorner \bigcup \mathscr{R}t'' \urcorner$ (11), (12)

(20) $PS_{w,L}(s'', t'', y', b')$ (11), (14), (17), (18)

(21) $\vdash \ulcorner \bigcup \mathscr{R}t'' \subset \bigcup \mathscr{R}t' \urcorner$ (15)

(22) $\ulcorner \bigcup \mathscr{R}t' \urcorner\ I_{w,L} \ulcorner \bigcup \mathscr{R}t'' \urcorner$ (11), (19), (20), (21)

(23) $b\ M_{w,L} \ulcorner \bigcup \mathscr{R}t' \urcorner$ (12)

(24) $\ulcorner \sim b \in \bigcup \mathscr{R}t'' \urcorner \in w$ (15)

(25) $\ulcorner 0 \neq 0 \urcorner \in L\text{-}Cn(w)$ (22), (23), (24); imp. 1, MD-9.3

(26) Case II: similar.

In general, if w is a statistically terminating rational corpus, it involves no reference terms known to denote infinite sets. Indeed, observe that the consequent of this theorem merely requires that we not *know of* a function mapping the natural numbers onto our reference class. It is conceivable that we should know that $\ulcorner \bigcup \mathscr{R}t \urcorner$ denotes an infinite class, without *knowing of* any 1-1 function f whose domain is $\bigcup \mathscr{R}t$ and whose range is ω; i.e., we could have $\ulcorner \bigvee_f (f \text{ is } 1-1 \wedge \mathscr{D}f = \bigcup \mathscr{R}t \wedge \mathscr{R}f = \omega) \urcorner$ in our body of knowledge while for no expression f of L do we have $\ulcorner f \text{ is } 1-1 \wedge \mathscr{D}f = \bigcup \mathscr{R}t \wedge \mathscr{R}f = \omega \urcorner$ in our body of knowledge. But even in this unlikely state of affairs, we would have no reason to think we could re-order the elements of $\bigcup \mathscr{R}t$, and thus no reason to suppose our rational corpus is not statistically terminating.

This is not implausible. It is hard to believe that there are useful empirical terms in the language of science that are known to denote infinite sets. It is one thing to know that a term denotes a finite class, without knowing how big it is, or even to know that it is finite, and a very different thing indeed to know that a term denotes an infinite set. Furthermore, if a reference term y is known to denote an infinite reference class, we must surely have some natural way of ordering the members of that class; and then, by the definition of the set of reference terms, any term denoting an initial segment (in relation to the natural ordering) will turn out also to be a reference term, and a rather more useful one.

We can now prove our fundamental theorem, the most important theorem of this section concerning randomness. It states that if a rational corpus w is statistically terminating, x is a random member of y with respect to having a z-value in b, x' is a random member of y' with respect to having a z' value in b', and $\ulcorner z\langle x\rangle\in b\urcorner\ B_{w,L}\ulcorner z'\langle x\rangle\in b'\urcorner$, then a strongest statistical statement of w about y, z, and b is exactly as strong as a strongest statistical statement of w about y', z', and b'.

MT-9.5 $RAN_L(x, y, z, b, w) \wedge RAN_L(x', y', z', b', w) \wedge w \text{ is } S\text{-}Term_L \wedge$
$\wedge \ulcorner z\langle x\rangle\in b\urcorner\ B_{w,L}\ulcorner z'\langle x'\rangle\in b'\urcorner \wedge PS_{w,L}(s, y, z, b) \wedge$
$\wedge PS_{w,L}(s', y', z', b') \rightarrow s\ E\text{-}STR_{w,L}\ s'$

Proof:

(1) Antecedent of 9.5 — ass.

(2) $s\ STR_{w,L}\ s' \vee s'\ STR_{w,L}\ s \vee s\ DIF_{w,L}\ s' \vee s\ E\text{-}STR_{w,L}\ s'$ — (1), MD-8.7, MT-7.13

(3) show $\sim s\ DIF_{w,L}\ s'$

(4) $\ulcorner x' \preccurlyeq x\urcorner \in w \wedge s\ DIF_{w,L}\ s'$ — ass. without loss of generality; MD-7.39, 7.40

(5) $[\bigvee_{y_0, z_0, s_0, b_0}]\ x\ M_{w,L}\ulcorner\bigcup \mathscr{R}y_0\urcorner \wedge PS_{w,L}(s_0, y_0, z_0, b_0) \wedge$
$\wedge\ s_0\ E\text{-}STR_{w,L}\ s \wedge \ulcorner z\langle x\rangle\in b\urcorner\ B_{w,L}\ulcorner z_0\langle x\rangle\in b_0\urcorner \wedge$
$\wedge \ulcorner\bigcup \mathscr{R}y_0\urcorner\ Incl_{w,L}\ulcorner\bigcup \mathscr{R}y\urcorner \wedge$

$\wedge \bigwedge_{y_1, z_1, s_1, b_1} (x\ M_{w,L} \ulcorner \bigcup \mathscr{R} y_1 \urcorner \wedge s_1\ E\text{-}STR_{w,L}\ s_0 \wedge$

$\wedge PS_{w,L}(s_1, y_1, z_1, b_1) \wedge$

$\wedge \ulcorner z_1 \langle x \rangle \in b_1 \urcorner\ B_{w,L} \ulcorner z_0 \langle x \rangle \in b_0 \urcorner \wedge$

$\wedge \ulcorner \bigcup \mathscr{R} y_1 \urcorner\ Incl_{w,L} \ulcorner \bigcup \mathscr{R} y_0 \urcorner \rightarrow$

$\rightarrow \ulcorner \bigcup \mathscr{R} y_1 \urcorner\ I_{w,L} \ulcorner \bigcup \mathscr{R} y_0 \urcorner)$ (1), MD-9.5, MD-9.6

(6) $s_0\ DIF_{w,L}\ s' \wedge \sim y_0\ D_L(x', y', z', b'w)$
(1), (4), (5) MT-7.10; MD-9.6

(7) $(x, y_0, z_0, b_0)\ SND_L(x', y', z', b', w) \vee$
$\vee (x, y_0, z_0, b_0)\ PND_L(x', y', z', b', w)$
(5), (6), MD-9.3

(8) $\sim (x, y_0, z_0, b_0)\ PND_L(x', y', z', b', w)$ (4), MD-9.2

(9) $[\bigvee_{y_2, z_2, b_2, s_2}]\ x\ M_{w,L} \ulcorner \bigcup \mathscr{R} y_2 \urcorner \wedge$
$\wedge \ulcorner z_2 \langle x \rangle \in b_2 \urcorner\ B_{w,L} \ulcorner z_0 \langle x \rangle \in b_0 \urcorner \wedge s_2\ E\text{-}STR_{w,L}\ s' \wedge$
$\wedge PS_{w,L}(s_2, y_2, z_2, b_2) \wedge [(\ulcorner \bigcup \mathscr{R} y_2 \urcorner\ Incl_{w,L} \ulcorner \bigcup \mathscr{R} y_0 \urcorner \wedge$
$\wedge \sim \ulcorner \bigcup \mathscr{R} y_2 \urcorner\ I_{w,L} \ulcorner \bigcup \mathscr{R} y_0 \urcorner) \vee \ulcorner \bigcup \mathscr{R} y_2 \urcorner\ I_{w,L} \ulcorner \{x\} \urcorner]$
(5), (6), MD-9.1

(10) $s_2\ DIF_{w,L}\ s \wedge \sim y_2\ D_L(x, y, z, b, w)$
(1), (7), MD-9.5, MT-7.10

(11) $(x, y_2, z_2, b_2)\ SND_L(x, y, z, b, w) \vee$
$\vee (x, y_2, z_2, b_2)\ PND_L(x, y, z, b, w)$ (10), (9), MD-9.3

(12) $\sim (x, y_2, z_2, b_2)\ PND_L(x, y, z, b, w)$ MD-9.2

(13) $[\bigvee_{y_1, z_1, b_1, s_1}]\ x\ M_{w,L} \ulcorner \bigcup \mathscr{R} y_1 \urcorner \wedge$
$\wedge \ulcorner z_1 \langle x \rangle \in b_1 \urcorner\ B_{w,L} \ulcorner z_2 \langle x \rangle \in b_2 \urcorner \wedge$
$\wedge PS_{w,L}(s_1, y_1, z_1, b_1) \wedge s_1\ E\text{-}STR_{w,L}\ s \wedge$
$\wedge [(\ulcorner \bigcup \mathscr{R} y_1 \urcorner\ Incl_{w,L} \ulcorner \bigcup \mathscr{R} y_2 \urcorner \wedge$
$\wedge \sim \ulcorner \bigcup \mathscr{R} y_1 \urcorner\ I_{w,L} \ulcorner \bigcup \mathscr{R} y_2 \urcorner) \vee \ulcorner \bigcup \mathscr{R} y_1 \urcorner\ I_{w,L} \ulcorner \{x\} \urcorner]$
(9), (11), (12), MD-9.1

(14) $s_2\ DIF_{w,L}\ s_1$ (4), (9), (13), MT-7.10

(15) $\ulcorner \bigcup \mathscr{R} y_1 \urcorner\ Incl_{w,L} \ulcorner \bigcup \mathscr{R} y_0 \urcorner \rightarrow \ulcorner \bigcup \mathscr{R} y_1 \urcorner\ I_{w,L} \ulcorner \bigcup \mathscr{R} y_0 \urcorner$
(5), (13)

(16) $\{\ulcorner\bigcup \mathscr{R}y_1 \subset \bigcup \mathscr{R}y_0 \rightarrow \bigcup \mathscr{R}y_1 = \bigcup \mathscr{R}y_0\urcorner,$
$\ulcorner(\bigcup \mathscr{R}y_1 \subset \bigcup \mathscr{R}y_2 \wedge \sim \bigcup \mathscr{R}y_1 = \bigcup \mathscr{R}y_2) \vee$
$\vee \bigcup \mathscr{R}y_1 = \{x\}\urcorner, \ulcorner(\bigcup \mathscr{R}y_2 \subset \bigcup \mathscr{R}y_0 \wedge$
$\wedge \sim \bigcup \mathscr{R}y_2 = \bigcup \mathscr{R}y_0) \vee \bigcup \mathscr{R}y_2 = \{x\}\urcorner\} \vdash_L \ulcorner 0 \neq 0\urcorner$ imp.

(17) show $\sim s\ STR_{w,L}\ s'$

(18) $s\ STR_{w,L}\ s'$ ass.

(19) $[\bigvee_{y''}]\ y''\ D_L(x, y, z, b, w)$ (1), MD-9.5, MD-9.4

(20) $\sim y''\ D_L(x, y, z, b, w)$ (1), MD-9.5

(21) $\sim s'\ STR_{w,L}\ s$ similarly

(22) $s\ E\text{-}STR_{w,L}\ s'$ (2), (3), (17), (21)

It is this theorem, ultimately, which will allow us to define probability in such a way that every member of a biconditional chain will be assigned the same probability. Rational corpora generated in a plausible manner will turn out generally to be statistically terminating, so that the theorem is in fact even stronger than it appears.

Finally, let us suppose that x is a term of L, y a random variable, and b a Borel term. We should like to claim that there exists a term x', a random variable y', and a Borel term b', such that x' is a random member of some reference sequence t', with respect to having a y'-value in b', relative to the body of knowledge w, and such that $\ulcorner y\langle x\rangle \in b\urcorner$ is connected by a biconditional chain in w to $\ulcorner y'\langle x'\rangle \in b'\urcorner$. In order to justify this claim, we need make only one special assumption: that there is no infinite sequence of disruptions of randomness due to stronger and stronger statistical knowledge. We shall write $P\text{-}Term_L(w)$ to mean that this does not happen in w:

MD-9.8 $P\text{-}Term_L(w) \leftrightarrow \sim \bigvee_{x,t,y,b} (x, t, y, b \text{ are } \omega\text{-sequences} \wedge$
$\wedge \bigwedge_{i\in\omega} (x_i\ M_{w,L} \ulcorner \bigcup \mathscr{R}t_i\urcorner \wedge$
$\wedge \ulcorner y_i\langle x_i\rangle \in b_i\urcorner\ B_{w,L} \ulcorner y_{i+1}\langle x_{i+1}\rangle \in b_{i+1}\urcorner \wedge$
$\wedge\ t_{i+1}\ S_L(x_i, t_i, y_i, b_i, w)))$

As usual, we suppose that w has a basis, is statistically closed, and is normal.

MT-9.6 w is $RC_L \wedge w$ *is statistically closed*$_L \wedge Normal_L\ w \wedge$
$\wedge P\text{-}Term_L(w) \wedge \bigvee_{n, A} (L\text{-}Cn(A) \cap St_L \subset w \wedge RV(n, y, \langle L, A\rangle) \wedge$
$\wedge \vdash_L \ulcorner b \in \mathbb{B}^n \urcorner) \wedge \ulcorner x \in \mathcal{D} y \urcorner \in w \wedge \sim \ulcorner 0 \neq 0 \urcorner \in L\text{-}Cn(w) \rightarrow$
$\rightarrow \bigvee_{x'', y'', t'', b''} (RAN_L(x'', t'', y'', b'', w) \wedge$
$\wedge \ulcorner y\langle x\rangle \in b \urcorner\ B_{w, L} \ulcorner y''\langle x''\rangle \in b'' \urcorner)$

Proof:

(1) Assume antecedent

(2) Let $t = \ulcorner \iota f\,(f \text{ is fun} \wedge \mathcal{D} f = \omega \wedge \bigwedge_{i \in \omega} f\langle i\rangle = \{x\}) \urcorner$

(3) $\ulcorner \bigcup \mathcal{R} t = \{x\} \urcorner \in w$ (2), MD-8.1

(4) $\ulcorner \{x\} \urcorner \in RT_L$ (1), MD-7.39

(5) $SS_{w, L} \ulcorner S\langle t, y, b, 0_r, 1_r\rangle \urcorner$ (1), (3), (4), MD-7.41

(6) $\{\ulcorner x \in \mathcal{D} y \urcorner\} \vdash_L \ulcorner S\langle t, y, b, 0_r, 1_r\rangle \urcorner$ (1), (2), D-7.2

(7) $\ulcorner S\langle t, y, b, 0_r, 1_r\rangle \urcorner \in w$ (1), (6), MD-8.1

(8) $[\bigvee_s]\ PS_{w, L}(s, t, y, b, w)$ (1), (5), (7), MT-8.28

(9) $s\ E\text{-}STR_{w, L} \ulcorner S\langle t, y, b, 1_r, 1_r\rangle \urcorner \vee$
$\vee s\ E\text{-}STR_{w, L} \ulcorner S\langle t, y, b, 0_r, 0_r\rangle \urcorner \vee$
$\vee s\ E\text{-}STR_{w, L} \ulcorner S\langle t, y, b, 0_r, 1_r\rangle \urcorner$ (2), MD-8.1

(10) Case I: $s\ E\text{-}STR_{w, L} \ulcorner S\langle t, y, b, 1_r, 1_r\rangle \urcorner$ ass.

(11) $\sim \bigvee_{t'} t'\ S_L(x, t, y, b, w)$ (1), (2), MD-9.4

(12) ~~show~~ $\bigwedge_{t'} \sim t'\ D_L(x, t, y, b, w)$

(13) $t'\ D_L(x, t, y, b, w)$ ass.

(14) $[\bigvee_{s', x', y', b'}]\ \ulcorner y'\langle x'\rangle \in b' \urcorner\ B_{w, L} \ulcorner y\langle x\rangle \in b \urcorner \wedge$
$\wedge x'\ M_{w, L} \ulcorner \bigcup \mathcal{R} t' \urcorner \wedge PS_{w, L}(s', t', y', b') \wedge s\ DIF_{w, L}\ s' \wedge$
$\wedge \sim (x', t', y', b')\ SND_L(x, t, y, b, w)$ (8), (13), MD-9.3

(15) $\bigwedge_{s'', t'', y'', b''} (\ulcorner y''\langle x'\rangle \in b'' \urcorner\ B_{w, L} \ulcorner y\langle x\rangle \in b \urcorner \wedge$
$\wedge x'\ M_{w, L} \ulcorner \bigcup \mathcal{R} t'' \urcorner \wedge \ulcorner \bigcup \mathcal{R} t'' \urcorner\ I_{w, L} \ulcorner \{x'\} \urcorner \wedge$
$\wedge PS_{w, L}(s'', t'', y'', b'') \rightarrow s\ E\text{-}STR_{w, L}\ s'')$
(8), (14), MD-9.1

(16) Let $t'' = \ulcorner \imath f\,(f\ is\ fun \wedge \mathscr{D}f = \omega \wedge \bigwedge_{i \in \omega} f\langle i \rangle = \{x'\})\urcorner$

(17) $\ulcorner \bigcup \mathscr{R}t'' = \{x'\}\urcorner \in w$ (1), (16), MD-8.1

(18) $\ulcorner \{x'\}\urcorner \in RT_L$ (14), MD-7.39

(19) $x'\ M_{w,L} \ulcorner \bigcup \mathscr{R}t''\urcorner$ (17), MD-8.1

(20) Let $y'' = \ulcorner \imath f\,(f\ is\ fun \wedge \mathscr{D}f = \{x'\} \wedge ((f\langle x'\rangle = \langle 1_r\rangle \wedge$
$\wedge\ y\langle x\rangle \in b) \vee (f\langle x'\rangle = \langle 0_r\rangle \wedge y\langle x\rangle \notin b)))\urcorner$

(21) $\bigvee_{n,A} (L\text{-}Cn(A) \cap St_L \subset w \wedge RV(n, y'', \langle L, A\rangle) \wedge$
$\wedge \vdash_L \{\langle 1_r\rangle\} \in \mathbb{B}^n)$ (1), (20)

(22) $\{\ulcorner S\langle t, y, b, 1_r, 1_r\rangle\urcorner\} \vdash_L \ulcorner y\langle x\rangle \in b\urcorner$ (2)

(23) $\ulcorner y\langle x\rangle \in b\urcorner \in w$ (8), (10), (22), MD-8.1

(24) $\ulcorner y''\langle x'\rangle \in \{\langle 1_r\rangle\}\urcorner \in w$ (20), (23), MD-8.1

(25) $\ulcorner S\langle t'', y'', \{\langle 1_r\rangle\}, 1_r, 1_r\rangle\urcorner \in w$ (16), (20), (24), MD-8.1

(26) $SS_{w,L} \ulcorner S\langle t'', y'', \{\langle 1_r\rangle\}, 1_r, 1_r\rangle\urcorner$
(17), (18), (21), MD-7.41

(27) $PS_{w,L}(\ulcorner S\langle t'', y'', \{\langle 1_r\rangle\}, 1_r, 1_r\rangle\urcorner, t'', y'', \ulcorner\{\langle 1_r\rangle\}\urcorner)$
(25), (26), MT-8.32

(28) $\ulcorner S\langle t'', y'', \{\langle 1_r\rangle\}, 1_r, 1_r\rangle\urcorner\ E\text{-}STR_{w,L}\ s$
(10), (8), (27), MD-7.42

(29) $\sim 0 = 0$ (15), (20), (19), (17), (27), (28)

(30) Case II: $s\ E\text{-}STR_{w,L} \ulcorner S\langle t, y, b, 0_r, 0_r\rangle\urcorner$ similar

(31) Case III: $s\ E\text{-}STR_{w,L} \ulcorner S\langle t, y, b, 0_r, 1_r\rangle\urcorner$ ass.

(32) $\sim \bigvee_{t'} t'\ D_L(x, t, y, b, w)$ (1), (31), MD-9.4

(33) sh~~o~~w $\bigvee_{x'', t'', y'', b''} (RAN_L(x'', t'', y'', b'', w) \wedge$
$\wedge \ulcorner y\langle x\rangle \in b\urcorner\ B_{w,L} \ulcorner y''\langle x''\rangle \in b''\urcorner)$

(34) $\bigwedge_{x'', t'', y'', b''} (\ulcorner y\langle x\rangle \in b\urcorner\ B_{w,L} \ulcorner y''\langle x''\rangle \in b''\urcorner \rightarrow$
$\rightarrow \sim RAN_L(x'', t'', y'', b'', w))$ ass.

Let m, n, p, q be ω-sequences constructed as follows:

(35) $[\bigvee_{t'}]\ t' S_L(x, t, y, b, w)$ (1), (32), MD-9.5

(36) $[\bigvee_{s, s', x', y', b'}]\ x'\ M_{w,L} \ulcorner \bigcup \mathscr{R}t'\urcorner \wedge PS_{w,L}(s, t, y, b) \wedge$

$\wedge \ulcorner y\langle x\rangle \in b\urcorner\ B_{w,L} \ulcorner y'\langle x'\rangle \in b'\urcorner \wedge$
$\wedge\ PS_{w,L}(s', t', y', b') \wedge s\ STR_{w,L}\ s' \wedge$
$\wedge \sim \bigvee_{y''} y''\ D_L(x', t', y', b', w)$ (36), MD-9.4

(37) $n_0 = x' \wedge p_0 = y' \wedge q_0 = b' \wedge m_0 = t'$

(38) $\sim RAN_L(n_0, m_0, p_0, q_0)$ (33), (36)

(39) $[\bigvee_{t''}]\ t''\ S_L(n_0, m_0, p_0, q_0, w)$ (38, (36), MD-9.3

(40) $m_1 = t''$ and so on.

(41) $m_k = t'''$

(42) $[\bigvee_{s''', x''', s'''', y''', b'''}]\ x'''\ M_{w,L} \ulcorner \bigcup \mathscr{R} m_k \urcorner \wedge$
$\wedge\ PS_{w,L}(s''', m_{k-1}, p_{k-1}, q_{k-1}) \wedge$
$\wedge\ p_{k-1}\langle n_{k-1}\rangle \in b_{k-1}\ B_{w,L}\ b'''\langle x'''\rangle \in b''' \wedge$
$\wedge\ PS_{w,L}(s'''', t''', y''', b''') \wedge s''''\ STR_{w,L}\ s''' \wedge$
$\wedge \sim \bigvee_{y''} y''\ D_L(x''', t''', y''', b''', w)$

(43) $n_k = x''' \wedge p_k = y''' \wedge q_k = b'''$

(44) m, n, p, and q contradict the last clause of the antecedent of MT-9.6, $P\text{-}Term_L(w)$

A few more theorems, none of them difficult, will be used in the next chapter.

MT-9.7 $\ulcorner b \in \mathbb{B}^n \urcorner \in w \wedge RAN_L(x, y, z, b, w) \rightarrow RAN_L(x, y, z, \ulcorner \mathbb{R}^n - b \urcorner, w)$

MT-9.8 $RAN_L(x, y, z, b, w) \wedge PS_{w,L}(s, y, z, b) \wedge$
$\wedge \ulcorner z\langle x\rangle \in b\urcorner\ B_{w,L} \ulcorner z'\langle x'\rangle \in b'\urcorner \wedge PS_{w,L}(s', y', z', b') \wedge$
$\wedge\ x'\ M_{w,L} \ulcorner \bigcup \mathscr{R} y' \urcorner \wedge s'\ E\text{-}STR_{w,L}\ s \rightarrow RAN_L(x', y', z', b', w)$

MT-9.9 $RAN_L(x, y, z, b, w) \wedge PS_{w,L}(s, y, z, b) \wedge y\ MIN_{w,L} \ulcorner z\langle x\rangle \in b\urcorner \rightarrow$
$\rightarrow \sim \bigvee_{s', y', z', b'} (\ulcorner \bigcup \mathscr{R} y' \urcorner\ Incl_{w,L} \ulcorner \bigcup \mathscr{R} y \urcorner \wedge$
$\wedge \sim \ulcorner \bigcup \mathscr{R} y' \urcorner\ I_{w,L} \ulcorner \bigcup \mathscr{R} y \urcorner \wedge \ulcorner z'\langle x\rangle \in b'\urcorner\ B_{w,L} \ulcorner z\langle x\rangle \in b\urcorner \wedge$
$\wedge\ PS_{w,L}(s', y', z', b') \wedge s\ DIF_{w,L}\ s')$

NOTES

[1] Reichenbach, Hans, *The Theory of Probability*, Berkeley and Los Angeles, 1949, p. 374. Italics Reichenbach's.

[2] *Ibid.*, p. 375.

[3] Hilpinen, Risto, *Rules of Acceptance and Inductive Logic*, Amsterdam, 1968, p. 74.

[4] *Probability and The Logic of Rational Belief*, Middletown, 1961.

CHAPTER TEN

PROBABILITY

1. It is already clear, from the discussion of the last chapter, how we are going to define probability: We shall say that the probability of a statement of the form $\ulcorner x \in y \urcorner$, relative to a strictly consistent body of knowledge *w* is the interval $[p, q]$, whenever relative to the rational corpus *w*, *x* is a random member of *z*, with respect to membership in *y*, and we know (in *w*) that the proportion of *y*'s among *z*'s lies in the interval $[p, q]$. Furthermore, again, in accordance with a remark made in the last chapter, we shall say that the probability of any statement *s* is $[p, q]$, relative to *w*, just when *s* is connected by a biconditional chain in *w* to $\ulcorner x \in y \urcorner$, and the probability of $\ulcorner x \in y \urcorner$, relative to *w*, is $[p, q]$. Finally, we take account of more general (not necessarily strictly consistent) rational corpora by stipulating that the probability of a statement *s*, relative to the rational corpus *w*, is $[p, q]$ just in case the probability of *s*, relative to every maximal consistent subset of *w* is $[p, q]$.

Probability will thus be understood as a relation between a statement and a body of knowledge, in a given language, and thus as relative to a language. The relativization of probability to a language has come in for considerable criticism. First of all, of course, the Carnap-Hintikka approach to logical probabilities has concerned itself only with very elementary languages. A more fundamental criticism, however, is that much of the important progress of science may be seen as the progressive refinement of the scientific vocabulary: not merely in the introduction of new terms ('electron', 'positron'), but in the refinement of terms already present ('length', 'temperature'). If probability is supposed to be a guide to what we believe, then it should not only take account of these changes in language, but should guide us in making them. Yet if probability is relative to a given language, we cannot sensibly compare the probability of a theory of employing one language to that of a theory employing another language.

There are two answers to this criticism. The short and perhaps not very satisfying one is that as long as we can define probability for the language of science we happen to use at the moment, we have at least a basis for assessing the probability of statements we make at this moment. The shortcoming of the logical range interpretations of probability is not that they are language dependent *per se*, but that they are dependent on languages far too simple to reflect the actual language of science, and at the same time that there is no plausible way of treating a part of this language in isolation from the rest. There is thus no way in which the logical range interpretations of probability can throw light on genuine controversies in statistics or in the applications of statistical knowledge. Philosophers who concern themselves with probability often act as if there were no such controversies; but a very shallow skimming of a fragment of statistical literature gives them the lie. The logical conception of probability offered here, however, is defined for very strong languages indeed (including set theory), and nothing depends on the number or character of the predicates and operation expressions of the language. In fact only those expressions that occur in the basis of a rational corpus (if it has one) are relevant to probabilities. It is thus quite possible (as we shall see in detail in due course) to apply quite directly the conception of probability developed here to real life problems of statistical inference.

The longer and philosophically more interesting answer is that we need not be able to compare directly the probability of a theory employing one language to the probability of a theory employing another language in order to receive guidance from probability considerations in the choice between the two theories. It might generally turn out that the probability of any theory is the whole interval [0, 1]. It might turn out that the best way to construe a theory is as a vocabulary and meaning postulates, so that all theories have the probability of one. But we might still be able to look at the bodies of knowledge that can be developed from the two theories. Since we assume that there is some difference in vocabulary, we cannot suppose simply that one body of knowledge is included in the other; but we may be able to give sense to the claim that one body of knowledge contains more, more interesting, or stronger statements than the other, or that one contains more information than the other. If we can employ probability to establish the contents of a rational corpus, then we can indirectly use probability to evaluate the difference in contents between the rational corpus devolving from one theory and that coming from the other. This is a very brief sketch of *one* direction the long answer might take. Filling out the details, or finding an alternative, is the most difficult and important problem a general theory of scientific inference must solve.

The relativization of probability to bodies of knowledge does not raise any novel problems. It has long been asserted by those who have offered logical range interpretations of probability, from Keynes to Hintikka, that if probability is to be taken as a guide in life, it must be relativized to a (total) body of knowledge. This is not at all to say that probability is not 'objective'; to relativize arithmetic to set theory is not to make it subjective. It is, however, to say that what is probable for us, at a given time, depends on what we know, and will be different at some other time if our body of knowledge is different at that time. There are those who have found this objectionable. It is argued that there are *true* probabilities, that are quite independent of what we know and don't know; that we often distinguish between true and false probability assignments not on the basis of what we do or don't know, but on the basis of the objective content of the world outside us; that it makes perfectly good sense to talk about hypothetical probabilities and unknown probabilities; and so on.

All of those ways of talking can be accommodated in the framework we have provided. To give literal sense to these locutions, we need to develop the semantics of our languages, and in particular the generalized semantics appropriate to rational corpora. Nevertheless it is clear that we can talk about rational corpora each of whose members is true in the real world. The 'true' probability that an a will be a b, then will be the probability of $\ulcorner x \in b \urcorner$ relative to a rational corpus whose basis consists of $\ulcorner x \in a \urcorner$, $\ulcorner a = \bigcup \mathcal{R} c \urcorner$, and an appropriate true statistical statement about c and b, where x is a constant foreign to L. (If there is another sequence, c', also satisfying the requirements but leading to a different probability we must take the question as ambiguous.) Let this rational corpus be supplemented by any other set of statements at all in the vocabulary of L, that are not inconsistent with these, and the probability will remain the same. (This theorem will be proven in due course in this chapter). The 'appropriate true statistical statement' depends of course on the objective world outside us, not at all on what we know and don't know. To say that a probability is unknown is to say roughly that there is nothing in our body of knowledge that we take to be a good approximation to the true statistical statement in question. To talk about a hypothetical probability is to talk about a rational corpus just like ours, except that it contains a certain statistical statement which, hypothetically, we suppose to be true.

Although the definition of probability contemplated is a logical one, and in fact depends only on the syntactical relation between a statement and a body of knowledge construed as a set of statements, it nevertheless has a component that is usually empirical. It follows from our conception of probability, in particular, that we cannot correctly say that the probability of a statement is $[p, q]$ unless we *know* a certain statistical statement mentioning these same parameters, or better, in order to avoid philosophical difficulties concerning knowledge (in particular, we should not identify knowledge and true belief), unless a certain statistical statement mentioning these same parameters is a member of our rational corpus. Thus it is that in general when we say that the probability of getting a six on the next roll of this loaded die is about $\frac{1}{3}$, the grounds for our assertion include the acceptance into our body of knowledge of a statement to the effect that about a third of the rolls of this die will produce sixes. We do not require that the statistical statement be *true* (for then we could hardly ever say anything: it would amount to requiring that our bodies of knowledge contain only true statements – a desideratum inappropriate to our human station in life), but that we have adequate grounds for accepting it. Furthermore it turns out (it will be established as a theorem later on) that if c is a constant not in L, then the probability of $\ulcorner c \in b \urcorner$, relative to the rational corpus $w \cup L\text{-Cn}(\{\ulcorner c \in \bigcup \bigcup \mathcal{R}a \urcorner\})$, is $[p, q]$ if and only if the corresponding statistical statements is in w.

Although most of the statistical statements in our rational corpora are empirical, and come to be members of our rational corpora in virtue of adequate empirical evidence having been marshalled for them, there are also some statistical statements that are true in every possible world. This includes not only rather uninteresting statistical statements of the form $\ulcorner S\langle w, x, y, 0, 1\rangle \urcorner$ but certain other, rather more interesting, statements. For example, we may accept formal counterparts of ‘practically all subsets of the set a that contain n members contain a proportion of members of b differing by at most ϵ_n from the proportion of members of a that are members of b’.

The value of a probability depends on the language we are speaking and the body of knowledge we have. This dependence operates through the notion of randomness. We cannot refer, except artificially by means of something like the introduction of a new name into the language, to a toss of a coin that is *merely* 'a' toss of the coin, nor to the weight of a Canadian goose that is *merely* 'a' Canadian goose. If we are referring to a real toss of the coin, or a real goose, we must refer to it by a description which reveals a number of things about the object described. It gives at least some indication of where and when the object described may be presumed to occur; it also gives, indirectly if not directly, some of the relations in which that object stands to other parts of the world (e.g., the toss was performed by John or by Sally, the goose is sighted by a hunter or by a birdwatching society). What we know about the object is relevant, as the logical interpreters of probability have all maintained, and as frequentists who have been concerned with the applications of probability have also maintained in their discussions of the selection of the appropriate reference class, to the assessments of probability we make. (Curiously, as observed earlier, this does not apply to subjectivists, for whom we need not take account of what we know, but only of the relations among our opinions.)

The second term of the randomness relation refers to an infinite sequence of finite sets; what is required is that we know that the union of all these sets be a set denoted by one of a special set of terms constructible in our language. We have here the classical reference set or fundamental probability set; but the possibilities are sharply limited by our language. Not only do we limit our attention to sets that can be named in the language, but we limit it to a very special class of sets that can be named in the language. It is here that the dependence of probability on a particular language comes in most strongly. It is not an arbitrary or metaphysical dependence: we *could* speak another language, but in order to do so, we would have to have good reasons for choosing to speak another language, and I claim that we can have good reasons, formulated in part in terms of probability, for choosing to speak one language rather than another.

There is nothing controversial or unclassical involved in the third and fourth terms of the randomness relation, except perhaps the distinction between a random variable (an expression of the language) and a random quantity (the function denoted by a random variable).

We come finally to the most delicate question, that of the ignorance stipulated by the definition of randomness. It turns out that a can be a random member of B only when certain things are *not* known. Is this the old principle of indifference in new guise? Not quite. For when we say that a is a random member of B, we do not at all mean that it is as likely to be one member of B as another; even if this were to make sense, which is doubtful, we cannot derive probabilities from relations of randomness alone. Every probability must be based on statistical knowledge. The randomness relation merely specifies what particular bit of statistical knowledge is relevant in our particular circumstances.

Given, then, that probability is going to be the sort of thing that can be taken as a guide in life, and that we are not going to push the conditions for the application of our statistical knowledge into a foggy limbo labeled 'pragmatics' and promptly forgotten about, there is nothing *prima facie* implausible about our program. Let us look at the properties of the probability relation defined in the manner suggested.

2. Let w be consistent. We shall ultimately wish to take account of rational corpora that are not strictly consistent, so we begin by defining *partial probability*. We say that the partial probability of s is $[p, q]$, relative to w, when there exist terms x, y, z, b such that $RAN_L(x, y, z, b, w)$, $PS_{w,L}(\ulcorner S\langle y, z, b, p, q\rangle\urcorner, y, z, b)$, and $\ulcorner s \leftrightarrow z\langle x\rangle \in b\urcorner \in w$. Observe that the '$p$' and '$q$' of this definition are expressions. Under the standard interpretation of our language, however, these expressions will denote real numbers, and thus we may just as well take the *interval* $[p, q]$ as the value of our probability as the *expression* $\ulcorner[p, q]\urcorner$. There are, of course, lots more intervals than there are names of them, but there are plenty of names for all practical purposes. We presuppose that we have a standard interpretation for the set-theoretical part of our language, so that the real number expression p denotes just the appropriate real number. This renders our definition a semantic one. We write '$p\text{-}prob_L(s, w)$' for the partial probability of s with respect to the strictly consistent set of statements w.

MD-10.1 $$p\text{-}prob_L = \iota f\,[f \text{ is } fun \wedge \mathscr{D}f = \prod(St_L, \mathscr{P}St_L) \wedge \bigwedge_{s,w}(s \in St_L \wedge$$
$$\wedge\, w \subset St_L \rightarrow [\bigvee_{x,y,z,b,p,q}(S\text{-}Term_L\, w \wedge RAN_L(x, y, z, b, w) \wedge$$
$$\wedge\, \ulcorner z\langle x\rangle \in b\urcorner\; B_{w,L}\; s \wedge \sim\ulcorner 0 \neq 0\urcorner \in L\text{-}Cn(w) \wedge$$

$$\wedge PS_{w,L}(\ulcorner S\langle y, z, b, p, q\rangle\urcorner, y, z, b) \wedge f(s, w)=[p, q]) \vee$$
$$\vee(\sim \bigvee_{x,y,z,b,p,q} (S\text{-}Term_L\, w \wedge RAN_L(x, y, z, b, w) \wedge$$
$$\wedge \ulcorner z\langle x\rangle \in b\urcorner\, B_{w,L}\, s \wedge \sim\ulcorner 0\neq 0\urcorner \in L\text{-}Cn(w) \wedge$$
$$\wedge PS_{w,L}(\ulcorner S\langle y, z, b, p, q\rangle\urcorner, y, z, b)) \wedge f(s, w)=0)])]$$

By way of explication of the definition, we have:

MT-10.1 $p\text{-}prob_L(s, w)=[p, q] \leftrightarrow \bigvee_{x,y,z,b} (RAN_L(x, y, z, b, w) \wedge$

$$\wedge \ulcorner z\langle x\rangle \in b\urcorner\, B_{w,L} s \wedge PS_{w,L}(\ulcorner S\langle y, z, b, p, q\rangle\urcorner, y, z, b) \wedge$$
$$\wedge\, w \text{ is } RC_L \wedge \sim\ulcorner 0\neq 0\urcorner \in L\text{-}Cn(w))$$

The first and most important theorem about the partial probability function states that the definite description of the definition is proper even in the interesting non-vacuous cases.

MT-10.2 $S\text{-}Term_L w \wedge \sim\ulcorner 0\neq 0\urcorner \in L\text{-}Cn(w) \wedge RAN_L(x, y, z, b, w) \wedge$
$$\wedge RAN_L(x', y', z', b', w) \wedge \ulcorner z\langle x\rangle \in b\urcorner\, B_{w,L}\, s \wedge$$
$$\wedge \ulcorner z'\langle x'\rangle \in b'\urcorner\, B_{w,L}\, s \wedge PS_{w,L}(\ulcorner S\langle y, z, b, p, q\rangle\urcorner, y, z, b) \wedge$$
$$\wedge PS_{w,L}(\ulcorner S\langle y', z', b', p', q'\rangle\urcorner, y', z', b') \rightarrow [p, q]=[p', q']$$

Proof: (1) Assume hypothesis

(2) $\ulcorner z\langle x\rangle \in b\urcorner\, B_{w,L}\, z'\langle x'\rangle \in b'\urcorner$ (1), MT-8.11

(3) $\ulcorner S\langle y, z, b, p, q\rangle\urcorner\, E\text{-}STR_{w,L}\, \ulcorner S\langle y', z', b', p', q'\rangle\urcorner$ (1), (2), MT-9.5

(4) $0 \vdash_L p=p' \wedge q=q'\urcorner$ (3), MD-7.42

(5) $[p, q]=[p', q']$ (4), assumed soundness

If s and t are connected by a biconditional in w (so that $\ulcorner s\leftrightarrow t\urcorner$ is among the logical consequences of w), then they have the same partial probability:

MT-10.3 $s\, B_{w,L}\, t \rightarrow p\text{-}prob_L(s, w)=p\text{-}prob_L(t, w)$

Proof: obvious.

Of course if w is not strictly consistent, the probability of any statement is the empty set:

MT-10.4 $\ulcorner 0\neq 0\urcorner \in L\text{-}Cn(w) \rightarrow p\text{-}prob_L(s, w)=0$

Proof: obvious.

On the other hand, if w is strictly consistent, then, except in odd circumstances, non-empty probabilities always exist.

MT-10.5 $\sim\ulcorner 0\neq 0\urcorner\in L\text{-}Cn(w)\wedge Normal_L\, w\wedge w$ *is statistically closed* $\wedge$
$\wedge\bigvee_z z$ *is basis*$_L$ *for* $w\wedge s\in St_L\wedge P\text{-}Term_L(w)\rightarrow$
$\rightarrow p\text{-}prob_L(s, w)\neq 0$

Proof:

(1) Assume hypothesis

(2) Let $y=\ulcorner\imath f\,(f \text{ is } fun\wedge\mathscr{D}f=\{0\}\wedge((f\langle 0\rangle=\langle 1_r\rangle\wedge s)\vee\vee(f\langle 0\rangle=\langle 0_r\rangle\wedge\sim s)))\urcorner$

(3) $RV(\ulcorner 1\urcorner, y, \langle L, 0\rangle)\wedge 0\vdash_L\ulcorner\{\langle 1_r\rangle\}\in\mathbb{B}^1\urcorner$ (1), (2), MD-7.29, MD-7.33

(4) $\ulcorner y\langle 0\rangle\in\{\langle 1_r\rangle\}\urcorner\, B_{w,L}s$ (1), (2), MD-8.1, MD-8.3

(5) $\bigwedge_{z,x,b}\ulcorner z\langle x\rangle\in b\urcorner\, B_{w,L}s\leftrightarrow\ulcorner y\langle 0\rangle\in\{1_r\}\urcorner\, B_{w,L}\ulcorner z\langle x\rangle\in b\urcorner$ (1), (4), MT-8.11

(6) $[\bigvee_{x'',y'',z'',b''}](RAN_L(x'', y'', z'', b'', w)\wedge\wedge s\, B_{w,L}\ulcorner z''\langle x''\rangle\in b''\urcorner)$ (1), (2), (3), (4), (5), MT-9.6

(7) $[\bigvee_{s''}]\, PS_{w,L}(s'', y'', z'', b'')$ (6), MD–9.3

(8) $[\bigvee_{p,q}]\, s''=\ulcorner S\langle y'', z'', b'', p, q\rangle\urcorner$ (7), MD-8.7

(9) $PS_{w,L}(\ulcorner S\langle y'', z'', b'', p, q\rangle\urcorner, y'', z'', b'')$ (7), (8)

(10) $p\text{-}prob_L(s, w)=[p, q]$ (1), (6), (9), MT-10.1

One of the shortcomings of our approach is that we do not have a neat 'calculus of probability'. There is, of course, a perfectly standard calculus of statistical statements, but all we have for probability is a partial and pale reflection of this underlying calculus.

If w is a set of statements meeting the requirements mentioned in the definition of partial probability, then if a statement s belongs to w, its probability relative to w is the degenerate interval $[1_r, 1_r]$; the probability of its denial relative to w is the degenerate interval $[0_r, 0_r]$.

MT-10.6 $s\in w\wedge S\text{-}Term_L\, w\wedge\sim\ulcorner 0\neq 0\urcorner\in L\text{-}Cn(w)\wedge w$ is $RC_L\rightarrow$
$\rightarrow p\text{-}prob_L(s, w)=[1_r, 1_r]$

Proof:

(1) $s \in w \wedge S\text{-}Term_L\, w \wedge \sim \ulcorner 0 \neq 0 \urcorner \in w \wedge w$ is RC_L *ass.*

(2) Let $t = \ulcorner \imath f\,(f \text{ is } fun \wedge \mathscr{D} f = \omega \wedge \bigwedge_{i \in \omega} f \langle i \rangle = \{0\}) \urcorner$

(3) Let $g = \ulcorner \imath f\,(f \text{ is } fun \wedge \mathscr{D} f = \{0\} \wedge [(f \langle 0 \rangle = \langle 0_r \rangle \wedge s) \vee$
$\vee (f \langle 0 \rangle = \langle 1_r \rangle \wedge \sim s)]) \urcorner$

(4) $\ulcorner g \langle 0 \rangle \in \{\langle 0_r \rangle\} \urcorner\, B_{w,L}\, s$ (1), (3), MD-8.1, 8.3

(5) $\ulcorner \bigcup \mathscr{R} t = \{0\} \urcorner \in w$ (2), MD-8.1

(6) $\ulcorner \{0\} \urcorner \in RT_L$ MD-7.39

(7) $\{s\} \vdash_L \ulcorner S \langle t, g, \{\langle 0_r \rangle\}, 1_r, 1_r \rangle \urcorner$ (2), (3)

(8) $SS_{w,L}\, \ulcorner S \langle t, g, \{\langle 0_r \rangle\}, 1_r, 1_r \rangle \urcorner$ (5), (6), MD-7.41

(9) $\sim y'\, S_L(\ulcorner 0 \urcorner, t, g, \ulcorner \{\langle 0_r \rangle\} \urcorner, w)$ (9), MD-9.4, MT-7.11

(10) $PS_{w,L}(\ulcorner S \langle t, g, \{\langle 0_r \rangle\}, 1_r, 1_r \rangle \urcorner, t, g, \ulcorner \{\langle 0_r \rangle\} \urcorner)$
(1), (7), (8), MT-8.32

(11) ~~show~~ $\sim y'\, D_L(\ulcorner 0 \urcorner, t, g, \ulcorner \{\langle 0_r \rangle\} \urcorner, w)$

(12) $y'\, D_L(\ulcorner 0 \urcorner, t, g, \ulcorner \{\langle 0_r \rangle\} \urcorner, w)$ ass.

(13) $[\bigvee_{z, s', x', t', g', b'}]\, x'\, M_{w,L}\, \ulcorner \bigcup \mathscr{R} t' \urcorner \wedge$
$\wedge PS_{w,L}(z, t, g, \ulcorner \{\langle 0_r \rangle\} \urcorner) \wedge$
$\wedge \ulcorner g \langle 0 \rangle \in \{\langle 0_r \rangle\} \urcorner\, B_{w,L}\, \ulcorner g' \langle x' \rangle \in b' \urcorner \wedge$
$\wedge PS_{w,L}(s', t', g', b') \wedge s'\, DIF_{w,L}\, z \wedge$
$\wedge \sim \bigvee_{s'', t'', g'', b''} (\ulcorner g'' \langle x' \rangle \in b'' \urcorner\, B_{w,L}\, \ulcorner g' \langle x' \rangle \in b' \urcorner \wedge$
$\wedge PS_{w,L}(s'', t'', g'', b'') \wedge s''\, E\text{-}STR_{w,L}\, z \wedge$
$\wedge x'\, M_{w,L}\, \ulcorner \bigcup \mathscr{R} t'' \urcorner \wedge [(\ulcorner \bigcup \mathscr{R} t'' \urcorner\, Incl_{w,L}\, \ulcorner \bigcup \mathscr{R} t' \urcorner \wedge$
$\wedge \sim \ulcorner \bigcup \mathscr{R} t' \urcorner\, I_{w,L}\, \ulcorner \bigcup \mathscr{R} t'' \urcorner) \vee$
$\vee \ulcorner \bigcup \mathscr{R} t'' \urcorner\, I_{w,L}\, \ulcorner \{x'\} \urcorner \in w])$ (12), (13), MD-9.1, MD-9.3

(14) Let $t'' = \ulcorner \imath f\,(f \text{ is } fun \wedge \mathscr{D} f = \omega \wedge \bigwedge_{i \in \omega} f \langle i \rangle = \{x'\}) \urcorner$

(15) Let $g'' = \ulcorner \imath f\,(f \text{ is } fun \wedge \mathscr{D} f = \{x'\} \wedge$
$\wedge [(f \langle x' \rangle = \langle 0_r \rangle \wedge s) \vee (f \langle x' \rangle = \langle 1_r \rangle \wedge \sim s)]) \urcorner$

(16) $\ulcorner g'' \langle x' \rangle \in \{\langle 0_r \rangle\} \urcorner\, B_{w,L}\, \ulcorner g' \langle x' \rangle \in b' \urcorner$
(15), (13), MT-8.11

(17) $\ulcorner \bigcup \mathscr{R} t'' = \{x'\} \urcorner \in w$ (14), MD-8.1

(18) $\ulcorner\{x'\}\urcorner \in RT_L$ MD-7.39

(19) $\{s\} \vdash_L \ulcorner S\langle t'', g'', \{\langle 0_r\rangle\}, 1_r, 1_r\rangle\urcorner$ (14), (15)

(20) $SS_{w,L} \ulcorner S\langle t'', g'', \{\langle 0_r\rangle\}, 1_r, 1_r\rangle\urcorner$ (17), (18), MD-7.41

(21) $PS_{w,L}(\ulcorner S\langle t'', g'', \{\langle 0_r\rangle\}, 1_r, 1_r\rangle\urcorner, t'', g'', \ulcorner\{\langle 0_r\rangle\}\urcorner)$
(1), (19), (20), MT-8.32

(22) $\ulcorner S\langle t'', g'', \{\langle 0_r\rangle\}, 1_r, 1_r\rangle\urcorner\ E\text{-}STR_{w,L}$
$\ulcorner S\langle t, g, \{0_r\rangle\}, 1_r, 1_r\rangle\urcorner$ MD-4.72

(23) $0 \neq 0$ (13), (16), (21), (22), (17)

(24) $RAN_L(\ulcorner 0\urcorner, t, g, \ulcorner\{\langle 0_r\rangle\}\urcorner, w)$ (1), (10), (11), MD-9.5

(25) $p\text{-}prob_L(s, w) = [1_r, 1_r]$ (1), (24), (4), (9), MT-10.1

MT-10.7 $s \in w \wedge S\text{-}Term_L\ w \wedge \sim\ulcorner 0 \neq 0 \in L\text{-}Cn(w) \wedge w$ is $RC_L \rightarrow$
$\rightarrow p\text{-}prob_L(\ulcorner\sim s\urcorner, w) = [0_r, 0_r]$

Proof: similar

In general, if the probability of s is $[p, q]$, then the probability of $\ulcorner\sim s\urcorner$ is $[1-q, 1-p]$:

MT-10.8 $p\text{-}prob_L(s, w) = [p, q] \rightarrow p\text{-}prob_L(\ulcorner\sim s\urcorner, w) = [1-q, 1-p]$

Proof:

(1) $p\text{-}prob_L(s, w) = [p, q]$ ass.

(2) $[\bigvee_{x,y,z,b}]\ S\text{-}Term_L\ w \wedge w$ is $RC_L \wedge \ulcorner z\langle x\rangle \in b\urcorner\ B_{w,L}\ s \wedge$
$\wedge PS_{w,L}(\ulcorner S\langle y, z, b, p, q\rangle\urcorner, y, z, b) \wedge RAN_L(x, y, z, b, w)$
(1), MT-10.1

(3) $[\bigvee_{A,n}]\ RV(n, z, \langle L, A\rangle)$ (2), MD-8.7, MD-7.41

(4) $PS_{w,L}(\ulcorner S\langle y, z, \mathbb{R}^n - b, 1-q, 1-p\rangle\urcorner, y, z, \ulcorner\mathbb{R}^n - b\urcorner)$
(2), (3), MD-8.7

(5) $\ulcorner z\langle x\rangle \in \mathbb{R}^n - b\urcorner\ B_{w,L}\ \ulcorner\sim s\urcorner$ (2), *MD*-8.3

(6) $RAN_L(x, y, z, \ulcorner\mathbb{R}^n - b\urcorner, w)$ (2), MT-9.7

(7) $p\text{-}prob_L(\ulcorner\sim s\urcorner, w) = [1-q, 1-p]$ (2), (4), (5), (6), MT-10.1

It would seem natural to have something corresponding to the intuitive judgement that if *P* entails *Q*, then the probability of *Q* must be at least as great as that of *P*. Subject to the condition that the upper bound for the probability of Q not be 1, and that the lower bound for the probability of P not be 0, we have not only this result, but the somewhat stronger result that if the conditional $\ulcorner P \rightarrow Q \urcorner$ belongs to w, either the lower bound or the upper bound of the probability of Q will be greater than (or equal to) the lower bound or the upper bound of the probability of P, respectively. To obtain this result we need a modest amount of deductive closure, as embodied in the first clause of the antecedent of the theorem.

MT-10.9 $\bigwedge_{x,y,s,t} ([\ulcorner x \prec \omega \urcorner \in w \wedge \ulcorner y \prec \omega \urcorner \in w \rightarrow \ulcorner x \times y \prec \omega \urcorner \in w] \wedge$
$\wedge [\ulcorner x \subset y \urcorner \in w \wedge \ulcorner s \subset t \urcorner \in w \rightarrow \ulcorner x \times s \subset y \times t \urcorner \in w] \wedge$
$\wedge [\ulcorner x = y \urcorner \in w \wedge \ulcorner s = t \urcorner \in w \rightarrow \ulcorner x \times s = y \times t \urcorner \in w]) \wedge$
$\wedge \textit{p-prob}_L(s, w) = [p, q] \wedge \textit{p-prob}_L(s', w) = [p', q'] \wedge$
$\wedge \ulcorner s' \rightarrow s \urcorner \in w \wedge Normal_L \; w \rightarrow \; \vdash_L \ulcorner p' \leqslant p \vee q' \leqslant q \urcorner$

Proof:

(1) Assume antecendent.

(2) $[\bigvee_{x,y,z,b,p,q}] RAN_L(x, y, z, b, w) \wedge$
$\wedge PS_{w,L}(\ulcorner S\langle y, z, b, p, q \rangle \urcorner, y, z, b) \wedge \ulcorner z\langle x \rangle \in b \urcorner \; B_{w,L} \; s$ (1), MT-10.1

(3) $[\bigvee_{x',y',z',b',p',q'}] RAN_L(x', y', z', b', w) \wedge$
$\wedge PS_{w,L}(\ulcorner S\langle y', z', b', p', q' \rangle \urcorner, y', z', b') \wedge$
$\ulcorner z'\langle x' \rangle \in b' \urcorner \; B_{w,L} \; s'$ (1), MT-10.1

(4) $S\text{-}Term_L \; w \wedge \sim \ulcorner 0 \neq 0 \urcorner \in w$ (1), MT-10.1

(5) Let $y^* = \ulcorner \imath f (f \text{ is } fun \wedge \mathscr{D} y^* = \omega \wedge \bigwedge_{i \in \omega} y_i^* = y_i \times y_i') \urcorner$

(6) Let $a = \ulcorner \imath f (f \text{ is } fun \wedge \mathscr{D} f = \bigcup \mathscr{R} y^* \wedge \bigwedge_{t \in \mathscr{D} f} f\langle t \rangle =$
$= z\langle \imath k(\bigvee_j \widehat{k\, j} = t) \rangle) \urcorner$

(7) Let $a' = \ulcorner \imath f (f \text{ is } fun \wedge \mathscr{D} f = \bigcup \mathscr{R} y^* \wedge \bigwedge_{t \in \mathscr{D} f} f\langle t \rangle =$
$= z'\langle \imath k(\bigvee_j \widehat{j\, k} = t) \rangle) \urcorner$

(8) $\ulcorner \widehat{x x'} \urcorner \; M_{w,L} \; \ulcorner \bigcup \mathscr{R} y^* \urcorner$ (2), (3), MD-9.5; MD-8.21

(9) $\vdash_L \ulcorner S\langle y, z, b, p, q \rangle \rightarrow S\langle y^*, a, b, p, q \rangle \ulcorner$ (2), (5), (6)

(10) $\vdash_L \ulcorner S\langle y', z', b', p', q' \rangle \rightarrow S\langle y^*, a', b', p', q' \rangle \urcorner$ (3), (5), (7)

(11) $PS_{w,L}(\ulcorner S\langle y^*, a, b, p, q\rangle\urcorner, y^*, a, b)$ (1), (2), (9)

(12) $PS_{w,L}(\ulcorner S\langle y^*, a', b', p', q'\rangle\urcorner, y^*, a', b')$ (1), (3), (10)

(13) $[\bigvee_{y^{*\prime\prime}, a'', b''}]\ulcorner\widehat{x}x'\urcorner M_{w,L}\ulcorner\bigcup\mathscr{R}y^{*\prime\prime}\urcorner\wedge$
$\wedge PS_{w,L}(\ulcorner S\langle y^{*\prime\prime}, a'', b'', p, q\rangle\urcorner, y^{*\prime\prime}, a'', b'')\wedge$
$\wedge\ulcorner a\langle\widehat{x}x'\rangle\in b\urcorner B_{w,L}\ulcorner a''\langle\widehat{x}x'\rangle\in b''\urcorner\wedge y^{*\prime\prime}\, MIN_{w,L}\, s$
(4), (6), (8), (11); *S-Term* MD-9.7

(14) $[\bigvee_{y^{*\prime\prime\prime}, a''', b'''}]\ulcorner\widehat{x}x'\urcorner M_{w,L}\ulcorner\bigcup\mathscr{R}y^{*\prime\prime\prime}\urcorner\wedge$
$\wedge PS_{w,L}(\ulcorner S\langle y^{*\prime\prime\prime}, a''', b''', p', q'\rangle\urcorner, y^{*\prime\prime\prime}, a''', b''')\wedge$
$\wedge\ulcorner a'\langle\widehat{x}x'\rangle\in b'\urcorner B_{w,L}\ulcorner a'''\langle xx'\rangle\in b'''\urcorner\wedge y^{*\prime\prime\prime}\, MIN_{w,L}\, s'$
(4), (7), (8), (10), (12); *S-Term* MD9.7

(15) Let $d=\ulcorner\imath f\ (f \text{ is } fun\wedge\mathscr{D}f=y^{*\prime\prime}\wedge$
$\wedge\bigwedge_{t\in y^{*\prime\prime}}([(a''\langle t\rangle\notin b''\vee\sim s')\wedge f\langle t\rangle=\langle 1_r\rangle]\vee$
$\vee[a''\langle t\rangle\in b''\wedge s'\wedge f\langle t\rangle=\langle 0_r\rangle]))\urcorner$

(16) $\ulcorner d\langle\widehat{x}x'\rangle\in\{\langle 1_r\rangle\}\urcorner B_{w,L}\ulcorner\sim s'\urcorner$ (1), (15), (13), MT-8.7

(17) $\{\ulcorner S\langle y^{*\prime\prime}, d, \{\langle 1_r\rangle\}, 1-r, 1\rangle\urcorner, \ulcorner S\langle y^{*\prime\prime}, a'', b'', p, q\rangle\urcorner\}\vdash_L$
$\ulcorner S\langle y^{*\prime\prime}, a'', b'', p, r\rangle\urcorner$

(18) $PS_{w,L}(\ulcorner S\langle y^{*\prime\prime}, d, \{\langle 1_r\rangle\}, 1-q, 1\rangle\urcorner, y^{*\prime\prime}, d, \ulcorner\{\langle 1_r\rangle\}\urcorner)$
(1), (17), (13), (15), MD-8.8

(19) sh~~o~~w $\vdash_L\ulcorner q'\leqslant q\vee p'\leqslant p\urcorner$

(20) $\sim\vdash_L\ulcorner q'\leqslant q\vee p'\leqslant p\urcorner$ ass.

(21) $\vdash_L\ulcorner q<q'\wedge p<p'\urcorner$ (20), Normality, MT-8.33

(22) $\vdash_L\ulcorner q<1_r\wedge 0_r<p'\urcorner$ (21)

(23) $\ulcorner S\langle y^{*\prime\prime}, d, \{\langle 0_r\rangle\}, 0_r, q\rangle\urcorner\, DIF_{w,L}\ulcorner S\langle y', z', b', p', q'\rangle\urcorner$
(1), (21), (22), MD-7.44

(24) $\sim y^{*\prime\prime}\, D_L(x', y', z', b', w)$ (3), MD-9.5

(25) $(\ulcorner\widehat{x}x'\urcorner, y^{*\prime\prime}, d, \ulcorner\{\langle 0\rangle\}\urcorner)\, SND_L(x', y', z', b', w)\vee$
$\vee(\ulcorner\widehat{x}x'\urcorner, y^{*\prime\prime}, d, \ulcorner\{\langle 0_r\rangle\}\urcorner)\, PND_L(x', y', z', b', w)$
(24), MD-9.3

(26) $\sim\ulcorner\widehat{x}x'\prec x'\urcorner\in w$ (1)

(27) $(\ulcorner\widehat{x}x'\urcorner.\ y^{*\prime\prime}, d, \ulcorner\{\langle 0_r\rangle\}\urcorner)\, SND_L(x', y', z', b', w)$
(25), (26), MD-9.2

(28) $[\bigvee_{y^{**}, a^{**}, b^{**}}] \ulcorner \widehat{xx'} \urcorner M_{w,L} \ulcorner \bigcup \mathscr{R}y^{**} \urcorner \wedge$
$\wedge PS_{w,L}(\ulcorner S\langle y^{**}, a^{**}, b^{**}, p', q'\rangle \urcorner, y^{**}, a^{**}, b^{**}) \wedge$
$\wedge \ulcorner a^{**}\langle \widehat{xx'}\rangle \in b^{**} \urcorner B_{w,L} \ulcorner a''' \langle \widehat{xx'}\rangle \in b''' \urcorner \wedge$
$\wedge [(\ulcorner \bigcup \mathscr{R}y^{**} \urcorner Incl_{w,L} \ulcorner \bigcup \mathscr{R}y^{*''} \urcorner \wedge$
$\wedge \sim \ulcorner \bigcup \mathscr{R}y^{**} \urcorner I_{w,L} \ulcorner \bigcup \mathscr{R}y^{*''} \urcorner) \vee$
$\vee \ulcorner \bigcup \mathscr{R}^{y**} I_{w,L} \{\widehat{xx'}\} \urcorner \in w]$ (21), MT-9.8, MD-9.1

(29) Let $d' = \ulcorner \imath f\ (f \text{ is } fun \wedge \mathscr{D}f = y^{**} \wedge$
$\wedge \bigwedge_{t \in \mathscr{D}f} (((s \vee a^{**}\langle t\rangle \in b^{**}) \wedge f\langle t\rangle = \langle 1_r\rangle) \vee$
$\vee (\sim s \wedge \sim a^{**}\langle t\rangle \in b^{**} \wedge f\langle t\rangle = \langle 0_r\rangle)))\urcorner$

(30) $\ulcorner d'\langle \widehat{xx'}\rangle \in \{\langle 1_r\rangle\} \urcorner B_{w,L}\, s$ (22), (23), (1), MT-8.11

(31) $PS_{w,L}(\ulcorner S\langle y^{**}, d', \{\langle 1_r\rangle\}, p', 1_r\rangle \urcorner, y^{**}, d', \ulcorner \{\langle 1_r\rangle\} \urcorner)$
(22), (23), (1)

(32) $\sim \ulcorner S\langle y^{**}, d', \{\langle 1_r\rangle\}, p', 1_r\rangle \urcorner DIF_{w,L} \ulcorner S\langle y^{*''}, a'', b'', p, q$
(22), (24), (25), (13), MT-9.9

(33) $\vdash_L \ulcorner p' \leqslant p \urcorner$ (32), MT-9.8, MT-9.3MD-7.44

The antecedent of this theorem we shall henceforth abbreviate as the cross-product condition, or *CPC*:

MD-10.2 $CPC(w) \leftrightarrow \bigwedge_{x,y,s,t} ([\ulcorner x \prec \omega \urcorner \in w \wedge \ulcorner y \prec \omega \urcorner \in w \rightarrow$
$\rightarrow \ulcorner x \times y \prec \omega \urcorner \in w] \wedge [\ulcorner x \subset y \urcorner \in w \wedge \ulcorner s \subset t \urcorner \in w \rightarrow$
$\rightarrow \ulcorner x \times s \subset y \times t \urcorner \in w] \wedge [\ulcorner x = y \urcorner \in w \wedge \ulcorner s = t \urcorner \in w \rightarrow$
$\rightarrow \ulcorner x \times s = y \times t \urcorner \in w])$

A number of special cases of thorem MT-10.8 turn out to be interesting and useful:

MT-10.10 $CPC(w) \wedge p\text{-}prob_L(s, w) = [p, p] \wedge p\text{-}prob_L(t, w) =$
$= [q, q] \wedge Normal_L\, w \wedge \ulcorner s \rightarrow t \urcorner \in w \rightarrow \vdash_L \ulcorner p \leqslant q \urcorner$

Proof: MT-10.9; the cases $p=0$ and $q=1$ are trivial.

MT-10.11 $CPC(w) \wedge p\text{-}prob_L(s, w) = [p, q] \wedge p\text{-}prob_L(\ulcorner s \vee s' \urcorner, w) =$
$= [p', q'] \wedge Normal_L\, w \rightarrow \vdash_L \ulcorner p \leqslant p' \vee q \leqslant q' \urcorner$

MT-10.12 $CPC(w) \wedge p\text{-}prob_L(s, w) = [p, q] \wedge p\text{-}prob_L(\ulcorner s \wedge s' \urcorner, w) =$
$= [p', q'] \wedge Normal_L\, w \rightarrow \vdash_L \ulcorner p' \leqslant p \vee q' \leqslant q \urcorner$

MT-10.13 $p\text{-}prob_L(s, w) = [p, q] \wedge \ulcorner s \rightarrow s' \urcorner \in w \rightarrow p\text{-}prob_L(\ulcorner s \wedge s' \urcorner, w) = [p, q]$

Proof: MT-10.3

MT-10.14 $\ulcorner x \in \bigcup \mathscr{R} y \urcorner \in w \wedge PS_{w,L}(\ulcorner S\langle y, z, b, p, q\rangle \urcorner, y, z, b) \wedge \wedge \sim \bigvee_t t\, D_L(x, y, z, b, w) \wedge p\text{-}prob_L(\ulcorner z\langle x\rangle \in b \urcorner, w) = [r, s] \rightarrow \vdash_L \ulcorner p \leqslant r \wedge s \leqslant q \urcorner$

The most important property of probabilities is their additivity. Our statistical measures are of course additive, but this additivity does not carry over to probabilities, except in a limited way. That is, in general it will not be the case that if the partial probability of s, relative to w, is $[p_0, p_1]$ and the partial probability of t, relative to w, is $[q_0, q_1]$, and if we know that statement $\sim (s \wedge t)\urcorner$, the partial probability of $\ulcorner s \vee t \urcorner$ will be $[p_0 + q_0, p_1 + q_1]$. To see that this is not implausible consider the following case:

CASE I. We have a die which we know to be perfectly balanced, except that either the one face turns up with a frequency of $\frac{1}{6} + e$ and the two face with a frequency of $\frac{1}{6} - e$, or the one face turns up with a frequency of $\frac{1}{6} - e$ and the two face with a frequency of $\frac{1}{6} + e$. The probability of a one on a future toss may plausibly be taken to be the interval $[\frac{1}{6} - e, \frac{1}{6} + e]$, and of a two on a future toss the same interval. But the probability of getting a one or a two is $[\frac{1}{3}, \frac{1}{3}]$.

We can, however, prove a theorem which asserts roughly the following: If the probability of s_1 is $[p_1, q_1]$ and the probability of s_2 is $[p_2, q_2]$, and $\ulcorner \sim (s_1 \wedge s_2) \urcorner$ is in our body of knowledge, then the probability of the disjunction $\ulcorner s_1 \vee s_2 \urcorner$ will fall within the interval $[p_1 + p_2, q_1 + q_2]$.

The importance of this modest result should not be underestimated. The proponents of the subjectivistic interpretation of probability make much of the fact that if an individual employs betting odds that correspond to his degrees of belief, and if these degrees of belief satisfy the axioms of the probability calculus, the individual will never have a book made against him. As has already been remarked there are a number of serious lacunae in this argument; in particular, the probability calculus provides the individual with nothing, beyond algorithms for computation, that straightforward deductive logic does not provide him with.

Nevertheless, if we suppose that the individual can be put into the position of having to make book on a finite number of statements – i.e., of posting odds on each of the statements, and then accepting any bets regarding those statements at those odds that anyone wants to offer him – then it is of interest to ask whether or not the individual *can* post odds that satisfy the constraints imposed by probabilities as here defined. It is not obvious that it need be the case: perhaps such a situation might be one in which rationality is no longer an available option. But it turns out that it is an option. If K is a finite set of statements, it turns out that (subject to our usual provisos concerning Normality and the like), there will exist a function P from the members of K to the real numbers in $[0, 1]$ such that (a) these numbers satisfy the probability calculus, and (b) for any statement s in K, $P(s) \in p\text{-}Prob(s, w)$.

We begin by proving the first theorem mentioned: if w is a rational corpus that is normal, statistically closed, probabilistically terminating, and has a basis and satisfies the cross-product closure condition, then if $\ulcorner \sim(s_1 \wedge s_2)\urcorner$ is a member of w, and the probabilities of s_1, s_2, and $\ulcorner s_1 \vee s_2 \urcorner$ are $[p_1, q_1]$, $[p_2, q_2]$ and $[p_0, q_0]$, respectively, it will be a theorem that $\ulcorner p_1 + p_2 \leqslant p_0 \wedge q_0 \leqslant q_1 + q_2 \urcorner$.

MT-10.15 $Normal_L\ w \wedge w$ *is statistically closed* $\wedge\ P\text{-}Term_L(w) \wedge$
$\wedge\ CPC(w) \wedge \bigvee_z z$ *is* $Basis_L$ *for* $w \wedge \ulcorner \sim(s_1 \wedge s_2)\urcorner \in w \wedge$
$\wedge\ p\text{-}prob_L(s_1, w) = [p_1, q_1] \wedge p\text{-}prob_L(s_2, w) = [p_2, q_2] \wedge$
$\wedge\ p\text{-}prob_L(\ulcorner s_1 \vee s_2 \urcorner, w) = [p_0, q_0] \rightarrow \vdash_L \ulcorner p_1 + p_2 \leqslant$
$\leqslant p_0 \wedge q_0 \leqslant q_1 + q_2 \urcorner$

Proof:

(1) Assume antecedent

(2) $[\bigvee_{r_0, t_0}]\ p\text{-}prob_L(\ulcorner \sim s_1 \wedge \sim s_2 \urcorner, w) = [r_0, t_0]$ (1), MT-10.5

(3) $[\bigvee_{r_1, t_1}]\ p\text{-}prob_L(\ulcorner s_1 \wedge \sim s_2 \urcorner, w) = [r_1, t_1]$ (1), MT-10.5

(4) $[\bigvee_{r_2, t_2}]\ p\text{-}prob_L(\ulcorner \sim s_1 \wedge s_2 \urcorner, w) = [r_2, t_2]$ (1), MT-10.5

(5) $[\bigvee_{r_3, t_3}]\ p\text{-}prob_L(\ulcorner s_1 \wedge s_2 \urcorner, w) = [r_3, t_3]$ (1), MT-10.5

(6) $\ulcorner \sim s_1 \wedge s_2 \urcorner B_{w,L}\, s_2$ (1), MD-8.3

(7) $\vdash_L \ulcorner r_2 = p_2 \wedge t_2 = q_2 \urcorner$ (1), (4), (6), MT-10.2

(8) $\ulcorner s_1 \wedge \sim s_2 \urcorner B_{w,L}\, s_1$ (1), MD-8.3

(9) $\vdash_L \ulcorner r_1 = p_1 \wedge t_1 = q_1 \urcorner$ (1), (3), (8), MT-10.2

(10) $\vdash_L \ulcorner r_3 = 0 \wedge t_3 = 0 \urcorner$ (1), (5), MT-10.3, MT-10.7

(11) $[\bigvee_{x_0, y_0, z_0, b_0}] RAN_L(x_0, y_0, z_0, b_0, w) \wedge$
$\wedge \ulcorner \sim s_1 \wedge \sim s_2 \urcorner B_{w,L} \ulcorner z_0 \langle x_0 \rangle \in b_0 \urcorner \wedge$
$PS_{w,L}(\ulcorner S\langle y_0, z_0, b_0, r_0, t_0\rangle \urcorner, y_0, z_0, b_0)$ (2), MT-10.1

(12) $[\bigvee_{x_1, y_1, z_1, b_1}] RAN_L(x_1, y_1, z_1, b_1, w) \wedge$
$\wedge \ulcorner s_1 \wedge \sim s_2 \urcorner B_{w,L} \ulcorner z_1 \langle x_1 \rangle \in b_1 \urcorner \wedge$
$\wedge PS_{w,L}(\ulcorner S\langle y_1, z_1, b_1, p_1, q_1\rangle \urcorner, y_1, z_1, b_1)$
(3), (9), MT-10.1

(13) $[\bigvee_{x_2, y_2, z_2, b_2}] RAN_L(x_2, y_2, z_2, b_2, w) \wedge$
$\wedge \ulcorner \sim s_1 \wedge s_2 \urcorner B_{w,L} \ulcorner z_2 \langle x_2 \rangle \in b_2 \urcorner \wedge$
$\wedge PS_{w,L}(\ulcorner S\langle y_2, z_2, b_2, p_2, q_2\rangle \urcorner, y_2, z_2, b_2)$
(4), (7), MT-10.1

(14) $[\bigvee_{x_3, y_3, z_3, b_3}] RAN_L(x_3, y_3, z_3, b_3, w) \wedge$
$\wedge \ulcorner s_1 \wedge s_2 \urcorner B_{w,L} \ulcorner z_3 \langle x_3 \rangle \in b_3 \urcorner \wedge$
$\wedge PS_{w,L}(\ulcorner S\langle y_3, z_3, b_3, 0, 0\rangle \urcorner, y_3, z_3, b_3)$
(5), (10), MT-10.1

(15) Let $x^* = \ulcorner x_0 \frown x_1 \frown x_2 \frown x_3 \urcorner$

(16) Let $y^* = \ulcorner \imath f\,(f \text{ is fun} \wedge \mathscr{D}f = \omega \wedge \bigwedge_{i \in \omega} f\langle i \rangle =$
$= y_0\langle i \rangle \times y_1\langle i \rangle \times y_2\langle i \rangle \times y_3\langle i \rangle) \urcorner$

(17) Let $z^* = \ulcorner \imath f\,(f \text{ is fun} \wedge \mathscr{D}f = \bigcup \mathscr{R}y^* \wedge \bigwedge_{t \in \mathscr{D}f} ((f\langle t \rangle =$
$= \langle 0 \rangle \wedge \bigvee_{x, y} (t = x \frown y \wedge x \in \bigcup \mathscr{R}y_0 \wedge z_0\langle x \rangle \in b_0)) \vee (f\langle t \rangle =$
$= \langle 1 \rangle \wedge \bigvee_{x, y, z} (t = x \frown y \frown z \wedge y \in \bigcup \mathscr{R}y_1 \wedge z_1\langle y \rangle \in b_1)) \vee$
$\vee (f\langle t \rangle = \langle 2 \rangle \wedge \bigvee_{x, y, z} (t = x \frown y \frown z \wedge y \in \bigcup \mathscr{R}y_2 \wedge$
$\wedge z_2\langle y \rangle \in b_2)) \vee (f\langle t \rangle = \langle 3 \rangle \wedge \bigvee_{x, y, z} (t = x \frown y \frown z \wedge$
$\wedge z \in \bigcup \mathscr{R}y_3 \wedge z_3\langle z \rangle \in b_3)))) \urcorner$

(18) $PS_{w,L}(\ulcorner S\langle y^*, z^*, \{\langle 0 \rangle\}, r_0, t_0\rangle \urcorner, y^*, z^*, \{\langle 0 \rangle\})$
1 (*CPC*), (11), (12), (13), (14), (16), (17)

(19) $PS_{w,L}(\ulcorner S\langle y^*, z^*, \{\langle 1\rangle\}, p_1, q_1\rangle\urcorner, y^*, z^*, \{\langle 1\rangle\})$
1 (CPC), (11), (12), (13), (14), (16), (17)

(20) $PS_{w,L}(\ulcorner S\langle y^*, z^*, \{\langle 2\rangle\}, p_2, q_2\rangle\urcorner, y^*, z^*, \{\langle 2\rangle\})$
1 (CPC), (11), (12), (13), (14), (16), (17)

(21) $PS_{w,L}(\ulcorner S\langle y^*, z^*, \{\langle 3\rangle\}, 0_r, 0_r\rangle\urcorner, y^*, z^*, \{\langle 3\rangle\})$
1 (CPC), (11), (12), (13), (14), (16), (17)

(22) $\{\ulcorner S\langle y^*, z^*, \{\langle 1\rangle\}, p_1, q_1\rangle\urcorner, \ulcorner S\langle y^*, z^*, \{\langle 2\rangle\}, p_2, q_2\rangle\urcorner,$
$\ulcorner S\langle y^*, z^*, \{\langle 3\rangle\}, 0, 0\rangle\urcorner\} \vdash_L$
$\ulcorner S\langle y^*, z^*, \{\langle 0\rangle\}, 1-q_1-q_2, 1-p_1-p_2\rangle\urcorner$
(17), (19), (20), (21); closure (1)

(23) $\ulcorner S\langle y^*, z^*, \{\langle 0\rangle\}, 1-q_1-q_2, 1-p_1-p_2\rangle\urcorner \in w$
(19), (20), (21), (22); (1)

(24) $\vdash_L \ulcorner 1-q_1-q_2 \leqslant r_0 \wedge t_0 \leqslant 1-p_1-p_2\urcorner$
(23), (18), MD-8.7

(25) $p\text{-}prob_L(\ulcorner s_1 \vee s_2\urcorner, w) = [1-t_0, 1-r_0]$ (2), MT-10.8

(26) $\vdash_L \ulcorner p_1 + p_2 \leqslant p_0 \wedge q_0 \leqslant q_1 + q_2\urcorner$ (1), (24), (25)

The general theorem now follows easily by induction:

MT-10.16 $\mathit{Normal}_L\ w \wedge w$ *is statistically closed* $\wedge\ P\text{-}Term_L(w) \wedge$
$\wedge \bigvee_z z$ *is basis*$_L$ *for* $w \wedge s$ *is n-sequence* $\wedge\ CPC(w) \wedge$
$\wedge \bigwedge_{i<n} (p\text{-}prob_L(s_i, w) = [p_i, q_i] \wedge \ulcorner s_0 \vee s_1 \vee \ldots \vee s_{i-1} \rightarrow \sim s_i\urcorner \in w)$
$\wedge\ p\text{-}prob_L(\ulcorner s_0 \vee s_1 \vee \ldots \vee s_{n-1}\urcorner, w) = [r, t] \rightarrow$
$\rightarrow \vdash_L \ulcorner \sum_{i<n} p_i \leqslant r \wedge t \leqslant \sum_{i<n} q_i\urcorner$

Employing the powerful antecedent of this theorem – the most powerful and important part of which is the cross-product condition combined with the condition of statistical closure – we can prove a stronger version of MT-10.9:

MT-10.17 $\mathit{Normal}_L\ w \wedge w$ *is statistically closed* $\wedge\ P\text{-}Term_L(w) \wedge CPC(w)$
$\wedge \bigvee_z (z$ *is Basis*$_L$ *for* $w) \wedge \ulcorner s_1 \rightarrow s_2\urcorner \in w \wedge p\text{-}prob_L(s_1, w) =$
$= [p, q] \wedge p\text{-}prob_L(s_2, w) = [p', q'] \rightarrow \vdash_L \ulcorner p \leqslant p' \wedge q \leqslant q'\urcorner$

Proof:

(1) Assume hypothesis.

(2) $[\bigvee_{r_0, t_0}]\ p\text{-}prob_L(\ulcorner s_1 \wedge s_2 \urcorner, w) = [r_0, t_0]$ (1), MT-10.5

(3) $[\bigvee_{r_1, t_1}]\ p\text{-}prob_L(\ulcorner \sim s_1 \wedge s_2 \urcorner, w) = [r_1, t_1]$ (1), MT-10.5

(4) $\ulcorner s_1 \wedge s_2 \urcorner B_{w,L}\, s_1$ (1)

(5) $\vdash_L \ulcorner r_0 = p \wedge t_0 = q \urcorner$ (1), (4), (2), MT-10.3

(6) $s_2\, B_{w,L} \ulcorner (s_1 \wedge s_2) \vee (\sim s_1 \wedge s_2) \urcorner$ MD-8.7

(7) $\vdash_L \ulcorner p + r_1 \leqslant p' \urcorner$ (2), (5), (6), (1), MT-10.15

(8) $\vdash_L \ulcorner p \leqslant p' \urcorner$ (7)

(9) $\ulcorner \sim s_2 \rightarrow \sim s_1 \urcorner \in w$ (1)

(10) $\vdash_L \ulcorner 1 - q' \leqslant 1 - q \urcorner$ (9), MT-10.8, argument of lines (1)–(8)

(11) $\vdash_L \ulcorner q \leqslant q' \urcorner$

3. These rather general relations among probabilities are not as familiar – nor as useful in ordinary circumstances – as the more specific relations we are all familiar with. These more familiar relations are essentially those that hold when there are no problems concerning randomness. Such problems are automatically avoided in two sorts of context: the abstract, hypothetical context in which most abstract talk about probabilities takes place; and the practical down to earth context in which a gambler or an engineer is concerned with uncontrollable apparatus having known stochastic properties. Both sorts of situation are easy to characterize in the present system, and in both we have both a classical point probability framework, and a slight generalization to interval probabilities.

In providing illustrations of this fact, we shall provide partial and informal interpretations of the lurking language L; we shall suppose, for example, that the term T of L is a reference term of L, and that it denotes the set of tosses of a particular coin, or the set of frogs of a particular species, or the set of balls currently in a particular urn. We shall throughout suppose, of course, that P_0^2, i.e., '$\in$', is given its normal interpretation so that '$\mathbb{R}$' does denote the set of real numbers, '$\mathbb{B}^n$' does denote the set of n-dimensional Borel sets, and so on.

Let us consider first the simplest ideal case, where we are speaking both precisely and hypothetically; where we say, for example, "Suppose that the weight of frogs of species S is distributed normally with mean m and standard deviation s; then the probability that a frog of this species will have weight in the interval $(m-s, m+s)$ is 0.6826...." In any such case we are concerned with a single (unspecified) object and a single distribution function assumed to characterize a random quantity in a reference set to which that object is assumed to belong. In ordinary language the object is referred to with the help of the indefinite article 'a': 'a frog'. It can be argued that the indefinite article generally has the sense of 'every random...'; for the present purpose we need merely add to the language in question a new name to denote the (otherwise unspecified) object in question, and an axiom asserting that that object belongs to the reference set in question. We take the reference set to be the union of an infinite sequence of finite sets. The manner in which that sequence is constructed is irrelevant if the union of its range is finite; in the (rare) event that we know that the reference set is infinite, we will (presumably) have grounds for supposing our distribution function to apply to some particular sequence of finite sets. A simple way of obtaining the appropriate finite sets is to consider temporal order in the case of events; in the case of objects, the temporal sequence of observations, or geographical locus. Thus if z is a reference class, $y(i)$ may be taken as the finite set of z's occurring before time t_i, or occurring between some primordial reference time t_0 and t_i, or the finite set of z's observed before t_i, or between t_0 and t_i, or the finite set of z's occupying a place within a distance d_i of some reference point (where of course $t_i < t_{i+1}$, $d_i < d_{i+1}$, etc.).

In short, let c be a term not occurring in L; let L' be the union of L and the singleton of c, let w be the (unique) rational corpus based on the single formula $\ulcorner \bigcup \mathscr{R} y = z \wedge \mathscr{R} x \subset \mathbb{R}^n \wedge D\langle y, x, \{s\}\rangle \wedge c \in z \urcorner$ and let $z \in RT_{L'}$. Under these circumstances, nothing exciting or unusual happens: we get the classical calculus of probability, as the following theorem shows:

MT-10.18 $\sim c \in L \wedge L' = L \cup \{c\} \wedge z \in RT_{L'} \wedge \vdash \ulcorner b_1, b_2 \in \mathbb{B}^n \urcorner \wedge$
$\wedge\, w = \imath r(r \textit{ is } RC_{L'} \wedge \{\ulcorner \bigcup \mathscr{R} y = z \wedge D\langle y, x, \{s\}\rangle \wedge \mathscr{R} x \subset \mathbb{R}^n \wedge$
$\wedge\, c \in z \urcorner\} \textit{ is basis}_{L'} \textit{ for } r) \rightarrow$

[Every statement of the right sort has a probability, which is a real number lying between 0 and 1.]

$$\bigvee_p (p\text{-}prob_{L'}(\ulcorner x\langle c\rangle \in b_1\urcorner, w)=[p,p] \wedge \vdash_{L'} \ulcorner 0 \leqslant p \leqslant 1\urcorner)$$

[If the probability of s is p, the probability of $\ulcorner \sim s\urcorner$ is $1-p$.]

$$p\text{-}prob_{L'}(s, w)=[p,p] \rightarrow p\text{-}prob_{L'}(\ulcorner \sim s\urcorner, w)=[1-p, 1-p]$$

[The special addition law: if s_1 and s_2 are jointly impossible, the probability of their disjunction is the sum of their probabilities.]

$$\begin{aligned}&0 \vdash_{L'} \ulcorner b_1 \cap b_2 = 0\urcorner \wedge p\text{-}prob_{L'}(\ulcorner x\langle c\rangle \in b_1\urcorner, w)=\\ &=[p,p] \wedge p\text{-}prob_{L'}(\ulcorner x\langle c\rangle \in b_2\urcorner, w)=\\ &=[q,q] \rightarrow p\text{-}prob_{L'}(\ulcorner x\langle c\rangle \in b_1 \cup b_2\urcorner, w)=[p+q, p+q]\end{aligned}$$

[The general addition law: the probability of the disjunction of s_1 and s_2 is the sum of their probabilities, less the probability of their conjunction.]

$$\begin{aligned}&p\text{-}prob_{L'}(\ulcorner x\langle c\rangle \in b_1\urcorner, w)=[p,p] \wedge p\text{-}prob_{L'}(\ulcorner x\langle c\rangle \in b_2\urcorner, w)=\\ &=[q,q] \wedge p\text{-}prob_{L'}(\ulcorner x\langle c\rangle \in b_1 \cap b_2\urcorner, w)=\\ &=[r,r] \rightarrow p\text{-}prob_{L'}(\ulcorner x\langle c\rangle \in b_1 \cup b_2\urcorner, w)=[p+q-r, p+q-r]\end{aligned}$$

[The probability of something impossible is 0, and the probability of something certain is 1.]

$$\begin{aligned}&p\text{-}prob_{L'}(\ulcorner x\langle c\rangle \in 0\urcorner, w)=\\ &=[0,0] \wedge p\text{-}prob_{L'}(\ulcorner x\langle c\rangle \in \mathbb{B}^n\urcorner, w)=[1,1]\end{aligned}$$

$$p\text{-}prob_{L'}(\ulcorner x\langle c\rangle \in b\urcorner, w)=\left[\underline{\int_b} ds, \overline{\int_b} ds\right]$$

Proof:

(1) Assume antecedent.

(2) $\bigwedge_b (b \in Math\ Tm_{L'} \rightarrow RAN_{L'}(c, y, x, b, w)$ — (1); poverty of w

(3) part (1) — (1), (2), MT-10.1

(4) part (2) — (1), (2), MT-10.8

(5) part (3) — (1), (2), MT-10.15

(6) part (4) — (1), (2), MT-10.15

(7) part (5) (1), (2), MT-10.6, MT-10.7
(8) part (6) (1), (2), MT-10.1

Note that one could as well associate the probabilities with the Borel sets themselves as with the corresponding statements, in this special case.

This is a rather more complicated way than usual of talking about probabilities, but in this case no confusion or errors are introduced by the usual technique of expressing probabilities in the object language. This is because it is built into our hypothetical body of knowledge that *all* we know relevant to the probabilities is expressed in our statistical knowledge about $\ulcorner \bigcup \mathscr{R}y \urcorner$, and in our knowledge that c belongs to this set. Under these special circumstances our assertions of probability are precisely paralleled by statistical assertions: hence, perhaps, the tendency that a large number of scholars have had to identify statistical assertions and probability assertions.

EXAMPLE I: Let T be an expression of L, denoting the set of balls in an urn; let Y be an expression of L, denoting a function whose domain is ω, and whose value for every natural number i is just the set denoted by T. Thus $\ulcorner \bigcup \mathscr{R}Y = T \urcorner$ is a statement in our body of knowledge. We suppose also that T is a reference term of L. Let Z also be an expression of L, denoting that function whose domain is the set denoted by T, and whose value for an object in its domain is $\langle 1_r \rangle$, $\langle 2_r \rangle$, or $\langle 3_r \rangle$ according as that object is Red, Blue, or Green. Finally, let X denote the function whose domain is $\mathbb{R}^1$, and whose value for every argument less than $\langle 1_r \rangle$ is 0; whose value for every argument in the interval $[1, 2)$ is $\frac{1}{4}$; whose value for every argument in the interval $[2, 3)$ is $\frac{3}{4}$, and whose value is 1 otherwise. Let us suppose that we know that X is the function characterizing the distribution of Z in the sequence Y. We would say then that the probability of drawing a red ball is $\frac{1}{4}$; the probability of drawing a red or a blue ball is $\frac{3}{4}$, and so on. Formally we represent this situation as follows: Let c be a constant not occuring in L; let L' be L supplemented by c. We consider the rational corpus whose basis is:

$$\{\ulcorner c \in \bigcup \mathscr{R}Y \wedge \bigcup \mathscr{R}y = T \wedge D\langle Y, Z, \{X\}\rangle \wedge R \urcorner\}$$

Where R is a meaning postulate reflecting the connection between the values of the random variable Z and the colors red, blue and green – or rather, the predicates in L denoting these colors.

As is evident from the preceding theorem, relative to this rational corpus, the probability that a ball is red is a quarter, that it is red or green is a half, and so on.

EXAMPLE II: Let P be an expression of L, among the potential reference terms RT_L, denoting the set of adult frogs of a certain species that are now alive. Let Q be an expression of L, denoting a function whose domain is ω, and whose value for every natural number i is the set of adult frogs of that species living on the earth's surface within i miles of the point marked with a star. Thus $\ulcorner \bigcup \mathscr{R}Q = P \urcorner$ is a statement in our body of knowledge. Let W be an expression of L denoting that function whose domain is the set of ponderable objects, and whose value for any argument is just the one-termed sequence whose term is the weight of that object in grams. Finally, let $N(m, s)$ denote the normal distribution function with mean m and standard deviation s. To suppose that $N(m, s)$ characterizes the distribution of weights among adult frogs in P is to consider a rational corpus whose basis is

$$\{\ulcorner d \in \bigcup \mathscr{R}P \wedge \bigcup \mathscr{R}P = Q \wedge D\langle P, W, \{N(m, s)\}\rangle \urcorner\}$$

where d denotes a new constant foreign to L. L' is L supplemented by the new constant d. Relative to this rational corpus, we have all the usual probability statements. It should be noted, however, that this hypothetical rational corpus cannot be regarded as one belonging to someone in the real world: in the real world we all know that the number of frogs of a certain species alive at any particular time (or even over an arbitrarily long, but bounded, period of time) is finite, and therefore that $\ulcorner D\langle P, W, \{N(m, s)\}\rangle \urcorner$ is false.

Let $R = \ulcorner \bigwedge_x W^*(x) = |m - W(x)|/s \urcorner$.

Conjoin R to the basis of the preceding rational corpus. Then, relative to this new rational corpus, we will be able to show that the probability that $W^*(d)$ is positive is $\frac{1}{2}$; the probability that it equals 0 is 0; etc.

The next stage of complication consists in considering a hypothetical rational corpus in which our statistical knowledge is not precise. Otherwise the situation is the same as before. We have as a metatheorem:

MT-10.19 $\sim c \in L \wedge L' = L \cup \{c\} \wedge z \in RT_{L'} \wedge \imath r\,(\{\ulcorner \bigcup \mathscr{R}y = z \wedge D\langle y, x, f\rangle \wedge c \in z \wedge \mathscr{R}x \subset \mathbb{R}^n \urcorner\}$ *is basis*$_{L'}$ *for* $r \wedge r$ *is* $RC_{L'}) = w \wedge \vdash_{L'} b_1, b_2 \in \mathbb{B}^n \rightarrow$

(1) $p\text{-}prob_{L'}(\ulcorner x\langle c\rangle \in b_1 \urcorner, w) =$
$$= \left[inf \left\{ \int_{b_1} ds : s \in f \right\}, sup \left\{ \int_{b_1} ds : s \in f \right\} \right]$$

(2) $p\text{-}prob_{L'}(\ulcorner x\langle c\rangle \in b_1 \urcorner, w) = [p, q] \to \vdash_{L'} \ulcorner 0 \leqslant p \leqslant q \leqslant 1 \urcorner$

(3) $p\text{-}prob_{L'}(\ulcorner x\langle c\rangle \in b_1 \urcorner, w) =$
$= [p, q] \to p\text{-}prob_{L'}(\ulcorner \sim x\langle c\rangle \in b_1 \urcorner, w) = [1-q, 1-p]$

(4) $\vdash_{L'} \ulcorner b_1 \cap b_2 = 0 \urcorner \wedge p\text{-}prob_{L'}(\ulcorner x\langle c\rangle \in b_1 \urcorner, w) =$
$= [p_1, q_1] \wedge p\text{-}prob_{L'}(\ulcorner x\langle c\rangle \in b_2 \urcorner, w) =$
$= [p_2, q_2] \wedge p\text{-}prob_{L'}(\ulcorner x\langle c\rangle \in b_1 \cup b_2 \urcorner, w) = [p_3, q_3] \to$
$\vdash_{L'} \ulcorner p_1 + p_2 \leqslant p_3 \wedge q_3 \leqslant q_1 + q_2 \urcorner$

EXAMPLE III: Just the same as example I, except that in lieu of the single distribution function X we consider two distribution functions X_1 and X_2; for example we might suppose that we were considering a hypothetical draw from an urn known to have one of two compositions.

Let $X_1 = \ulcorner \imath f(\mathscr{D}f = \mathbb{R}^1 \wedge f \text{ is fun} \wedge \bigwedge_{x \in \mathbb{R}^1} ((x < 1 \wedge fx = 0) \vee (x \in [1, 2) \wedge \wedge fx = \frac{1}{4}) \vee (x \in [2, 3) \wedge fx = \frac{3}{4}) \vee (3 \leqslant x \wedge fx = 1))) \urcorner$

Let $X_2 = \ulcorner \imath f(\mathscr{D}f = \mathbb{R}^1 \wedge f \text{ is fun} \wedge \bigwedge_{x \in \mathbb{R}^1} ((x < 1 \wedge fx = 0) \vee (x \in [1, 2) \wedge \wedge fx = \frac{1}{3}) \vee (x \in [2, 3) \wedge fx = \frac{2}{3}) \vee (3 \leqslant x \wedge fx = 1))) \urcorner$

Let c be a constant not occuring in L and L' be L supplemented by c. Consider the rational corpus w whose basis is

$$\{\ulcorner c \in \bigcup \mathscr{R}y \wedge \bigcup \mathscr{R}y = T \wedge D\langle y, z, \{X_1, X_2\}\rangle \wedge R \urcorner\}$$

For every borel term b we have $RAN(c, y, z, b, w)$, thus the probability that c is red is $[\frac{1}{4}, \frac{1}{3}]$, since

$$p\text{-}prob_{L'}(\ulcorner z\langle c\rangle \in \{\langle 1\rangle\} \urcorner, w) = [\tfrac{1}{4}, \tfrac{1}{3}];$$

the probability that c is red or blue is $[\frac{2}{3}, \frac{3}{4}]$, since

$$p\text{-}prob_{L'}(\ulcorner z\langle c\rangle \in \{\langle 1\rangle, \langle 2\rangle\} \urcorner, w) = [\tfrac{2}{3}, \tfrac{3}{4}];$$

the probability that c is blue is $[\frac{1}{3}, \frac{1}{2}]$, since

$$p\text{-}prob_{L'}(\ulcorner z\langle c\rangle \in \{\langle 2\rangle\} \urcorner, w) = [\tfrac{1}{3}, \tfrac{1}{2}];$$

Since $\vdash_{L'} \ulcorner\{\langle 1\rangle\}\cap\{\langle 2\rangle\}\urcorner=0$ part (4) of theorem 19 applies; we observe that $\frac{1}{4}+\frac{1}{3}\leqslant\frac{2}{3}\leqslant\frac{3}{4}\leqslant\frac{1}{3}+\frac{1}{2}$, or $\frac{7}{12}\leqslant\frac{8}{12}\leqslant\frac{9}{12}\leqslant\frac{10}{12}$ as required.

Finally, the probability that c is either blue or red, or else blue or green is [1, 1]. We have

$$p\text{-}prob_{L'}(\ulcorner z\langle c\rangle\in\{\langle 1\rangle,\langle 2\rangle\}\cap\{\langle 2\rangle,\langle 3\rangle\}\urcorner, w)=[\tfrac{1}{3},\tfrac{1}{2}]$$
$$p\text{-}prob_{L'}(\ulcorner z\langle c\rangle\in\{\langle 1\rangle,\langle 2\rangle\}\cup\{\langle 2\rangle,\langle 3\rangle\}\urcorner, w)=[1,1]$$
$$p\text{-}prob_{L'}(\ulcorner z\langle c\rangle\in\{\langle 1\rangle,\langle 2\rangle\}\urcorner, w)=[\tfrac{2}{3},\tfrac{3}{4}]$$
$$p\text{-}prob_{L'}(\ulcorner z\langle c\rangle\in\{\langle 2\rangle,\langle 3\rangle\}\urcorner, w)=[\tfrac{2}{3},\tfrac{3}{4}]$$

We observe that $\frac{2}{3}+\frac{2}{3}-\frac{1}{3}\leqslant 1\leqslant 1\leqslant\frac{3}{4}+\frac{3}{4}-\frac{1}{2}$
$1\leqslant 1\leqslant 1\leqslant 1$

EXAMPLE IV: Except for the family of distribution functions, this is the same as the preceding example: here we suppose the family to include *any* distribution function bounded by X_1 and X_2. This would be appropriate if our knowledge of the contents of the urn were *approximate* – if we knew, not as before that the urn had one of two precisely known compositions, but merely that between $\frac{1}{4}$ & $\frac{1}{3}$ of the balls were blue, between $\frac{1}{3}$ & $\frac{1}{2}$ were red, and between $\frac{1}{4}$ and $\frac{1}{3}$ were green.

Let $X=\ulcorner\{x: x \text{ is } Dist.fun \wedge \mathscr{D}x=\mathbb{R}^1\wedge \bigwedge_{y\in\mathbb{R}^1}(X_1y\leqslant xy\leqslant X_2y\vee$
$\vee X_2y\leqslant xy\leqslant X_1y)\}\urcorner$

Let the basis of the rational corpus w be

$$\{\ulcorner c\in\bigcup\mathscr{R}y\wedge\bigcup\mathscr{R}y=T\wedge D\langle y,z,X\rangle\wedge R\urcorner\}$$

The probabilities in this example are precisely those of the preceding example, since for every real number s, $inf\left\{\int_s dt: t\in X\right\}=Min(X_1s-$
$-X_1(s-\varepsilon), X_2s-X_2(s-\varepsilon))$ and $sup\left\{\int_s dt: t\in X\right\}=Max(X_1s-X_1(s-\varepsilon),$
$X_2s-X_2(s-\varepsilon))$

EXAMPLE V: This is the same as example II, except that for the supposition that the distribution is $N(m,s)$, we suppose that it is either

$N(m_1, s_1)$ or $N(m_2, s_2)$. The probabilities are easily computable: for any borel expression b, we have

$$p\text{-}prob_{L'}(\ulcorner W^*(c)\in b\urcorner, w) = \left[Min\left(\int_b dN(m_1, s_1), \int_b dN(m_2, s_2)\right), \right.$$
$$\left. Max\left(\int_b dN(m_1, s_1), \int_b dN(m_2, s_2)\right)\right]$$

EXAMPLE VI: As before, now let us suppose, not that we know that the weight is distributed according to one of two precise distributions, but that it is distributed according to one of an infinite number of distributions sandwiched between these two distributions. To be precise, suppose we know that the distribution is normal, with a mean between m_1 and m_2, and a variance of exactly s.

Let $X_m = \{x : \bigvee_m (m_1 \leqslant m \leqslant m_2 \wedge x = N(m, s))\}$

Then let w be the rational corpus based on $\{\ulcorner c \in \bigcup \mathscr{R}y \wedge \bigcup \mathscr{R}y = T \wedge \wedge D\langle y, W^*, X_m\rangle \wedge R\urcorner\}$.

For any borel term b, we have $RAN_{L'}(c, y, W^*, b, w)$ and thus

$$p\text{-}prob_L{}'(\ulcorner W^*\langle c\rangle \in b\urcorner, w) = \left[inf\left\{\int_b dz : z \in X_m\right\}, sup\left\{\int_b dz : z \in X_m\right\}\right]$$

The graphs of the frequency functions in this case look like this (see Figure 7).

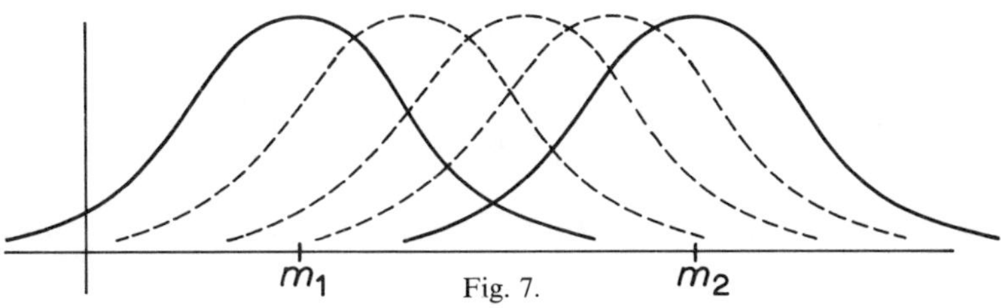

Fig. 7.

If every point of the set denoted by b lies to the left of m_1, we have

$$inf\left\{\int_b dz : z \in X_m\right\} = \int_b dN(m_2, s) \text{ and } sup\left\{\int_b dz : z \in X_m\right\} = \int_b dN(m_1, s)$$

corresponding relations hold if every point of the set denoted by b lies to the right of m_2.

In general, however, things are not so simple. Let b_n denote an *interval* whose midpoint n is in (m_1, m_2), then $sup\left\{\int_{b_n} dz : z \in X_m\right\} = \int_{b_n} dN(n, s)$ and

$$inf\left\{\int_{b_n} dz : z \in X_m\right\} = \int_{b_n} dN(m_1, s) \text{ if } n - m_1 \geq m_2 - n \text{ and}$$

$$inf\left\{\int_{b_n} dz : z \in X_m\right\} = \int_{b_n} dN(m_2, s) \text{ if } n - m_1 \leq m_2 - n.$$

In the case of more general families of distributions, things get yet more complicated. If $\ulcorner D\langle Y, S, Z\rangle\urcorner$ belongs to our rational corpus and the set denoted by Z contains only a finite number of members, we have no problems in principle. If that set of distribution functions has an infinite number of members, they must all fall into a finite number of parametric families. We may suppose (also on expistemological grounds?) that no more than a finite number of parameters are involved. We may then often differentiate with respect to these parameters to find maxima and minima for certain subsets of distributions, and thence ultimately arrive at the probabilities we seek. But sometimes an explicit solution may not be attainable. Even if it is not, however, all is not lost – we may still be able to produce useful limits.

EXAMPLE VII: Now let X_v be $\{N(m, s) : s_1 < s < s_2\}$ and let w be the rational corpus based on $\{\ulcorner c \in \bigcup \mathscr{R} y \wedge \bigcup \mathscr{R} y = T \wedge D\langle y, W^*, X_v\rangle\urcorner\}$.

As before, for any borel term b, we have $RAN_{L'}(d, Y, W^*, b, w)$ and thus

$$p\text{-}prob_{L'}(\ulcorner W^*\langle c\rangle \in b\urcorner, w) = \left[inf\left\{\int dz : z \in X_v\right\}, sup\left\{\int dz : z \in X_v\right\}\right]$$

The graphs of the frequency functions in this case look like this (see Figure 8).

If every point of the set denoted by b lies to the left of $m - s_2$, then

$$inf\left\{\int_b dz : z \in X_v\right\} = \int_b dN(m, s_1) \text{ and } sup\left\{\int_b dz : z \in X_v\right\} = \int_b dN(m, s_2).$$

Corresponding relations hold if every point of the set denoted by b lies to the right of $m + s_2$.

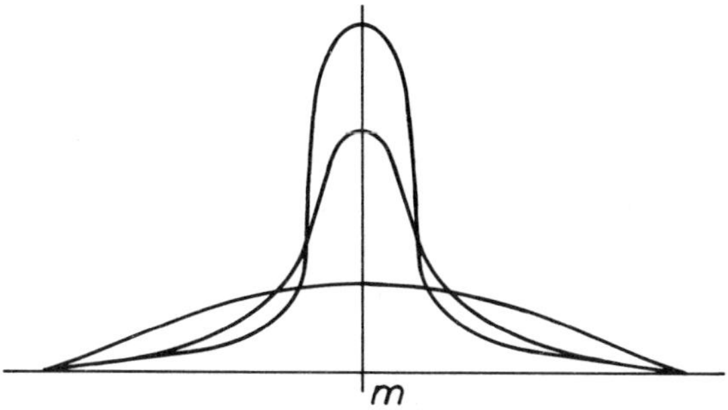

Fig. 8.

The second class of simple illustrative cases is also familar from introductory texts on probability; these are cases having to do with well-adjusted gambling apparatus, with the application of statistical laws in physics, and the like. This is the sort of case that Karl Popper and Ian Hacking have taken as paradigmatic. Hacking, for example, takes probability as a physical characteristic of what he calls a chance set up. A chance set up might consist of a coin and an apparatus for tossing it. Characteristic of such a set up might be a certain propensity *p* for the coin to land heads. One will then say that the probability that a trial of this chance set up will yield heads is *p*. This is perfectly plausible as regards trials whose outcome is not known. It is also plausible if one wishes to say, as one might, that the probability of heads on the toss of the coin I just performed which landed tails, is a half. If one does decide to talk in this way, however, it must be observed that 'probability' as used in that way is no longer a guide in life. In life there is an asymmetry, in tossing coins and in many other things, between the past and the future. Even money is a fair bet with respect to the next toss, but not with respect to a toss whose outcome is known to be tails. I prefer to use 'probability' in such a way that it is always a guide in life, and thus to take account of this asymmetry.

We can easily characterize bodies of knowledge in which the Hacking-Popper approach to probability applies perfectly to future (or in general, unknown) trials of a chance set up. We need merely stipulate of every term *t*, that if it is known to denote an object belonging to the reference set in question, then either it is a random member of its unit set with respect to the property being considered – this comprises the case of known in--

stances – or it is a random member of that reference set with respect to the property being considered – this is the case of the unknown instance. What is ruled out here are just the states of partial knowledge other than those in which just that particular statistical statement being considered is relevant. We do not, if we are going to use probability as a guide in life, want to rule these states out always: If I am told that ten tosses of a certain coin were performed yesterday and their outcomes recorded, and that seven of the ten resulted in heads, then 0.7/0.3 represents the appropriate odds for me to offer on tails on the third of those tosses, even if I know for sure that the coin and the tossing apparatus are perfectly fair. But *in general* when confronted with gambling aparatus or a physical set up embodying a statistical law of physics, it is true that with respect to a given trial, at a given time, one of two epistemological states obtains: either we know the outcome of that trial, or thereis nothing we know that keeps that trial from being a random member of the general class of such trials.

EXAMPLE VIII: Let Y denote a sequence of sets of tosses of a certain coin; let X denote the random quantity whose domain is the union of the range of Y, and whose value for any member of that range is $\langle 0\rangle$ if the coin lands heads on that toss, and $\langle 1\rangle$ otherwise; let Z denote the function whose domain is the set of 1-sequences of reals, and whose value is 0 for 1-sequences less than $\langle 0\rangle$, whose value is $\frac{1}{2}$ for 1-sequences less than $\langle 1\rangle$ and greater than or equal to $\langle 0\rangle$, and whose value is 1 otherwise. Let w_6 be a rational corpus containing $\ulcorner D(Y, X, \{Z\})\urcorner$. Furthermore, let us suppose that it is true of w_6, whatever else it may contain, that for any term t and any borel term b, if t is known in w_6 to belong to $\ulcorner\bigcup\mathscr{R}Y\urcorner$, i.e., if $\ulcorner t\in\bigcup\mathscr{R}y\urcorner\in w_6$ then either $RAN_L(t, Y, X, b, w_6)$ or $RAN_L(t, \ulcorner\{\langle i, \{t\}\rangle : i\in\omega\}\urcorner, X, b, w_6)$. Then the classical calculus of probability applies without difficulty: the probability of heads is a half, of non-heads is a half, of heads or tails is 1, etc., in the case that $RAN_L(t, Y, X, b, w_6)$ holds; and correspondingly the probability of heads is one (or zero), of non-heads zero (or one), etc., if the other alternative holds.

Formally, the condition imposed on the rational corpus w_6 is this:

(1) $$\bigwedge_{t\in Tm_L} (\ulcorner t\in\bigcup\mathscr{R}Y\urcorner\in w_6\rightarrow RAN_L(t, Y, X, b, w_6)\vee RAN_L(t, \ulcorner\{\langle i, \{t\}\rangle : i\in\omega\}\urcorner, X, b, w_6))$$

This is a perfectly reasonable condition for real bodies of knowledge. Furthermore, even if the coin in question is the one mentioned earlier, we can apply this technique to terms known to denote some toss *other* than one of the ten recorded yesterday about which we have special knowledge. Let $\ulcorner\bigwedge_{i\in\omega}(Y^*(i)=Y(i)-T)\urcorner$ belong to w_7, where T denotes the ten tosses recorded yesterday. It is not hard to imagine that we may still have $\ulcorner D\langle Y^*, X, \{Z\}\rangle\urcorner$ in our rational corpus w_7. If this is so, it is not hard to imagine that the condition $\bigwedge_{t\in Tm_L}(\ulcorner t\in\bigcup\mathscr{R}Y^*\urcorner\in w_7\rightarrow RAN_L(t, Y^*, X, b, w_7)\vee RAN_L(t, \ulcorner\{\langle i, \{t\}\rangle : i\in\omega\}\urcorner, X, b, w_7))$ holds for w_7. (Note that if $\bigcup\mathscr{R}Y$ is known to be identical to a reference term of L, and T is known to be identical to a reference term of L, then $\ulcorner\bigcup\mathscr{R}Y^*\urcorner$ will be known to be identical to a reference term of L.)

EXAMPLE IX: Let Y denote a sequence of sets of molecules of hydrogen gas at a certain temperature t. Let T be a fragment of kinetic theory, according to which the velocity of the molecules in a given direction is normally distributed with mean $m(t)$ and variance $[s(t)]^2$. Let V denote the random quantity whose domain is the set of hydrogen molecules, and whose value for any member of that domain at a given time is the velocity of that molecule in the given direction at that time. Let $Z=$ $=N(m(t), s(t))$. Finally, let us suppose that our rational corpus, containing the theory T, satisfies the condition

$$(2)\qquad \bigwedge_{c\in Tm_L}(\ulcorner c\in\bigcup\mathscr{R}Y\urcorner\in w_8\rightarrow RAN_L(c, Y, X, b, w_8)\vee \vee RAN_L(c, \ulcorner\{\langle i, \{c\}\rangle : i\in\omega\}\urcorner, X, b, w_8))$$

Just as we considered before certain bodies of knowledge that contained approximate rather than precise statistical knowledge, so we may take another step toward reality by introducing the same generalization here.

EXAMPLE X: Suppose that this is the same as example VIII, except that in lieu of claiming the precise statistical knowledge that the sequence of coin tosses has a limiting frequency of heads of $\frac{1}{2}$, we claim only that this limiting frequency lies between 0.4 and 0.6. This is certainly more like the kind of knowledge we do have about coins. And it is still not implausible to suppose that condition (1) holds of this body of knowledge (say w_9), so that we get the same sort of slight generalization of the standard probability calculus we discovered in Example IV Finally, in this example, we are considering just the sort of body of knowledge that people might really and plausibly be supposed to have.

EXAMPLE XI: Suppose that this is the same as example IX, except that an element of error is introduced yielding an approximate distribution, rather than a precise one. There are various sources for this error. The theory itself may apply exactly only under some limiting conditions. In particular, of course, to say that the velocity of the molecules is distributed normally entails that there are an infinite number of molecules present, while we know perfectly well that there are only a finite number. Furthermore, in applying this theory, we suppose the sample of gas we are talking about is at temperature t; but we don't *know* that it is at exactly that temperature; we know only that it is practically certain (as a result of our measurements) that the temperature is in a small interval about t. Furthermore, we can be sure that any molecules inside our container are moving with speeds well under the speed of light, and so that the distribution of velocities is not even theoretically normal, but rather truncated at both ends. We have, in the last analysis, a family of distributions Z^*, none of which is normal, but all of which are, in a sense that can be made as precise as one pleases, 'close' to the distribution $N(m(t), s(t))$. It is even more natural in this example than in the preceding one that condition (2) should apply.

4. So far, we have defined partial probability, that is, a probability function that has values relative only to strictly consistent sets of statements. It was earlier argued that strict consistency is an unrealistic and implausible condition to impose on bodies of knowledge, and it is relative to bodies of knowledge that we want to define probability in order that it should be a proper guide in life. We shall get our final and finished notion of probability by stipulating that the probability of a statement, relative to a body of knowledge, is $[p, q]$ just in case the probability of that statement, relative to every maximal consistent subset of that body of knowledge is $[p, q]$. It will no longer turn out, as it did before, that every statement of our language will have a probability. But the statements that lack probabilities will turn out (at least, given a plausible rule of acceptance) to be just those puzzling statements that give rise to the lottery paradox.

We shall see that relative to a body of knowledge in which statements of the form 'ticket i will not win the lottery', are accepted, there will be no probabilities for these statements. And conversely, in those bodies of knowledge relative to which the probability of 'ticket i will not win the lottery' has a probability at all, it will be the probability determined by the number of tickets in the lottery, and no such statement will be accepted.

We begin by defining a *strand* of a rational corpus: we say that c is a strand of w if and only if w is a rational corpus, and c is a subset of statements of w, c is consistent, and if any statement of w that is not a member of c is added to c, the resulting set of statements is inconsistent.

MD-10.3 $$c \text{ is } Strand_L \text{ of } w \leftrightarrow w \text{ is } RC_L \wedge c \subset w \wedge \sim \ulcorner 0 \neq 0 \urcorner \in L\text{-}Cn(c) \wedge \wedge \bigwedge_z (z \in w - c \rightarrow \ulcorner 0 \neq 0 \urcorner \in L\text{-}Cn(c \cup \{z\}))$$

Of course it turns out that if a rational corpus is itself deductively closed, it is its own only strand. We do have cause to speak of such rational corpora, as the rational corpora of the preceding examples indicate.

MT-10.20 $$LCn(w) \cap St_L \subset w \subset St_L \rightarrow \bigwedge_c (c \text{ is } Strand_L \text{ of } w \rightarrow c = w)$$

Furthermore, in those cases just discussed in which c is a constant foreign to L, and we consider a rational corpus w with the statement $\ulcorner c \in \bigcup \mathscr{R} y \urcorner$ added to it, it will be the case that $RAN_L(c, y, z, b, s)$ for every strand s, if it holds for any strand s.

MT-10.21 $$w \text{ is } RC_L \wedge \sim c \in L \wedge L' = L \cup \{c\} \wedge L' \text{ is language} \rightarrow \rightarrow [\bigvee_s (s \text{ is } Strand_{L'} \; w \cup L'\text{-}Cn(\{\ulcorner c \in \bigcup \mathscr{R} y \urcorner\}) \wedge \wedge RAN_{L'}(c, y, z, b, s)) \rightarrow \bigwedge_s (s \text{ is } Strand_{L'} \; w \cup L'\text{-}Cn(\{\ulcorner c \in \bigcup \mathscr{R} y \urcorner\}) \rightarrow RAN_{L'}(c, y, z, b, s))]$$

Thus all of the preceding theorems concerning the special cases in which randomness is guaranteed by either the abstractness of the discussion or the use of the indefinite article 'a', may be directly generalized to hold for our general notion of probability to be defined now.

MD-10.4 $$PROB_L = \imath f (f \text{ is } fun \wedge \mathscr{R} f = \prod_{i<2} \langle St_L, \mathscr{P} St_L \rangle \wedge \bigwedge_{x \in \mathscr{D} f} fx = = \imath k (\bigwedge_w (w \text{ is } strand_L \; x_1 \rightarrow p\text{-}prob_L(x, w) = k)))$$

If everything works out all right, then the probability of a statement of L, relative to a rational corpus of L, will be the appropriate interval:

MT-10.22 $PROB_L(s, w)=[p, q] \leftrightarrow w \text{ is } RC_L \wedge \bigwedge_x (x \text{ is } Strand_L\ w \rightarrow$

$\rightarrow p\text{-}prob_L(s, x)=[p, q])$

On the other hand, if something goes wrong, the probability is the empty set:

MT-10.23 $PROB_L(x)=0 \leftrightarrow \sim L \text{ is } language \vee \sim \bigvee_{z,y} (x=\langle y, z\rangle \vee y \in St_L \wedge$

$\wedge z \in \mathscr{P} St_L \wedge y \text{ is } RC_L) \vee \imath k (\bigwedge_w (w \text{ is } strand_L \text{ of } x_1 \rightarrow$

$\rightarrow p\text{-}prob_L(x_0, w))=k)=0$

Although we have not yet said much of anything about the way in which items come to be accepted into our rational corpora, it seems clear that the only circumstances that can be responsible for the acceptance of an inconsistent set of statements s are those analogous to the lottery case: each member of s, relative to another (higher level) rational corpus is so probable that we should accept it; and the denial of the conjunction of the mebers of s, relative to that other rational corpus of higher level, is also so probable that we should accept it. What happens in such cases? Let r be the sequence of statements 'ticket 0 will not win', 'ticket 1 will not win', ... 'ticket n will not win', and let t be the statement 'There are exactly n tickets of which exactly 1 will win'. Each strand of the (inconsistent) rational corpus containing statements from the sequence r will yield a partial probability for 'ticket k will not win'. But this probability will be [1, 1] or [0, 0] according as the statement in question happens to be a member of that strand of the rational corpus or not. Thus there will be no overall probability for 'ticket k will not win'.

Relative to the higher level rational corpus, we may perfectly well suppose that the probabilities of these statements do exist (i.e., aren't the empty set). There is no reason why statement t should not be a member of every strand of our rational corpus, even if that rational corpus does have more than one strand. Furthermore, there is no reason why it should not be the case that ticket i is a random member of the set of tickets, relative to each strand of this higher level rational corpus. Thus, relative to the higher level rational corpus we may have appropriate probabilities for the statements in question, and the conventional description of the lottery applies.

5. A much cited stumbling block for logical interpretations of probability has been the difficulty of explaining the connection between the apparent factual content of certain probability statements and the fact that on such an interpretation, a probability statement, if true at all, must be regarded as logically true. A related difficulty is that of finding some plausible way to construe talk about unknown probabilities, true as opposed to apparent probabilities, hypothetical probabilities, and so on.

Under the interpretation of probability offered in this chapter, logical though it be, these difficulties are easily met.

Let us begin with the most common connection between probability and factual content. I say that the probability of getting a head on the next toss of this coin is a half, or that the probability of the next frog we see of species X being less than two inches long is 0.1584. Surely, one might say, such assertions cannot be made *a priori*; surely there must be some empirical basis for them. I agree. As I have analysed probability, the first statement amounts to: (a) relative to my body of knowledge, the next toss of this coin is a random member of its tosses with respect to yielding heads and (b) the statement that half (more plausibly, close to half) of the tosses of this coin yield heads is a part of that body of knowledge. The second statement amounts to: (a) The next frog we see is, relative to my body of knowledge, a random member of the set of frogs of this species with respect to being less than two inches long, and (b) I know (believe rationally) that the proportion of frogs of that species less than two inches long is (close to) 0.1584. Thus in the analysis of probability offered above, a statistical statement is *mentioned*, and furthermore mentioned as belonging to my rational corpus, my body of knowledge. To make this assertion about the statistical statement is, implicitly, to assert that I have adequate grounds for accepting it. It is to say that a certain possibly (usually) empirical statement is worthy of my belief, and indeed that it ought, in general, to be worthy of your belief too. In general terms, to say that the probability of a statement S is p, is to say roughly that S is equivalent to some statement of the form $\ulcorner x \in y \urcorner$, that x is a random member of z with respect to belonging to y, relative to my body of knowledge, and that I include in my body of knowledge a statistical statement to the effect that the proportion of *y*'s among *z*'s is *p*.

There are two very distinct grounds, both amenable to reasonable adjudication, on which you and I might disagree about a probability. One is more straight-forwardly factual than the other. If I claim that the probability that the next frog of species *X* that we see will be less than two inches long is 0.1584, while you claim that it is 0.3547, it may in the first place be that we both agree on the distribution of lengths in frogs of that species, but that one of us knows something about that frog that the other doesn't: for example, you may have heard it croak, which enabled you, before seeing it, to place it in a special subspecies. This is not a case in which your statistical knowledge need differ from mine: we may perfectly well both have the same general statistical knowledge. But you are able to put the frog in a smaller reference class than I am. Of course, as soon as you give me the information that it is not merely of species *X*, but of subspecies *B*, then we shall agree on the probability. We will have shared an item of your rational corpus that was lacking in mine, and that will render the same reference class appropriate for both of us. This is not a question, really, of factual disagreement – you and I did not *disagree* about whether the frog belonged to subspecies *B* or not; it was merely that you knew something of which I was ignorant.

But it may be that our disagreement about the probability stems from a disagreement in our statistical knowledge. It may well be that both in the case of your rational corpus and in the case of mine, the next frog of species *X* is simply going to be a random member of species *X* with respect to having a length under two inches, but that I include in my rational corpus the statistical statement that 0.1584 of those frogs have a length under 2 inches, while you include in your rational corpus the statistical statement that 0.3547 of those frogs have a length under two inches. We now have a real disagreement: one of us must clearly be wrong. Since we have supposed that these items are in our bodies of knowledge, our rational corpora, we are supposing that we each have 'adequate evidence' for accepting the statistical statements in question. It is therefore clear that your evidence differs from mine. If, however, we are willing, as true scientists, to share our evidence, we may yet come to agreement. Perhaps one of us is right – e.g., my frog manual is out of date, and yours is not; or perhaps the combined evidence we have available, will show that we are both wrong: if your statistics are based on the study of a thousand frogs, and mine on the study of two thousand frogs, it may be that when our data are combined, (together with auxilliary hypotheses) we will obtain a statistical parameter of 0.2987. Observe that in this case, although what sparked the disagreement was a claim about probability, it was a divergence in factual beliefs underlying the probability assertions that required adjudication; and thus it was entirely appropriate to look at the world and collect data about frogs. This no more entails that probability statements are factual than disagreement about the mortality of Socrates entails that *modus ponens* is a factual mode of inference. I can doubt that Socrates is mortal, even when you tell me that you believe both that all men are mortal and that Socrates is a man, without impugning the validity of your logic.

The factual content underlying most probability statements comes even more clearly into view when we consider the very common locution: The probability that a B is a C is p. I suggested earlier that the indefinite article 'a' was often used in the sense of 'every random'. A convenient formal way of representing such locutions is through the introduction of a new constant c to the language, together with the addition to the rational corpus in question of the statement $\ulcorner c \in B \urcorner$. Since c is a new constant we will in general be assured that c is a random member of B with respect to C. If this is so, then a necessary and sufficient condition for the assertion that the probability that c is a C is that the rational corpus in question contain the statistical statement, the proportion of B's that are C's is p. Hence the extraordinary ease with which statisticians and philosophers, who are typically concerned with the probability of survival of 'a' male of 42, rather than the probability of survival of Tom Jones, have slipped into the error of identifying such assertions with statistical assertions. More formally, we may say that if w is a Normal, and Statistically terminating, rational corpus, T is a reference term of the language L, Y is an ω-sequence the union of whose range is known in w to be T, and c is a constant not occuring in L, and if w' is the union of w with the logical consequences of $\ulcorner c \in \bigcup \mathscr{R} Y \urcorner$, then the probability of $\ulcorner z \langle c \rangle \in b \urcorner$, relative to w', will be $[p, q]$ if and only if the strongest statement in w about $\ulcorner \bigcup \mathscr{R} Y \urcorner$, z, and b is $\ulcorner S \langle Y, z, b, p, q \rangle \urcorner$. We have as a theorem:

MT-10.24 $\quad$ *Normal*$_L$ $w \wedge w$ *is statistically closed* $\wedge \bigvee_z$ *z is Basis*$_L$ *for* $w \wedge$

$$\wedge P\text{-}Term_L(w) \wedge \sim c \in L \wedge L' = L \cup \{c\} \wedge w' =$$

$$= w \cup (L'\text{-}Cn(\ulcorner c \in \bigcup \mathscr{R} y \urcorner) \cap St_{L'}) \rightarrow \bigwedge_{p,q} (PROB_{L'}(\ulcorner z \langle c \rangle \in b \urcorner, w')$$

$$= [p, q] \rightarrow PS_{w,L}(\ulcorner S \langle y, z, b, p, q \rangle \urcorner, y, z, b)$$

Proof: Obvious.

If we suppose, as we plausibly may, that our epistemological acceptance rules are such that if $\ulcorner S \langle y, z, b, p, q \rangle \urcorner$ is acceptable in w, then it is acceptable in every $strand_L$ of w, the consequent of the theorem may be strengthened into a biconditional.

A puzzling difficulty cited in connection with both the Bayesian subjectivistic approach to probability and in connection with logical interpretations of probability, was the implausible precision of certain of our items of probabilistic knowledge. Recall, for example, the instance in which we assigned a prior probability of a ball being purple of 0.01 (according to subjective estimate, or according to some logical measure), and a conditional probability of 0.02 for the probability of a ball being purple, given that some other ball is purpole.

We found that it followed that we would assign a probability of 0.99 to the hypothesis that no more than 11 percent pf the balls would turn out to be purple in the arbitrarily long run. Does the present approach alleviate this puzzlement?

The answer is that it does, and in two ways. First of all, the *a priori* probability – the probability of getting a purple ball from the bag, given no information at all – may plausibly be expected to be the whole interval [0, 1], and this will therefore also be the probability of any long run frequency of purple balls. On the other hand, it is a logical truth that at least 99% of the samples of 10000 from a finite population of draws of balls from the bag will contain a proportion r of purple balls that differs by no more than 0.1 from the parameter p representing the proportion of purple balls in the whole population. In our state of pristine ignorance, the first ten thousand draws is a random member of the set of all samples of 10000, with respect to having this property. We may thus say that the probability is [0.99, 1.00] that the set of 10000 draws will 'reveal' the proportion of purple balls within a margin of error of 0.01. But there is nothing in appropriate about this, since we have supposed that we are ignorant of the value of p, and this probability statement says nothing either about the value of p, or about the value of r. It says something only about the reasonable epistemological attitude to take regarding a relation between them, in the absence of any useful empirical knowledge.

Suppose, now, that we do have some vague knowledge about p – say that it lies between 0.1 and 0.5. We now can make a computation like that performed earlier, and find that the probability is at least 0.99 that the frequency of purple balls in the long run will lie between 0.09 and 0.51. But there is nothing at all surprising about this: indeed, the probability is 1 that the frequency in the indefinitely long run will lie between these limits, since this is a deductive consequence of the assumed knowledge that the frequency in the indefinitely long run lies between 0.1 and 0.5!

To assert, then, that the probability that 'a' B is a C is p, is simply to assert that the strongest statistical statement in our body of knowledge concerning B's and C's incorporates the parameter p. This indeed is a factual statement, but it is a statement about your body of knowledge, not about B's and C's. In turn, this statement does *mention* a statement about B's and C's. Furthermore, the probability assertion itself, spelled out in full, mentions the body of knowledge explicitly, and thus is devoid of *any* factual content. '$PROB_L(s, w) = [p, q]$' if true at all, is true once and for all, regardless of the nature of the world, of the frequency with which B's and C's, regardless of whether w is, or was or will be, my body of knowledge or anybody's body of knowledge. To say that, relative to the body of knowledge w, the probability that a B is a C is p, is to say something that is either logically true or logically false, as they say, eternally.

This conventional suppression of the reference to a body of knowledge in a probability statement is one of the things that has made it awkward to deal with unknown, hypothetical, apparent as opposed to true, probabilities within the framework provided by a logical interpretation. Examples I and II provide an indication of how one might treat such probability assertions. Thus to say that the TRUE probability that a B is a C is p, would be to say that relative to a rational corpus w containing only true statements, and containing a precise statistical statement about B and C, together with the statement $c \in B$, where c is a constant new to the language, the probability that c will be a C is p, or, what is the same thing, by MT-10.23, that that rational corpus will contain the statement $\ulcorner S\langle B, z, \{\langle 1\rangle\}\, p, p\rangle\urcorner$, where z is the characteristic function of C. To say that Tom does not know the true probability of a *B* being a *C*, that for him it is *p* when in fact it is *q*, is just to say something like the following. Let *L* be a language in which Tom's rational corpus can be expressed. Let *w* be his rational corpus, *c* a constant new to *L*. To say that for Tom the probability of a *B* being a *C* is *p*, is to say that, relative to the rational corpus $(L\text{-}Cn(\ulcorner c \in b \urcorner) \cap St_L) \cup w$ the probability of $\ulcorner c \in C \urcorner$ is (p, p). Tom might, of course, in fact believe that this probability was *r*, or *s*, or even *q*; in so believing, however, he would be committing a *logical* blunder. To say that in fact the 'true' probability is *q*, is to say that a rational corpus w^* in the language *L*, containing only true statements, and containing a true, precise statement about the frequency of *C* in *B* is such that relative to the logical consequences of the union of w^* and $\ulcorner c \in B \urcorner$, the probability of $\ulcorner c \in C \urcorner$ is (q, q). To say all this is not to say that Tom is culpable; it is only to say that he is ignorant. To be culpable, he must think that something is probable relative to his rational corpus when it is not, or to have a rational corpus which is arrived at in violation of the principles of inductive logic. It is to make a mistake of inductive logic in forming a rational corpus, or a mistake of deductive logic in evaluating a probability relative to that rational corpus.

Similarly, to say that someone does not know what the probability of *S* is may mean any of three things. It may mean that he lacks the logical acumen to compute the probability of *S*, relative to his rational corpus. It may also mean that relative to his rational corpus, the probability of *S* is *p*, but relative to a rational corpus containing only true statistical statements, including a precise one of appropriate subject matter, (which his does not), the probability is q. Or finally (and most usually, when one professes oneself ignorant of a probability) it may mean that relative to one's own rational corpus the probability of *S* is the whole interval [0, 1], or almost that whole interval, so that one's knowledge of the probability is relatively useless as a guide in life. (We shall see why such probabilities are relatively useless in Chapter XII.)

Finally, to deal with hypothetical probabilities is to deal with hypothetical rational corpora. To say, 'Suppose that the probability that a *B* is a *C* is *p*', is to say, 'Let us consider the rational corpus whose basis is so and so, including, perhaps among other things, $\ulcorner S\langle B, z_C, \{\langle 1\rangle, p, p\rangle \urcorner$'. Note that even in this case we must have a specific language in mind; for to say that $\ulcorner S\langle B, z_C, \{\langle 1\rangle\}, p, p\rangle \urcorner$ is in a rational corpus is to say (among other things) that it is known in that rational corpus that the union of the range of B is equal to T, where T is a reference term of L.

The upshot of this chapter is this: a purely logical, and indeed mainly syntactical, definition of probability has been offered. The definition is in terms of rational corpora, construed as sets of statements, and the notion of randomness, also construed as a syntactical relation holding among four expressions and a set of statements. It turns out that under some mild restrictions, every statement of the language has a probability (although sometimes this will be the useless probability $[0, 1]$), relative to a strictly consistent rational corpus. It turns out that under some less mild restrictions some of the analogues of the usual quantitative relations of probability turn out to hold. In certain special cases, namely, when dealing with artificially simple rational corpora, and in dealing with the special kinds of events that fit the Popper-Hacking paradigm of probability, we get essentially the full classical calculus of probability, or the slight generalization of that calculus required to deal with intervals. Finally, we have seen how the definition offered makes sense of the confusing relation between the logical and empirical aspects of probability, and also provides a framework in which contexts allegedly inaccessible to a logical interpretation of probability can be analysed.

CHAPTER ELEVEN

CONDITIONAL PROBABILITY

1. There are two completely distinct ways in which conditional probability has been thought of. The most common way of thinking of conditional probability is to take as basic the locution, 'The probability that an *A* is a *C*, given that it is a *B*'. This probability will correspond in a natural way to the measure of *C*'s among objects belonging to the intersection of *A* and *B*. It is the way conditional probability is usually introduced in textbooks. It is what I shall call here *conditional measure*. In discussions of logical probability, and in some other places, where probability is taken to be a function whose domain is sentences or propositions or facts, conditional probability is introduced by means of a simple generalization of the forgoing definition: we define the conditional probability of *s*, given *t*, as the quotient of the probability of both *s* and *t*, and the probability of *t*. It will be claimed in due course that in those cases in which such a definition is plausible, it is because those are cases in which the relation to conditional measure is direct and straightforward. (There are also cases in which it seems just wrong to take the probability of *s* given *t* as the probability of the conjunction divided by the probability of *t*.)

Corresponding to this concept of conditional probability, that is, to the concept of conditional measure – there is a concept of independence that I shall call *stochastic* independence. It is this sort of independence we have in mind when we say that the tosses of a coin – even an oddly bent one – are independent. If it were not more obfuscatory than enlightening, we could call this concept of independence 'causal'. More formally, we could say, as most texts do, that being a *C* is independent of being a *B*, among *A*'s, when the conditional measure of *C*'s among *A*'s, given that they are also *B*'s, is just the same as the unconditional measure of *C*'s among *A*'s. In a somewhat more general way, one may define *s* and *t* as independent when the probability of *s*, conditional on *t*, is the same as the unconditional probability of *s*. Again, this definition *works* when there are underlying measures, satisfying the first relationship, on which these probabilities are directly based.

There is also another way of thinking of conditional probability, and a corresponding notion of independence. If probability is understood as an epistemic concept, the conditional probability that an *A* is a *C*, given that it is a *B*, will be interpreted as the probability that an *A* is a *C*, relative to a rational corpus that embodies the information that *that A* is also a *B*. The corresponding notion of independence, epistemological independence, is also to be explicated by reference to bodies of knowledge: we say that *s* is independent of *t*, relative to a body of knowledge, just when the probability of *s*, relative to that body of knowledge, is the same as the probability of *s*, relative to that body of knowledge augmented by the logical consequences of *t*. In this sense the tosses of a symmetrical coin about those whose behavior we suppose ourselves to have complete knowledge are independent (knowledge of a string of heads will not change the probability we assign to another head) while the tosses of a bent coin, about which we suppose ourselves to be ignorant, will not be independent. In the latter case, a string of heads may well cause us reasonably to change our assessment of the probability of heads on a succeeding toss.[1]

One striking difference between stochastic independence and epistemological independence is that the former is symmetrical (assuming that none of the relevant measures is 0), while the latter is not. It is perfectly possible for *t* to be epistemologically independent of *s*, while *s* is not epistemologically independent of *t*. To see this, we observe first that stochastic independence is symmetrical:

$$p(s)=p(s \mid t)=\frac{p(s \wedge t)}{p(t)} \leftrightarrow$$

$$p(s) \cdot p(t)=p(s \wedge t) \leftrightarrow$$

$$p(t)=\frac{p(s \wedge t)}{p(s)}=p(t \mid s), \text{ provided } p(t) \neq 0 \text{ and } p(s) \neq 0.$$

On the other hand, suppose that *s* is 'the fifth ball drawn from *U* is green', and *t* is 'the second toss of *c* lands heads'. Further, let us suppose that our rational corpus contains the statements: '10% of the balls drawn from *U* are green', '20% of the lead balls drawn from *U* are green', '50% of the tosses of *c* yield heads', and 'if *c* lands heads on the second toss, the fifth ball drawn from *U* is lead'.

Now it is perfectly clear that s is not independent of t: adding t to our rational corpus will change the epistemological probability of s from 0.1 and 0.2. On the other hand, it is perfectly plausible to suppose that t is independent of s: knowing nothing about the proportion of green balls that are lead, or any connection between the color of the fifth ball and the occurrence of heads or tails on the second toss of the coin c, it is hard to imagine what grounds there could be for attributing anything but the probability $\frac{1}{2}$ to the occurrence of heads on the second toss of c. Our information provides an upper limit for the proportion of lead balls drawn from the urn, but this doesn't lead us anywhere.

A small budget of examples will help to illustrate the difference between conditional measure and conditional probability.

EXAMPLE I: The simplest illustration of the difference between conditional measure and conditional probability is given by a long sequence of tosses of a coin with unknown bias. *Stochastically speaking*, the tosses are independent (we may suppose) with unknown but constant probability of yielding heads. The conditional probability of heads on the nth toss, given a certain sequence of results on the first $n-1$ tosses is just the unconditional probability of heads on the nth toss – i.e., the unknown probability of heads. *Epistemologically speaking*, the tosses are not at all independent: What could better serve to guide our expectations of heads on the nth toss than sequence of heads and tails on the first $n-1$ tosses? The conditional probability of heads on the nth toss, conditioned on the addition to our body of knowledge of the result that most of the first $n-1$ tosses have yielded heads will quite properly be different from – and larger than – the unconditional probability of heads on the nth toss. Exactly how this will work will depend, of course, on what else we have in our rational corpus.

EXAMPLE II: Let a die and a method of tossing it be absolutely fair and known to be so. Then the conditional measure of tosses yielding aces, given the restriction to tosses yielding an odd number, is $\frac{1}{3}$. Under most circumstances, this will also be the conditional probability. But suppose we know of a particular toss that it was performed yesterday by Peter, who rolled an odd number only once, that odd number being an ace; then the conditional probability that that particular toss yielded an ace, given the addition to our body of knowledge of the fact that it yielded an odd number, is one.

EXAMPLE III: Let there be a set of k urns, of which j contain 50% white balls, and the rest 10% white balls. Let an urn be selected by an appropriate randomizing procedure, and then let a sample of s balls be drawn (with replacement) of which r turn out to be white. We may readily use Bayes' theorem to compute the conditional measure of such experiments in which the urn is one containing 50% white balls. But we might yet be informed that the urn was in fact one of those containing 50% white balls (For example, those urns might have the legend '$\frac{1}{2}$' embossed on their sides. Or we might be informed that although the randomizing procedure was perfectly fair in general, on the day we are talking about it yielded m/n urns with the even mixture of white and black balls). Furthermore, the probability that the next ball to be drawn is white, conditional on the sample, will in the first case precisely match the conditional measure; in the other two cases, it will not – or to speak more precisely, it will match a quite different conditional measure.

EXAMPLE IV: Consider an urn containing six red and four black balls. Let a black ball be drawn. The conditional measure of the instances in which the second ball is black will be $\frac{1}{3}$. But if we don't know (or remember) what color the first ball is, the probability that the second ball is black is $\frac{2}{5}$.

EXAMPLE V: Let E be an experiment designed to have a level of significance of 0.05. In view of the nature of the subject matter, it turns out that E is a randomized experiment, consisting of a mixture of E_1 and E_2, where E_1 and E_2 are mixed in proportions 3:2, the level of significance of E_1 is 0.01, and the level of significance of E_2 is 0.11. The 'conditional probability', given E, of committing an error of the first kind is precisely the conditional measure, 0.05. But in general we will know how the randomizing device came out, and thus we will know, in general, which of the experiments, E_1 or E_2 we *actually* performed. The conditional *probability* of having committed an error of the first kind, after the experiment has been performed, is therefore either 0.01 or 0.11, according as the first or the second version of the experiment was performed, and thus in no case equal to the conditional measure 0.05.

What underlies all these examples is the tension – conflict, in fact – between an objectivistic or empirical notion of probability (what I call 'measure'), that depends upon the state of the world and not upon our epistemological position in it, and a logical or epistemological or even subjectivistic concept of probability which is designed to function as a guide in life. The resolution of the conflict in each example is straightforward enough from our point of view – it is a question of finding the appropriate reference class, for on this point of view every valid probability assertion has some appropriate statistical underpinning. But this reference class may perfectly well not be the one which would be appropriate for the unconditional probability, nor even necessarily any subset of that class.

Nevertheless, within the framework provided by an empirical interpretation of probability, no resolution is possible, for within this framework no principle concerning the selection of an appropriate reference class can be formulated. For that the concept of a body of knowledge is clearly central. From a traditional subjectivistic or logical point of view, the frequentistic conditional probabilities are simply wrong; thus de Finetti regards the tosses of a coin as exchangeable, but never, never, independent. This seems unsatisfactory, however, because the frequentistic assertions of our examples do seem to make sense, to be useful, and even, in some contexts, to be true. In a certain sense they seem to be true (as conditional measure statements) in *any* context, though irrelevant in some.

In the framework provided by the preceding chapters, we can do justice to both intuitions. Corresponding to the frequentist intuition, there is in the object language a definition of conditional measure and of stochastic independence that are perfectly straight-forward and conventional. In a number of familiar (generally abstract) contexts, this calculus is reflected directly in metalinguistic statements of conditional probability. Corresponding to the logical or subjectivistic intuitions, there is an epistemological concept of conditional probability. This concept corresponds directly to the stochastic concept only in special cases; but the special cases are so pervasive and important that it is not surprising that confusion has been generated. Indeed it is this very confusion that underlies the controversy concerning the universality of the application of Bayes' theorem.

2. Let us begin by considering a slightly more complicated example than we considered in the last section. Let a trial consist of the roll of a die, followed by the toss of a coin as many times as there were dots uppermost on the die. Let x be an increasing sequence of sets of trials. We let y be a (two dimensional) random quantity defined on the union of the range of x, having the value $\langle i, j \rangle$ when i dots showed up on the roll of the die and j heads resulted from the tossing of the coin. We suppose that in general the die and coin are both fair, so that we know the distribution of y. Now let us add to our knowledge one item consistent with the forgoing: namely, we suppose that we know that if, on the seventh roll, the die yielded a three, the coin with which the tosses were subsequently performed was biassed two to one in favor of heads. In general, of course, the conditional probability of getting one head, given that a three has been rolled, is 0.375; but the probability of getting one head, given a three on the seventh toss, will be $\frac{2}{9}$. Can we construe this probability as a conditional measure? There would be no difficulty if the biassed coin were used whenever a three was rolled. But all we know about is the 7th toss; the reference set would have to be construed as something like:

> $\{\langle u, v \rangle$:(u is 7th roll of die $\wedge$ v is throws of biassed coin $\wedge$ u results in 3) $\vee$ ((u is not 7th roll of die $\vee$ u does not result in 3) $\wedge$ v is throws of fair coin)$\}$

But this is precisely the disjunctive kind of set description that led us into trouble before; and furthermore it is difficult to see how to construct the appropriate measure. The general conditional measure will be the measure of $\{\langle 3, 1 \rangle\}$ in the entire set, which, as already mentioned is 0.375.

The only subset of this general set in which the conditional probability might be correct is the set consisting of the unit set of the seventh trial of the experiment. But this subset having only one member, can only yield probabilities (0, 0), (0, 1), and (1, 1) – none of which seem appropriate. The attempt to find a conditional measure that can be construed as a conditional probability thus seems to be frustrated in this case.

On the other hand, we have no trouble at all interpreting the conditional probability in question epistemologically: the probability that the coin will yield one head, given that the die has yielded a three on the seventh toss, is given just by the probability that that coin will land heads exactly once on the seventh trial, when we have added to our body of knowledge the statement that the die landed three on the seventh trial, i.e., when we have added to our body of knowledge what we assume to be a consequence of this, that the coin is biassed two to one in favor of heads. The probability that such a coin will yield exactly one head on three tosses is, just as we knew all along, 0.222; so this is the probability that we will get exactly one head on the seventh trial, *given* that the seventh trial yielded a three on the throw of the die.

Does this mean that the theory of conditional measures (developed in the last part of the appendix) is of relatively little importance for computing conditional probabilities? Hardly. For the kind of situation described above is clearly a very special – indeed, peculiar – one. Usually, it will be the case that the conditional measure will be the conditional probability. Even more often the limits set to the conditional measure will set limits within which the probability interval must fall. Only under extremely special circumstances will the conditional probability actually differ (in our strong sense) from the 'natural' conditional measure.

First of all, let us note that this 'natural' conditional measure often exists. Indeed, let w be any consistent and statistically closed set of statements, and s and t any statements such that they and their conjunction have probabilities provided by the same reference class and the same random quantity. Then there exists a potential reference sequence for t conditional on s which has parameters appropriately limited by the probabilities of $s \wedge t$ and s.

MT-11.1 $$\begin{aligned}&\mathit{Normal}_L\, w \wedge \bigvee_{z'} z' \textit{ is basis}_L \textit{ for } w \wedge \\ &\wedge \sim\ulcorner 0 \neq 0 \urcorner \in L\text{-}Cn(w) \wedge w \textit{ is statistically closed} \wedge \\ &\wedge s B_{w,L} \ulcorner z\langle x\rangle \in b_1 \urcorner \wedge RAN_L(x, y, z, b_1, w) \wedge \\ &\wedge PS_{w,L}(\ulcorner S\langle y, z, b_1, p_1, q_1\rangle \urcorner, y, z, b_1) \wedge t B_{w,L} \ulcorner z\langle x\rangle \in b_2 \urcorner \wedge \\ &\wedge RAN_L(x, y, z, b_2, w) \wedge PS_{w,L}(\ulcorner S\langle y, z, b_2, p_2, q_2\rangle \urcorner, y, z, b_2) \wedge \\ &\wedge \ulcorner s \wedge t \urcorner B_{w,L} \ulcorner z\langle x\rangle \in b_1 \cap b_2 \urcorner \wedge RAN_L(x, y, z, \ulcorner b_1 \cap b_2 \urcorner, w) \wedge \\ &\wedge PS_{w,L}(\ulcorner S\langle y, z, b_1 \cap b_2, p_3, q_3\rangle \urcorner, y, z, \ulcorner b_1 \cap b_2 \urcorner) \wedge \\ &\wedge \vdash_L \ulcorner 0 < p_1 \urcorner \rightarrow \bigvee_{y', r_1, r_2} (PS_{w,L}(\ulcorner S\langle y', z, b_2, r_1, r_2\rangle \urcorner, y', z, b^2) \wedge \\ &\wedge \vdash_L \ulcorner \left(\frac{p_3}{q_1} \leqslant r_1 \wedge r_2 \leqslant \frac{q_3}{p_1}\right) \urcorner\end{aligned}$$

Proof:

(1) Assume antecedent

(2) Let $y' = \ulcorner \imath f(f \text{ is } \omega\text{-sequence} \wedge \bigwedge_{i \in \omega} f\langle i \rangle = = y\langle i \rangle \cap \{u : z\langle u \rangle \in b_1\})\urcorner$

(3) $\vdash_L \ulcorner \bigcup \mathscr{R}y' = \bigcup \mathscr{R}y \cap \{u : z\langle u \rangle \in b_1\}\urcorner$ (2)

(4) $\bigvee_{z'} (z' \in RT_L \wedge \ulcorner z' = \bigcup \mathscr{R}y' \urcorner \in w)$ (1), (3), MD-7.40

(5) $SS_{w,L} \ulcorner S\langle y', z, b_2, p_3/q_1, q_3/p_1\rangle\urcorner$ (1), (4), MD-7.41

(6) $[\bigvee_{r_1, r_2}]\ PS_{w,L}(\ulcorner S\langle y', z, b_2, r_1, r_2\rangle\urcorner, y', z, b_2)$ (1), (5), MT-8.28, MD-8.7

(7) $\ulcorner S\langle y', z, b_2, p_3/q_1, q_3/p_1\rangle\urcorner \in w$ (1), closure

(8) $\vdash_L \ulcorner p_3/q_1 \leqslant r_1 \wedge r_2 \leqslant q_3/p_1 \urcorner$ (5), (7), MD-8.7 MD-7.42, MD-7.43

Two things should be noted about this theorem, aside from its extreme weakness. First, it should be observed that there need be no connection at all between $[p_2, q_2]$ and $[r_1, r_2]$; this should be obvious, in view of the fact that one represents an unconditional probability and the other a conditional probability. Second, the conditional measure represented by $[r_1, r_2]$ may be very much more precise than the limits represented by $[p_3/q_1, q_3/p_1]$. To see this, suppose that these probabilities are derived from the family of distribution functions

$$z = \{F_1, F_2\}, \quad \text{where} \quad \int_{b_1 \cap b_2} dF_1 = 0.2, \quad \int_{b_1 \cap b_2} dF_2 = 0.4,$$

$$\int_{b_1} dF_1 = 0.4, \quad \text{and} \quad \int_{b_1} dF_2 = 0.8; \text{ we then have } p_3/q_1 = 0.25, \quad q_3/p_1 = 1, \quad r_1 = r_2 = 0.5.$$

The form of conditional probability that interests us is the epistemological form. The analysis whose possibility is established by the preceding theorem is a bastard cross between epistemological and stochastic conditionality (but it is often appropriate in the sense that it yields appropriate values; we shall provide necessary and sufficient conditions for its appropriateness in due course). If we take probability epistemologically, we want to interpret the conditional probability of *s*, given *t* (relative to our body of knowledge w) as being simply the probability of s, relative to a body of knowledge just like ours, except that it contains t and its consequences; it is thus the probability of s, relative to the logical consequences of t union w. Now if this probability is *not* the same as the conditional measure of the preceding theorem, certain conditions must be met. If it is simply more precise than the conditional measure, there is not much to be said. But if it differs, if it is in actual conflict with the conditional measure, then in the expanded rational corpus, we must know that there is a subset of the original reference set agreeing with the conditional probability (and thus at variance with the conditional measure in the original reference set). We shall state the theorem first, and then show how the original example of this section illustrates it.

MT-11.2
$$\ulcorner z\langle x\rangle \in b_0 \urcorner B_{w,L}\, s \wedge \ulcorner z\langle x\rangle \in b_1 \urcorner B_{w,L} \ulcorner s\wedge t\urcorner \wedge$$
$$\wedge\, RAN_L(x, y, z, b_0, w) \wedge w^* = w \cup (St_L \cap L\text{-}Cn(\{s\}) \wedge$$
$$\wedge\, y^* = \ulcorner \iota f(f \text{ is } \omega\text{-sequence} \wedge \bigwedge_{i\in\omega} f\langle i\rangle = y\langle i\rangle \cap$$
$$\cap \{u : z\langle u\rangle \in b_0\})\urcorner \wedge PS_{w,L}(\ulcorner S\langle y^*, z, b_1, r_0, r_1\rangle\urcorner, y^*, z, b_1) \wedge$$
$$\wedge \sim \bigvee_{y'} y'\, D_L(x, y^*, z, b_1, w^*) \wedge p\text{-}prob_L(t, w^*) =$$
$$= [p, q] \rightarrow\ \vdash_L \ulcorner r_0 \leqslant p \wedge q \leqslant r_1 \urcorner$$

Proof:

(1) Assume antecedent.

(2) The consequent follows from 10.2 and 10.14.

In the preceding example, we supposed that x was an increasing sequence of sets of trials of an experiment consisting of a roll of die followed by a set of tosses of a coin, and that y was a two dimensional random quantity defined on the union of the range of x, having the value $\langle i, j\rangle$ when i dots show up on the roll of the die, and j heads result from tossing the coin i times. The general conditional probability of getting one head, given that a three has been rolled on the die is 0.375. But we have supposed ourselves also to know that if a three resulted on the seventh roll of the die, the coin used for the tosses was biassed two to one in favor of heads. Thus the appropriate conditional probability for heads occuring once in the seventh trial, given that the die resulted on three in that trial, is 0.222. We deal here with degenerate intervals, and it is clear that the two statements of conditional measure are in conflict. We may accept that in point of fact the appropriate conditional probability is 0.222. But according to our definition of randomness, we must be able to find a subset – a proper subset – of the set of throws of a coin following the rolling of a three, in which the measure of heads is just that in our chosen reference set. And sure enough, we can: namely the set of coin-throwings following a three rolled on the die, in which the two-to-one coin was used.

Observe that this is a perfectly good reference set, despite the fact that we know nothing about how often the two-to-one coin is used. (That fact is what prevents the joint distribution for a roll of a die, the selection (choice?) of a coin, and throws of a coin, from being an appropriate reference set.) There is nothing in *this* reference set involving the seventh toss explicitly – nor need there be. The knowledge we have supposed ourselves to possess about the seventh toss dictates the selection of a certain reference class, but it does not itself figure in the description of that class.

It is now fairly easy to see what the necessary and sufficient conditions are for a conditional measure to be a real live conditional probability: The conditional measure will be a conditional probability when relative to the body of knowledge constructed by including as part of that knowledge the conditioning statement, our original item x is a random member of the appropriate part of the original reference class. Thus if x is a random member of y with respect to having a z-value in b_1, relative to w; and if w' is the body of knowledge constructed by adding 'the z-value of x is in b_1' to w; and if, relative to this new rational corpus w', x is a random member of that part of y consisting of objects with z-values in b_1 with respect to having a z-value in b_2; *then* the conditional measure of b_2 in b_1 is just the conditional probability of 'the z-value of x is in b_2' given 'the z-value of x is in b_1'. The converse relation also holds in virtue of the fact that if the appropriate measures are known to correspond in two classes, then something is a random member of one in the appropriate sense if and only if it is a random member of the other in that sense.

Formally, we have:

MT-11.3 $RAN_L(x, y, z, b_1, w) \wedge \ulcorner z\langle x\rangle \in b_1 \urcorner\, B_{w,L}\, s \wedge w^* =$
$= (St_L \cap L\text{-}Cn(\{s\})) \cup w \wedge \sim\ulcorner 0 \neq 0 \urcorner \in L\text{-}Cn(w^*) \wedge$
$\wedge\, Normal_L\, w^* \wedge S\text{-}Term_L\, w^* \wedge \ulcorner z\langle x\rangle \in b_2 \urcorner\, B_{w^*,L}\, t \wedge$

$$\wedge\, y' = \ulcorner \imath f\,(f \textit{ is fun} \wedge \mathscr{D}f = \omega \wedge \bigwedge_{i \in \omega} f\langle i\rangle =$$
$$= y\langle i\rangle \cap \{u : z\langle u\rangle \in b_1\})\urcorner \wedge \ulcorner D\langle y, z, v\rangle\urcorner \in w \,\wedge$$
$$\wedge \bigwedge_{v' \in \mathit{Math\ Tms}_L} (\ulcorner D\langle y, z, v'\rangle\urcorner \in w^* \rightarrow \ulcorner v \subset v'\urcorner \in w) \,\wedge$$
$$\wedge \vdash_L \ulcorner \mathit{inf}\left\{\int_{b_1} df : f \in v\right\} > 0\urcorner \rightarrow (\textit{p-prob}_L(t, w^*) =$$
$$= \left[\mathit{inf}\left\{\int_{b_1 \cap b_2} df \Big/ \int_{b_1} df : f \in v\right\}, \mathit{sup}\left\{\int_{b_1 \cap b_2} df \Big/ \int_{b_1} df : f \in v\right\}\right] \leftrightarrow$$
$$\leftrightarrow \mathit{RAN}_L(x, y', z, b_2, w^*))$$

Proof:

(1) Assume antecedent

(2) Let $k = \left\{\int_{b_1 \cap b_2} df \Big/ \int_{b_1} df : f \in v\right\}$

(3) Sh~~o~~w $PS_{w^*, L}(\ulcorner S\langle y', z, b_2, \mathit{inf}\, k, \mathit{sup}\, k\rangle\urcorner, y', z, b_2)$

(4) $\ulcorner S\langle y', z, b_2, \mathit{inf}\, k, \mathit{sup}\, k\rangle\urcorner \in w^*$ (1)

(5) Sh~~o~~w $\bigwedge_{p, q} \ulcorner S\langle y', z, b_2, p, q\rangle\urcorner \in w^* \rightarrow$
$\rightarrow \vdash_L \ulcorner p \leqslant \mathit{inf}\, k \wedge \mathit{sup}\, k \leqslant q\urcorner$

(6) $\ulcorner S\langle y', z, b_2, p, q\rangle\urcorner \in w^*$ ass.

(7) $\sim\vdash_L \ulcorner p \leqslant \mathit{inf}\, k\urcorner$ ass.

(8) $\vdash_L \ulcorner \mathit{inf}\, k < p\urcorner$ (1), (7) (Normality)

(9) $\vdash_L \ulcorner 0 < p - \mathit{inf}\, k\urcorner$ (8)

(10) $\vdash_L \ulcorner D\langle y, z, v\rangle \wedge \mathit{inf}\left\{\int_{b_1} df : f \in v\right\} > 0 \wedge 0 < r \rightarrow$
$\rightarrow \bigvee_{f' \in v} \left(\int_{b_1 \cap b_2} df' \Big/ \int_{b_1} df' - \mathit{inf}\, k < r\right)\urcorner$

(11) $\ulcorner \bigvee_{f' \in v} \int_{b_1 \cap b_2} df' \Big/ \int_{b_1} df' < p\urcorner \in w^*$ (9), (10)

(12) $\ulcorner \bigvee_{f' \in v} D\langle y, z, v\text{-}\{f'\}\rangle\urcorner \in L\text{-}Cn(w)$ (6), (11)

(13) $\ulcorner \bigvee_{v'} (D\langle y, z, v'\rangle \wedge \sim v \subset v')\urcorner \in L\text{-}Cn(w)$ imp. by (1)

(14) $SS_{w^*, L} \ulcorner S\langle y', z, b_2, inf\ k, sup\ k\rangle\urcorner$ (1), MD-7.40, MT-7.9 MD-7.41

(15) $PS_{w^*, L}(\ulcorner S\langle y', z, b_2, inf\ k, sup\ k\rangle\urcorner, y', z, b_2)$ (4), (5), (14), (1), MD-8.7

(16) Shøw $RAN_L(x, y', z, b_2, w^*) \rightarrow p\text{-}prob_L(t, w^*) = [inf\ k, sup\ k]$

(17) $RAN_L(x, y', z, b_2, w^*)$ ass.

(18) $p\text{-}prob_L(t, w^*) = [inf\ k, sup\ k]$ (17), (3), (1), MT-10.1

(19) Shøw $p\text{-}prob_L(t, w^*) = [inf\ k, sup\ k] \rightarrow RAN_L(x, y', z, b_2, w^*)$

(20) $p\text{-}prob_L(t, w^*) = [inf\ k, sup\ k]$ ass.

(21) $[\bigvee_{x'', y'', z'', b''}] RAN_L(x'', y'', z'', b'', w^*) \wedge \ulcorner z''\langle x''\rangle \in b''\urcorner B_{w^*, L}\, t \wedge PS_{w^*, L}(\ulcorner S\langle y'', z'', b'', inf\ k, sup\ k\rangle\urcorner, y'', z'', b'')$ (20), MT-10.1

(22) $x\, M_{w^*, L} \ulcorner \bigcup \mathcal{R} y'\urcorner$ (1), MT-8.25

(23) $RAN_L(x, y', z, b_2, w^*)$ (21), (22), MT-8.11 MT-9.8, MD-7.42

3. One of the most important concepts with which we have to deal is that of *independence*. Earlier we distinguished between stochastic and epistemological independence. Stochastic independence we defined in the appendix: the random quantities x and y are independent according to a distribution function F, just in case, for every value of y, the conditional distribution of x given y is the same as the marginal distribution of x. Let s and t be increasing sequences of indices, the terms of which are disjoint and comprise the natural numbers less than n. We take s to be an m sequence, t to be an $n-m$ sequence, and the terms of the s-sequence to be indices corresponding to the m-dimensional random quantity x, and the terms of the t-sequence to correspond to the $n-m$ dimensional random quantity y. We combine x and y in order to get the n-dimensional random quantity z. Let F be the distribution of z, sF the

marginal distribution of x, tF the marginal distribution of y, sFy' the conditional distribution of x, given a specific set of y-values y', and tFx' the conditional distribution of y, given a specific set of x-values x'. We have defined the independence of x and y (D-60, appendix) by the stipulation that for every $x \in \mathbb{R}^m$ and every $y \in \mathbb{R}^{n-m}$, $Fz = sFy \cdot tFx$, where z is an n-sequence of reals put together from x and y in the appropriate way. It is obvious that this is a symmetrical relation.

When dealing directly with the set function P_F independence is often defined in the following way: membership in the set b_1 is independent of membership in the set b_2, just when the conditional measure of b_1 in b_2, $P_F(b_1/b_2) = P_F(b_1 \cap b_2)/P_F(b_2)$, is equal to the unconditional measure of b_1, $P_F(b_1)$ – provided that the measure of b_2 is not zero. This definition also yields the symmetry of independence, subject to the condition that the measures of b_1 and b_2 are not 0:

$$P_F(b_1) = P_F(b_1/b_2) = P_F(b_1 \cap b_2)/P_F(b_2) \leftrightarrow$$
$$\leftrightarrow P_F(b_1) \cdot P_F(b_2) = P_F(b_1 \cap b_2) \leftrightarrow$$
$$\leftrightarrow P_F(b_2) = P_F(b_1 \cap b_2)/P_F(b_1) = P_F(b_2/b_1)$$

Epistemological independence receives a definition quite analogous to that of stochastic independence; we say that the statement s is independent of the statement t, relative to the body of knowledge w, just in case the partial probability of s relative to w is the same as the partial probability of the statement s relative to the set of statements that are consequences of t or are already in w.

MD-11.1 $s\ Indep_{w,L}\ t \leftrightarrow p\text{-}prob_L(s, w) =$
$= p\text{-}prob_L(s, (w \sim \{x : \ulcorner \sim t \urcorner \in L\text{-}Cn(\{x\})\} \cup$
$\cup (L\text{-}Cn(\{t\}) \cap St_L)))$

It follows from this definition that if w is inconsistent, or fails to support probabilities for some other reason, everything is independent of everything. But why not?

A somewhat more powerful and general notion of independence is also useful: We say that the y_1 values of a_1 are independent of the y_2 values of a_2, just when, for every borel term b_1 and every borel term b_2, $\ulcorner y_1 \langle a_1 \rangle \in b_1 \urcorner$ is independent of $\ulcorner y_2 \langle a_2 \rangle \in b_2 \urcorner$

MD-11.2 y_1, a_1 *G-Indep*$_{w,L}$ $y_2, a_2 \leftrightarrow \bigwedge_{b_1, b_2} (\bigvee_{n_1, n_2} (\vdash_L \ulcorner b_1 \in \mathbb{B}^{n_1} \wedge$
$\wedge\, b_2 \in \mathbb{B}^{n_2} \urcorner) \rightarrow \ulcorner y_1 \langle a_1 \rangle \in b_1 \urcorner\, Indep_{w,L} \ulcorner y_2 \langle a_2 \rangle \in b_2 \urcorner)$

It is interesting to observe that while stochastic independence is symmetrical, epistemological independence is not. This conforms to the intuitions of a number of people who have written on inductive logic; in particular it is sometimes maintained that *s* may 'confirm' *t*, so that *t* is not independent of *s*, while *s* is quite independent of *t*. The following example exhibits this lack of symmetry.

Let a die be thrown, and then a coin tossed. If the die lands with a 1 up, we throw a coin biassed 2:1 in favor of heads; if it lands with a 2 up, we throw a coin biassed 2:1 against heads; and otherwise, we throw a fair coin. We suppose the die to be fair. Now let *s* be 'the coin lands heads on the 5th trial' and let *t* be 'the die yields 1 on the 5th trial'. Let *E* be an increasing sequence of sets of trials; and let E' be an increasing sequence of sets of trials, whose union is *properly* included within the union of the range of *E*, and to which the fifth trial belongs. We suppose (what is not outlandish) that we happen to know that $\frac{1}{6}$ of the trials in E' that yield heads also yield a 1 on the throw of the die.

Then, intuitively, we have,

$$p\text{-}prob_L(s) = \tfrac{1}{2} \neq \tfrac{2}{3} = p\text{-}prob_L(s, w \cup (St_L \cap L\text{-}Cn(\{t\})))$$
$$p\text{-}prob_L(t) = \tfrac{1}{6} = \tfrac{1}{6} = p\text{-}prob_L(t, w \cup (St_L \cap L\text{-}Cn(\{s\})))$$

These relations hold because the fifth trial is a random member of the set of trials with respect to yielding heads (and we know that half the trials, in general yield heads); when we add *t* to our body of knowledge, the fifth trial becomes a random member of the set of those trials resulting in a one thrown with the die, and of these trials we know that $\frac{2}{3}$ yield heads. On the other hand, the fifth trial is a random member of *E* with respect to yielding a 1 on the throw of the die, and we know that a sixth of these trials yield a 1; when we add *s* to our body of knowledge the fifth trial becomes a random member of the trials in E' in which the coin lands heads, and we know that $\frac{1}{6}$ of these produced a 1. To be sure, there is something just slightly unexpected about this, for in general we would expect that the proportion of 1's among trials in which the coin landed heads would be somewhat higher than $\frac{1}{6}$. But there is nothing peculiar or even anomolous about our knowing that the fifth trial happened to fall into a subclass in which $\frac{1}{6}$ of the trials yielding H also yielded 1. It becomes less peculiar still if we suppose that the biassed coins are only slightly biassed.

A more complicated but completely unpeculiar example can be provided by considering a case in which our knowledge is not assumed to be precise. We consider the same experiment as before, but we suppose that what we know of the die is that the frequency with which each side turns up lies between 0.15 and 0.18. The coin tossed when the one turns up we know to land heads between 0.6 and 0.7 of the time; that tossed when the two turns up we know to land heads between 0.3 and 0.4 of the time; and the 'normal' coin we suppose to exhibit a frequency of heads between 0.45 and 0.55. We also suppose ourselves to know that the tosses of the coin and the rolls of the die are stochastically independent, so that the frequency with which an experiment yields a 1 on the die and a head on the coin is (for example) between $0.6 \times 0.15 = 0.090$ and $0.7 \times 0.18 = 0.126$. On this basis we may suppose that our body of knowledge w contains the following statistical statements. (We let H, and $\bar{H}$ be the sets of trials on which H occurs and fails to occur, respectively, and the digits *1*, *2*, *3*, *4*, *5* and *6* represent the sets of trials embodying the corresponding outcome on the roll of the die. We represent the statistical statement that the proportion of trials of E that belong to the class C lies between p and q by '$S\langle E, C, p, q\rangle$'):

$$S\langle E, H \cap 1, 0.090, 0.126\rangle$$
$$S\langle E, H \cap 2, 0.045, 0.072\rangle$$
$$S\langle E, H \cap i, 0.0675, 0.099\rangle \quad (2<i<7)$$
$$S\langle E, i, 0.15, 0.18\rangle \quad (i<7)$$
$$S\langle E, H, 0.4455, 0.5545\rangle$$

Again, we consider a particular toss – say the fifth. This time, however, we suppose that we know nothing about the fifth toss that prevents it from being a random member of E with respect to any of the classes with which we are concerned. Then, in accordance with this discussion of paragraph 2, we may take the conditional measures as imposing constraints on the conditional probabilities. These conditional measures (derived from the foregoing statistical statements, together with the description of the experiment) include:

$$S\langle E \cap H, \mathit{3}, 0.1217, 0.222\rangle$$
$$S\langle E \cap \mathit{3}, H, 0.375, 0.660\rangle$$

For probabilities, we thus have (s is 'the fifth trial results in H'; t is 'the fifth trial results in $\mathit{3}$')

$$p\text{-}prob_L(s, w) = [0.4455, 0.5545]$$ (since the fifth trial is a random member of E with respect to H)

$$p\text{-}prob_L(t, w) = [0.15, 0.18]$$ (similarly, with respect to $\mathit{3}$)

Let us now compute the conditional probabilities:

$$p\text{-}prob_L(s, (w \cup (St_L \cap L\text{-}Cn(\{t\})))) = \\ = [0.45, 0.55] \neq [0.4455, 0.5545]$$

Although the conditional measure of H in $E \cap \mathit{3}$ can be computed to be in the interval [0.375, 0.660] (by considering the lower measure of $H \cap 3$, divided by the upper measure of 3, and the upper measure of $H \cap 3$ divided by the lower measure of 3), in point of fact we know more: we know that if a roll of the die lands 3, we shall be tossing a coin with the property that between 0.45 and 0.55 of the tosses yield heads. Since this is a stronger statistical statement than the statement that applies to the whole series of experiments, and since there is nothing to which the fifth toss belongs which differs in proportion of heads – we have assumed that all of the statistical information we have is stated above – it, rather than the general probability, represents the appropriate conditional probability. Since this probability differs from the corresponding unconditional probability, we can assert that s is not independent of t: knowing t makes a difference in the probability we assign.

$$p\text{-}prob_L(t, w \cup (St_L \cap L\text{-}Cn(\{s\}))) = [0.15, 0.18] = p\text{-}prob_L(t, w)$$

The only information we have concerning the measure of $\mathit{3}$ in $E \cap H$ is that given by $S\langle E \cap H, \mathit{3}, 0.1217, 0.2222\rangle$. But we also have information concerning the measure of $\mathit{3}$ in E as a whole, namely, $S\langle E, \mathit{3}, 0.15, 0.18\rangle$. Since nothing prevents the fifth toss from being a random member of E with respect to $\mathit{3}$ by difference, it is a random member of this set with respect to $\mathit{3}$, even though we also happen to know (in the conditioning body of knowledge) that it belongs to H. Thus the probability of t, given knowledge of s, is still [0.15, 0.18], and we may assert that t is independent of s. We have: t is independent of s, relative to w, but s is not independent of t, relative to w.

Let us consider another, slightly different example: The experiment is similar: we throw a die, about which our knowledge is the same as before. If it lands 1, we toss a coin exhibiting a general frequency of heads lying between 0.3 and 0.4. Otherwise, we toss a coin with a frequency of heads between 0.45 and 0.55. We have in w a description of the experiment. We let s'' be: the fifth trial results in heads, and t'' be: the fifth trial results in *2*. In w we have:

$$S\langle E'', i, 0.15, 0.18\rangle \qquad i<7$$
$$S\langle E'', H\cap 1, 0.045, 0.072\rangle$$
$$S\langle E'', H\cap i, 0.0675, 0.099\rangle \quad 1<i<7$$
$$S\langle E'', H, 0.4230, 0.5275\rangle$$

for probabilities, we have:

$$p\text{-}prob_L(s, w) = [0.4320, 0.5275]$$
$$p\text{-}prob_L(t, w) = [0.15, 0.18]$$

The conditional measure of *2* in $E''\cap H$ may be computed to lie between 0.1137 and 0.2553. If we suppose, as before, that the statistical knowledge cited is all we have, then, since this is weaker than our knowledge of the measure of *2* in E'' as a whole, the fifth toss will be a random member of E'' with respect to *2*, even relative to the body of knowledge that includes s, and thus we will have:

$$p\text{-}prob_L(t, w\cup(L\text{-}Cn(\{s\})\cap St_L)) = [0.15, 0.18]$$

and thus t is independent of s, relative to w.

But the conditional measure of H in $E''\cap 2$ is known to be [0.45, 0.55]. This actually differs from the overall measure of H in E''; but since $E''\cap 2$ is known to be properly included in E'', and since in $w\cup(L\text{-}Cn(\{s\})\cap St_L)$, the fifth toss is known to belong to $E''\cap 2$, it will be a random member of $E''\cap 2$, rather than of E'', and thus

$$p\text{-}prob_L(t, w\cup(St_L\cap L\text{-}Cn(\{s\}))) = [0.45, 0.55] \neq$$
$$\neq [0.4230, 0.5275]$$

4. We may now define conditional probability and independence generally. Conditional probability will be a function whose domain is the product of the set of statements of *L*, the set of statements of *L*, and the set of subsets of the set of statements of *L*. For given *s*, *t*, and *w*, its value will be the common value (if any) of the partial probability of *s*, relative to each strand of *w* supplemented by the logical consequences of *t*. Otherwise, of course, it will be the empty set.

MD-11.3 $C\text{-}PROB_L = \imath f(f \text{ is } fun \wedge \mathscr{D}f = \prod_{i<3} \langle St_L, St_L, \mathscr{P}St_L\rangle \wedge$
$\wedge \bigwedge_{x\in\mathscr{D}f} (\bigvee_{s,t,z} ((x = \langle s, t, z\rangle \wedge fx = \imath k(\bigwedge_{w} (w \text{ is } strand_L \text{ of }$
$z \cup (St_L \cap L\text{-}Cn(\{t\})) \rightarrow p\text{-}prob_L(s, w) = k)) \wedge z \text{ is } RC_L) \vee$
$\vee (\sim z \text{ is } RC_L \wedge fx = 0)))$

MT-11.4 $C\text{-}PROB_L(s, t, w) = [p, q] \leftrightarrow \bigwedge_{z} (z \text{ is } strand_L \text{ of }$
$w \cup (St_L \cap (L\text{-}Cn(\{t\}))) \rightarrow p\text{-}prob_L(s, z) = [p, q]) \wedge w \text{ is } RC_L$

Similarly, we may now define independence and general independence quite generally.

MD-11.4 $s\ INDEP_{w,L}\ t \leftrightarrow PROB_L(s, w) = C\text{-}PROB_L(s, t, w)$

MT-11.5 $s\ INDEP_{w,L}\ t \wedge PROB_L(s, w) \neq 0 \rightarrow \bigwedge_{z,z'} (z \text{ is } strand_L \text{ of }$
$w \wedge z' \text{ is } strand_L \text{ of } w \cup (St_L \cap L\text{-}Cn(\{t\})) \rightarrow$
$\rightarrow p\text{-}prob_L(s, z) = p\text{-}prob_L(s, z'))$

MT-11.6 $s\ INDEP_{w,L}\ t \wedge PROB_L(s, w) \neq 0 \rightarrow$
$\rightarrow \bigwedge_{z} (z \text{ is } strand_L \text{ of } w \rightarrow s\ Indep_{z,L}\ t)$

MT-11.7 $\bigwedge_{z,z'} (z \text{ is } strand_L \text{ of } w \wedge z' \text{ is } strand_L \text{ of } w \cup (St_L \cap L\text{-}Cn(\{t\})) \rightarrow$
$\rightarrow p\text{-}prob_L(s, z) = p\text{-}prob_L(s, z')) \rightarrow s\ INDEP_{w,L}\ t$

Observe that neither MT-11.6 nor MT-11.7 can be strengthened into a biconditional.

MD-11.5 $y, a\ G\text{-}INDEP_{w,L}\ y', a' \leftrightarrow \bigwedge_{b_1,b_2} (\bigvee_{n_1,n_2} (\vdash_L \ulcorner b_1 \in \mathbb{B}^{n_1} \wedge b_2 \in \mathbb{B}^{n_2}\urcorner) \rightarrow$
$\rightarrow \ulcorner y\langle a\rangle \in b_1\urcorner\ INDEP_{w,L} \ulcorner y'\langle a'\rangle \in b_2\urcorner)$

MT-11.8 $\bigwedge_{z} (z \textit{ is strand}_L \textit{ of } w \cup (St_L \cap L\text{-}Cn(\{r\})) \rightarrow s\ B_{z,L}\ t) \rightarrow$
$\rightarrow C\text{-}PROB_L(s, r, w) = C\text{-}PROB_L(t, r, w)$

MT-11.9 $\bigwedge_{z} (z \textit{ is strand}_L \textit{ of } w \cup (St_L \cap L\text{-}Cn(\{r\})) \rightarrow s\ B_{z,L}\ t) \rightarrow$
$\rightarrow (s\ INDEP_{w,L}\ r \leftrightarrow t\ INDEP_{w,L}\ r)$

We are particularly concerned with conditional probabilities that match corresponding conditional measures. We can construct a general condition for this to happen on the basis of MT-11.3: namely, a condition concerning randomness. Let *s* and *t* be two statements of *L*, *w* a rational corpus. We know that if the probability of *s* relative to *w* exists, it exists (and has the same value) as the partial probability of *s* relative to every strand of *w*. We know furthermore that under these circumstances we can often find a term *x*, a pair of functors *y* and *z*, and a pair of borel terms b_1 and b_2 such that in a given strand, *x* is a random member of *y* with respect to having a *z*-value in b_1, *s* is known to be equivalent to $\ulcorner z\langle x\rangle \in b_1 \urcorner$, *x* is a random member of *y* with respect to having a *z*-value in b_2, *t* is known to be equivalent to $\ulcorner z\langle x\rangle \in b_2 \urcorner$. MT-11.3 informs us that a necessary and sufficient condition for the partial probability to be the conditional measure of the items having *z*-values in b_2, in the subset of the reference set in which things have *z*-values belonging to b_1, is that *x* be a random member of this subset, with respect to having a *z*-value in b_2, relative to the logical consequences of *s*, union *w*. If this happens in the case of every strand of *w*, then the conditional probability of *s*, given *t*, relative to *w*, will be the conditional measure in that basic reference set. For this to happen, *x* must be a random member of the *same* reference set, relative to every strand. Of sourse if *w* is identical with its only strand, this theorem becomes merely a special case of MT-11.3, since then *p-prob* = PROB.

MT-11.10 $\bigwedge_{w, w'} (w \textit{ is strand}_L \textit{ of } w^* \wedge w' \textit{ is strand}_L \textit{ of}$
$w^* \cup (St_L \cap L\text{-}Cn(\{t\})) \rightarrow RAN_L(x, y, z, b_1, w) \wedge$
$\wedge \ulcorner z\langle x\rangle \in b_1 \urcorner\ B_{w,L}\ t \wedge \sim \ulcorner 0 \neq 0 \urcorner \in L\text{-}Cn(w') \wedge Normal_L\ w' \wedge$
$\wedge S\text{-}Term_L\ w' \wedge \ulcorner z\langle x\rangle \in b_2 \urcorner\ B_{w',L}\ s \wedge y' = \ulcorner \imath f (f \textit{ is fun} \wedge$
$\wedge \mathscr{D}f = \omega \wedge \bigwedge_{i \in \omega} f\langle i\rangle = y\langle i\rangle \cap \{u : z\langle u\rangle \in b_1\}) \urcorner \wedge$
$\wedge \ulcorner D\langle y, z, v\rangle \urcorner \in w \wedge \bigwedge_{v' \in Math\ Tms_L} (\ulcorner D\langle y, z, v'\rangle \urcorner \in w' \rightarrow$

$$\rightarrow \ulcorner v \subset v' \urcorner \in w') \wedge \vdash_L \ulcorner inf \left\{ \int_{b_1} df : f \in v \right\} > 0 \urcorner) \rightarrow$$

$$\rightarrow (C\text{-}PROB_L(s, t, w^*) = \left[inf \left\{ \int_{b_1 \cap b_2} df / \int_{b_1} df : f \in v \right\}, \right.$$

$$\left. sup \left\{ \int_{b_1 \cap b_2} df / \int_{b_1} df : f \in v \right\} \right] \leftrightarrow \bigwedge_w (w \text{ is } strand_L \text{ of}$$

$$w^* \cup (L\text{-}Cn(\{t\}) \cap St_L) \rightarrow RAN_L(x, y', z, b_2, w))$$

The following theorem shows that conditional probability, as defined in MD-11.3, is a probability: the conditional probability of t given s, relative to w is just the probability of t relative to the union of w and the statement consequences of s.

MT-11.11 $RC_L\, w \rightarrow C\text{-}PROB_L(t, s, w) = PROB_L(t, w \cup (L\text{-}Cn(\{s\}) \cap St_L))$

Proof:

(1) $RC_L\, w$ ass.

(2) $\bigvee_{p,q} C\text{-}PROB_L(t, s, w) = [p, q] \vee C\text{-}PROB_L(t, s, w) = 0$ MD-11.3

(3) Case I $[\bigvee_{p,q}]\ C\text{-}PROB_L(t, s, w) = [p, q]$ ass.

(4) $\bigwedge_z (z \text{ is } strand_L \text{ of } w \cup (L\text{-}Cn(\{s\}) \cap St_L) \rightarrow$
$\rightarrow p\text{-}prob_L(s, z) = [p, q])$ MT-11.4, (3)

(5) $PROB_L(t, w \cup (L\text{-}Cn(\{s\}) \cap St_L)) = [p, q]$ (1), (4), MT-10.11

(6) Case II $C\text{-}PROB_L(t, s, w) = 0$ ass.

(7) $\sim \bigvee_{p,q} (\bigwedge_z (z \text{ is } strand_L \text{ of } w \cup (L\text{-}Cn(\{s\}) \cap St_L) \rightarrow$
$\rightarrow p\text{-}prob_L(t, z) = [p, q]))$ (1), MD-11.3

(8) $\imath k (\bigwedge_z (z \text{ is } strand_L \text{ of } w \cup (L\text{-}Cn(\{s\}) \cap St_L) \rightarrow$
$\rightarrow p\text{-}prob_L(t, z) = k)) = 0$

(9) $PROB_L(t, w \cup (L\text{-}Cn(\{s\}) \cap St_L)) = 0$ (8), MD-10.4

It should be observed that the definition given of conditional probability is slightly broader than the conventional definition, since it does not depend on conditional measures. Thus suppose that we know that heights and weights are distributed in a bivariate normal manner $N(m_w, s_w, m_h, s_h, \varrho)$ in a population P. Suppose, as we may hypothetically, that we *know* that Tom's weight is exactly 200 lbs., and suppose that relative to what we know Tom is a random member of the 200 lbs. items in this population with respect to having a height over six feet. The conditional measure is *not* defined, since the measure of the items in P weighing exactly 200 lbs. is zero. But the conditional probability is defined perfectly well: It is equal in value to the integral of the conditional frequency function (with $w=200$) from 6′ to infinity.

Suppose, to be concrete, that in P height h and weight w are distributed according to the joint density function

$$\frac{1}{2\pi\sigma_n\sigma_w\sqrt{1-\varrho^2}}\times$$

$$\times \exp\left[-\frac{(x-\mu_h)^2\,\sigma_w^2-2\varrho\sigma_h\sigma_w(x-\mu_h)\,(y-\mu_w)+(y-\mu_w)^2\sigma_h^2}{2\sigma_h^2\sigma_w^2(1-\varrho^2)}\right].$$

Let $\sigma_h=\frac{1}{2}$, $\sigma_w=30$, $\varrho=0.8$, $\mu_h=5.9$, $\mu_w=160$. Then the conditional density for h, given $w=200$, is

$$\frac{1}{2\pi\sigma_h\sqrt{1-\varrho^2}}\exp\left[-\frac{((\mu_h+(200-\mu_w)\,\varrho\sigma_h/\sigma_w)-x)^2}{2\sigma_h^2(1-\varrho^2)}\right].$$

A little computation reveals that the distribution of heights among 200 pounders in P is normal (6.43, 0.09). Thus the probability that Tom is over 6 feet tall is 0.96 and the probability that he is under six feet tall, (given that he weighs exactly 200 lbs.) is 0.04.

We may deal with intervals, too: suppose we know that he weighs between 195 and 210 pounds. Then the distribution appropriate for assessing the probability that Tom is over six feet is one of the set of distributions $\{N(m, 0.09): 6.37 \leqslant m \leqslant 6.66\}$, and the probability is approximately [0.890, 0.986].

Finally, let us suppose that we have weighed Tom on a scale whose error is normally distributed with mean 0 and variance 1. Then if we have read '200' on the scale, the density appropriate to assessing Tom's height is arrived at by taking the scale error to be stochastically independent of Tom's height.

5. Some examples may help to illustrate the intimate connection between the foregoing analysis of conditional probability and more conventional analyses. The simplest situation is that corresponding to the classical frequentist treatment (although, as already observed, our analysis is slightly broader even in this strictly limited case.)

EXAMPLE I: Let F denote the distribution function given by the following table:

$$
\begin{array}{llll}
F(x, y)=0 & \text{if} \quad x<1 & \text{or} & y<1 \\
F(x, y)=\frac{1}{8} & \text{if} \quad x\in[1, 2) & \text{and} & y\in[1, 2) \\
F(x, y)=\frac{1}{4} & \text{if} \quad x\in[1, 2) & \text{and} & y\in[2, \infty) \\
F(x, y)=\frac{1}{8} & \text{if} \quad x\in[2, 3) & \text{and} & y\in[1, 2) \\
F(x, y)=\frac{3}{4} & \text{if} \quad x\in[2, 3) & \text{and} & y\in[2, \infty) \\
F(x, y)=\frac{1}{4} & \text{if} \quad x\in[3, \infty) & \text{and} & y\in[1, 2) \\
F(x, y)=1 & \text{if} \quad x\in[3, \infty) & \text{and} & y\in[2, \infty)
\end{array}
$$

Let x denote the random quantity which has the value 1_r if its argument is red, the value 2_r if its argument is green, and the value 3_r if its argument is blue; let y denote the random quantity which has the value 1_r if its argument is lead, and the value 2_r if its argument is gold.

Let s be a function whose domain is omega, and whose value for every $i\in\omega$, is simply the set of balls in a certain urn. We suppose that F, s, x, and y are all expressions in L, as well as z, where z denotes that function whose value for any argument a is the ordered pair $\langle x\langle a\rangle, y\langle a\rangle\rangle$; we suppose that b denotes the set of balls in the urn, and that $b\in RT_L$.

Let c be a constant foreign to L, and let $L\cup\{c\}$ be L'.

Let $w=St_{L'}\cap L'\text{-}Cn(\{\ulcorner z \textit{ is fun} \wedge \mathscr{R}z\subset\mathbb{R}^2 \wedge b=\bigcup \mathscr{R}s \wedge c\in b \wedge$
$\wedge D\langle s, z, \{F\}\rangle\urcorner\})$. Then:

(1) $w \textit{ is } RC_{L'}$

(2) $\bigwedge_{w'} (w' \textit{ is strand}_{L'} \textit{ of } w\rightarrow w=w')$

(3) $PS_{w, L'}(\ulcorner S\langle s, z, (\frac{1}{2}]\times\mathbb{R}, \frac{1}{4}, \frac{1}{4}\rangle\urcorner, s, z, \ulcorner(\frac{1}{2}, 1]\times\mathbb{R}\urcorner)$

(4) $RAN_{L'}(c, s, z, \ulcorner(\frac{1}{2}, 1]\times\mathbb{R}\urcorner, w)$

(5) $\bigwedge_{b} (\vdash_{L'}\ulcorner b\in\mathbb{B}^{2}\urcorner\rightarrow PS_{w, L'}(\ulcorner S\langle s, z, b, \int_b dF, \int_b dF\rangle\urcorner, s, z, b))$

$$(6)\ \bigwedge_{b} (\vdash_{L'} \ulcorner b \in \mathbb{B}^2 \urcorner \rightarrow RAN_{L'}(c, s, z, b, w))$$

From (1), (3), and (4), we have

$$(7)\ p\text{-}prob_{L'}(\ulcorner z\langle c\rangle \in (\tfrac{1}{2}, 1] \times \mathbb{R} \urcorner, w) = [\tfrac{1}{4}, \tfrac{1}{4}]$$

And, taking account of (2), this yields:

$$(8)\ PROB_{L'}(\ulcorner z\langle c\rangle \in (\tfrac{1}{2}, 1] \times \mathbb{R} \urcorner, w) = [\tfrac{1}{4}, \tfrac{1}{4}]$$

In general, we have, from (1), (2), (5), and (6):

$$(9)\ \bigwedge_{b} (\vdash_{L'} \ulcorner b \in \mathbb{B}^2 \urcorner \rightarrow PROB_{L'}(\ulcorner z\langle c\rangle \in b \urcorner, w) = \left[\int_b dF, \int_b dF\right]$$

All this is to be expected. Interpreting the locution 'a ball from the urn' as c, it is merely to say that relative to this hypothetical rational corpus, the probability that the quantity z of c will fall in the borel set b, is just the integral of dF over b.

Let us now look at the conditional probabilities of this example. We note that the conditions stipulated in the antecedent of MT-11.3 are satisfied here; in view of the fact that c is a constant new to the language L, we have

$$\begin{aligned}(10)\ \bigwedge_{b_1, b_2} & (\vdash_{L'} \ulcorner b_1, b_2 \in \mathbb{B}^2 \urcorner \rightarrow RAN_{L'}(c, \ulcorner \imath f\,(f \textit{ is fun} \wedge \\ & \wedge \mathscr{D}f = \omega \wedge \bigwedge_{i \in \omega} f\langle i\rangle = s\langle i\rangle \cap \{u : z\langle u\rangle \in b_1\})\urcorner, \\ & z, b_2, L\text{-}Cn(w \cup \{z\langle c\rangle \in b_1\}) \cap St_{L'}))\end{aligned}$$

In particular, where b_1 is '$(\tfrac{1}{2}, 1] \times \mathbb{R}$' and b_2 is '$\mathbb{R} \times (1, 2]$', we have

$$(11)\ C\text{-}PROB_{L'}(\ulcorner z\langle c\rangle \in b_2 \urcorner, \ulcorner z\langle c\rangle \in b_1 \urcorner, w) = [\tfrac{1}{2}, \tfrac{1}{2}]$$

Interpreted in ordinary language: given that a ball is red, the probability that it is gold is a half. In general, we have, for any $b_1, b_2 \in \mathbb{B}^2$,

$$\begin{aligned}(12)\ & PROB_{L'}(\ulcorner z\langle c\rangle \in b_1 \urcorner, w) = \\ & = [q, q] \wedge C\text{-}PROB_{L'}(\ulcorner z\langle c\rangle \in b_2 \urcorner, \ulcorner z\langle c\rangle \in b_1 \urcorner, w) = \\ & = [p, p] \rightarrow PROB_{L'}(\ulcorner z\langle c\rangle \in b_1 \cap b_2 \urcorner, w) = [p \cdot q, p \cdot q]\end{aligned}$$

EXAMPLE II: Let G denote the distribution function given by the following table:

$$G(x, y)=0 \quad \text{if} \quad x<0 \quad \text{or} \quad y<0$$
$$G(x, y)=2xy-y^2 \quad \text{for} \quad 0\leqslant x\leqslant 1, 0\leqslant y\leqslant x$$
$$G(x, y)=x \quad \text{for} \quad 0\leqslant x\leqslant 1, x<y$$
$$G(x, y)=1 \quad \text{for} \quad 1<x, 1<y$$

Let s be a sequence of sets of rectangles; let x be the random quantity which has as its value the length of a rectangle, and y the random quantity whose value is the width of a rectangle, where all of the rectangles in the union of the range of s are inside the unit square. We suppose that G, s, x, y, and z are all expressions in L, where z denotes that function whose value for any argument a is the ordered pair $\langle x\langle a\rangle, y\langle a\rangle\rangle$; and we suppose that $r\in RT_L$.

Let c be a constant foreign to L, and let $L\cup\{c\}=L'$.

Let $w=St_{L'}\cap L'\text{-}Cn(\{\ulcorner z \textit{ is fun} \wedge \mathscr{R}z\subset\mathbb{R}^2\wedge r=\bigcup\mathscr{R}s\wedge c\in r\wedge D\langle s, z, \{G\}\rangle\urcorner\})$.

Then:

(1) w *is* $RC_{L'}$

(2) $\bigwedge_{w'}(w' \textit{ is strand}_{L'} \textit{ of } w\rightarrow w'=w)$

(3) $\bigwedge_{b}\left(\vdash_L\ulcorner b\in\mathbb{B}^2\urcorner\rightarrow PS_{w,L'}\left(\ulcorner S\left\langle s, z, b, \int_b dG, \int_b dG\right\rangle\urcorner, s, z, b\right)\right)$

(4) $\bigwedge_{b}(\vdash_L\ulcorner b\in\mathbb{B}^2\urcorner\rightarrow RAN_{L'}(c, s, z, b, w))$

Hence,

(5) $\bigwedge_{b}\left(\vdash_L\ulcorner b\in\mathbb{B}^2\urcorner\rightarrow PROB_{L'}(\ulcorner z\langle c\rangle\in b\urcorner, w)=\left[\int_b dG, \int_b dG\right]\right)$

Since c is merely a new constant, we also have

(6) $\bigwedge_{b_1, b_2}(\vdash_L\ulcorner b_1, b_2\in\mathbb{B}^2\urcorner\rightarrow RAN_{L'}(c, \ulcorner\imath f(f \textit{ is fun}\wedge\mathscr{D}f=$
$=\omega\wedge\bigwedge_{i\in\omega} f\langle i\rangle=s\langle i\rangle\cap\{u{:}z\langle u\rangle\in b_1\})\urcorner, z, b_2, w\cup$
$\cup(St_{L'}\cap L'\text{-}Cn(\{\ulcorner z\langle c\rangle\in b_1\urcorner\}))))$

Thus we have in general, for any b_1 and b_2 in $\mathbb{B}^2$:

$$\begin{aligned}(7)\ & PROB_{L'}(\ulcorner z\langle c\rangle \in b_1\urcorner, w) = \\ & = [p, p] \wedge C\text{-}PROB_{L'}(\ulcorner z\langle c\rangle \in b_2\urcorner, \ulcorner z\langle c\rangle \in b_1\urcorner, w) = \\ & = [q, q] \rightarrow PROB_{L'}(\ulcorner z\langle c\rangle \in b_1 \cap b_2\urcorner, w) = [p\cdot q, p\cdot q]\end{aligned}$$

Numerical example:

$$PROB_{L'}(\ulcorner z\langle c\rangle \in (0.3, 0.4] \times \mathbb{R}\urcorner, w) = [0.035, 0.035]$$
$$C\text{-}PROB_{L'}(\ulcorner z\langle c\rangle \in \mathbb{R} \times (0.3, 0.4]\urcorner, z\langle c\rangle \in (0.3,/4] \times \mathbb{R}, w) = \\ = [0.143, 0.143]$$
$$PROB_{L'}(\ulcorner z\langle c\rangle \in (0.3, 0.4] \times (0.3, 0.4]\urcorner, w) = [0.005, 0.005]$$

The principle embodied in these examples can be stated as general theorem.

MT-11.12 $\sim c \in L \wedge L' = L \cup \{c\} \wedge Normal_L\, w \wedge S\text{-}Term_L\, w \wedge \wedge \ulcorner D\langle y, z, \{f\}\rangle\urcorner \in w \wedge \sim \ulcorner 0 \neq 0\urcorner \in w \wedge \bigvee_n (num\, n \wedge \wedge \vdash_L \ulcorner b_1, b_2 \in \mathbb{B}^n\urcorner) \wedge \vdash_L \ulcorner 0 < \int_{b_1} df\urcorner \wedge w' = = [L'\text{-}Cn(w) \cup L'\text{-}Cn(\{\ulcorner c \in \bigcup \mathscr{R}y\urcorner\})] \cap St_{L'} \rightarrow \rightarrow C\text{-}PROB_{L'}(\ulcorner z\langle c\rangle \in b_2\urcorner, \ulcorner z\langle c\rangle \in b_1\urcorner, w) = = \left[\int_{b_2 \cap b_1} df \Big/ \int_{b_1} df, \int_{b_2 \cap b_1} df \Big/ \int_{b_1} df\right]$

As in the discussion of probability in Chapter X, we see that relative to a normal, consistent set of statements, containing a distribution statement about y and z, the conditional probability of $\ulcorner z\langle c\rangle \in b_1\urcorner$, given $\ulcorner z\langle c\rangle \in b_2\urcorner$, has a given value if and only if there is a conditional measure statement in that set of statements corresponding to that value.

In addition to abstract or general discussions in which we make essential use of the locution involving the indefinite article 'a' (*a* trial of the experiment, *a* point in the plane, *a* forty year old male), there are other contexts in which the requirements of randomness are met more or less automatically. Typical examples occur in experiments involving well adjusted gambling apparatus, and in certain kinds of physical experiments.

EXAMPLE III: Suppose that we have before us a well-tested die and a well-tested coin, and a standardized die throwing and coin tossing apparatus. Let y denote the increasing sequence of trials of the following experiment performed with this apparatus: the die is rolled, and then the coin is tossed a number of times corresponding to the number of spots showing on the die. Let z denote that function (whose range is in $\mathbb{R}^2$) whose value for a member of y is the ordered pair consisting of the number of spots on the die and the number of heads on the tosses of the coin. Let a denote the very next trial of the experiment. Let w be our very own body of knowledge.

Consider an arbitrary strand w' of w. If f denotes the orthodox distribution function for this experiment, it is hard to envisage ordinary circumstances under which $\ulcorner D\langle y, z, \{f\}\rangle\urcorner$ will not appear in w'. For it *not* to appear there, we would have to have good reasons for accepting something in w which was inconsistent with $\ulcorner D\langle y, z, \{f\}\rangle\urcorner$, and under the circumstances described, it is hard to imagine what that might be. Similarly, it is hard to imagine that there is anything in w that contradicts $\ulcorner a \in \bigcup \mathscr{R}y\urcorner$. Thus *if*, relative to w', a is a random member of y with respect to having a z-value in b, the conventional measure will be the appropriate probability. Similarly, if (for any b_1 and b_2), relative to $w \cup (L\text{-}Cn(\{\ulcorner z\langle a\rangle \in b_1\urcorner\}) \cap St_L)$, a is a random member of $\ulcorner \imath f\,(f \text{ is fun} \wedge \wedge \mathscr{D}f = \omega \wedge \bigwedge_{i\in\omega} f\langle i\rangle = y\langle i\rangle \cap \{u : z\langle u\rangle \in b_1\})\urcorner$, with respect to having a z-value in b_2, the conventional conditional measure will be the appropriate conditional probability. Since we are talking about the *next* toss, and since we have carefully described the situation in such a way that we have no special information about the 'next' toss that distinguishes it from the rest of the trials of the experiment, both stipulations concerning randomness are satisfied, and we have:

$$p\text{-}prob_L(\ulcorner z\langle a\rangle \in \{4_r\} \times \mathbb{R}\urcorner, w') = [\tfrac{1}{6}, \tfrac{1}{6}]$$
$$p\text{-}prob_L(\ulcorner z\langle a\rangle \in \mathbb{R} \times \{3_r\}\urcorner, w' \cup (L\text{-}Cn(\{\ulcorner z\langle a\rangle \in \{4_r\} \times \mathbb{R}\urcorner\}) \cap \cap St_L)) = [\tfrac{1}{4}, \tfrac{1}{4}]$$

Since this is true for an arbitrary strand w', it is true for all strands of w, and we also have:

$$PROB_L(\ulcorner z\langle a\rangle \in \{4_r\} \times \mathbb{R}\urcorner, w) = [\tfrac{1}{6}, \tfrac{1}{6}]$$
$$C\text{-}PROB_L(\ulcorner z\langle a\rangle \in \mathbb{R} \times \{3_r\}\urcorner, \ulcorner z\langle a\rangle \in \{4_r\} \times \mathbb{R}\urcorner, w) = [\tfrac{1}{4}, \tfrac{1}{4}]$$

EXAMPLE IV: Let us suppose we are involved in a physical experiment on certain small but not microscopic particles, and that we know (from some fancy physical theory) that the joint distribution of their velocities and masses is (very close to!) bivariate normal with velocity normal (50, 5), mass normal (100, 4), and correlation coefficient of -0.7. We suppose that one of the particles carries a radioactive marker so that it can be identified; call it '*a*'. In the circumstances described, *a* will be a random member of the set of particles (the sequence of finite sets consisting of all the particles, to be precise) relative to having a *z*-value in any 2-dimensional borel set (*z* being the two dimensional random quantity given by velocity and mass), relative to every strand of our rational corpus. Thus

$$PROB_L(\ulcorner\text{velocity of } a \in (50, 60]\urcorner, w) = [0.47725, 0.47725]$$

$$C\text{-}PROB_L\ulcorner\text{mass of } a \in (98, 99.5]\urcorner, \ulcorner\text{velocity of } a = 55\urcorner, w) = \\ = [0.19, 0.19]$$

Observe that the latter conditional probability could not be identified with a conditional measure on a strictly frequentistic view, since the measure of the set of particles with velocity exactly equal to 0.55 is zero.

One warning should be sounded now: In such circumstances as this – as well as in the circumstances of the previous example – it is something of an idealization to claim that we 'know' a distribution statement of the form $\ulcorner D\langle y, z, \{f\}\rangle\urcorner$. What we may plausibly claim to *know* (at least actually, as distinct from hypothetically) is that the actual distribution with which we are concerned is *very close* to *f*. We know, of course, that there are no negative masses and no negative velocities, so that we know perfectly well that the bivariate normal is not strictly correct. We therefore know better than to pay attention to the probability of the mass of *a* conditional on its having a velocity of -5. But we get into implausibilities long before we get into the range of negative velocities. To say that *f* is very close to the true distribution may be to say, for example, that for every pair of real numbers *x* and *y*, the difference between the long run proportion of objects having velocities less than *x* and masses less than *y* and $f\langle x, y\rangle$, is less than ε. But the conditional probabilities concerning a value of one variable given a value of the other may be considerably more in error, and indeed may be whole magnitudes off when we deal with extreme values. Thus suppose that *f* is never in error by more than 0.001. The value of $\int_{b_1} df$ for an interval b_1 may be 0.002. Then the frequency with which objects have z-values falling in the interval $b_1 \cap b_2$ will lie between $\int_{b_1 \cap b_2} df - 0.002$ and $\int_{b_1 \cap b_2} df + 0.002$; and the frequency with which objects have z values falling in the interval b_1 will lie between $\int_{b_1} df - 0.002$ and $\int_{b_1} df + 0.002$. If the value of $\int_{b_1} df$ is less than 0.002, however, the quotient may have any value at all between 0 and 1; and if $\int df$ is close to 0.002, the value of the quotient is correspondingly uncertain. This is one reason, among others, that it seems important to bear in mind that our statistical knowledge is generally approximate rather than precise; and particularly important to bear it in mind when dealing with conditional probabilities.

NOTE

[1] For a similar distinction between 'probability given an event' and 'conditional probability', see J. S. Williams, 'The Role of Probability in Fiducial Inference', *Sankhyā Series A* **28** (1966) 271–296.

CHAPTER TWELVE

INTERPRETATIONS OF PROBABILITY

The interpretation of probability offered in the preceding chapters is unorthodox in a number of respects. It is a logical interpretation, but it does not satisfy the axioms of the conventional probability calculus (if only because that calculus deals with real numbers, rather than with intervals of them). It involves limiting frequencies, yet probability statements are taken to be logically true, and thus to be devoid of empirical content. Probability is taken to be legislative for rational belief, and thus as a guide in life, but the usual constraints embodied in the requirements of coherence are not satisfied. Since one can hardly claim that those who have spent all or a large part of their lives contemplating the complexities of the use and meaning of probability have been totally blind, it behooves anyone who wants to claim that they are all wrong to show either how they have been plausibly led astray, or how the interpretations they have offered may be construed as special cases of a more general and pervasive interpretation.

1. The oldest and most classical interpretation of probability, takes probability to be the ratio of favorable to the total number of possible cases. Thus the probability of a one on a throw of a die is $\frac{1}{6}$, because there are six possible cases, of which one is favorable to the appearance of a 'one'. It has long been recognized that this does not by itself give us a viable interpretation of probability. There are various ways of counting possible cases (for example, why not regard 'one' and 'not-one' as the two possible cases for rolling a die?) How does one treat a loaded die, of which one wishes to say that the probability of a 'one' is $\frac{1}{10}$? What are the ten possible cases? It would be impossible to speak of a loaded die for which the probability of a 'one' was $1/\sqrt{3}$, for that probability cannot be represented as a ratio at all. This becomes an even knottier problem when one wishes to speak of the *continuous* probability distribution of a random variable. What on earth can the possible cases be in this situation? An answer is suggested by what is taken to be another problem for the clas-

sical interpretation of probability: how do you treat the probability of drawing a black ball from an urn of balls? The possibilities here must (?) be interpreted as balls in the urn, not as possible colors. A similar interpretation of 'possibility' must be provided for probabilities based on actuarial statistics. (It is easy to see how the logical conception of probability developed from the former interpretation of 'possibility', and frequentist conceptions from the latter.)

There are circumstances under which this classical interpretation of probability can be directly (and even usefully) applied: when we have in our body of knowledge a theory according to which in the long run over time each of a certain finite set of possibilities will be realized approximately equally often. Thus in roulette our knowledge of physics tells us that only the tiniest difference in angular momentum imparted to the balanced wheel initially will make all the difference in which number comes up; our knowledge of human physiology and perception tells us that these differences are undetectable and uncontrollable by a wide margin, so that we may expect each number to come up equally often in the long run. Thus the probability that any specified trial in the future will have any particular property *of the right sort* (will yield a '7', will yield a prime number, will yield an odd number, etc.) will be just the ratio of the cases exhibiting that property to the total number of cases. It also makes sense to speak of the probability that a trial that we know yielded a '7' yielded a '7' – we say 'It's true that you won, but the probability was still only $\frac{1}{37}$'. What this means is that, relative to what we knew *before* the event, that trial was a random one with respect to yielding a '7', and, by the preceding analysis, there was just one chance out of 37 that it would yield a '7'.

It is clear that far from requiring an equal distribution of ignorance regarding the possibilities, this mode of analysis requires – not an equal distribution of knowledge – but knowledge of an equal distribution of frequencies among the possibilities.

Can we be quite sure that something amounting to the bare counting of possibilities won't arise under the interpretation of probability that has been offered? For example, one could consider the increasing sequence of sets of rolls with a die as the increasing sequence of sets of numbers yielded by the die. (Or, the increasing sequence of values of the random quantity representing the number of spots on the die.) The union of the range of this sequence is just the finite set $\{1, 2, 3, 4, 5, 6\}$. This is a mathematical object, and '$\{1, 2, 3, 4, 5, 6\}$' is a mathematical term. But scrutiny of MD-7.39 and MD-7.40 will reveal that although mathematical terms may enter intersectively into reference terms (as in 'the set of all n-fold subsets of S') they do not by themselves qualify as reference terms. Thus we cannot have as a reference term, a term which denotes simply a set of 'possible outcomes', unless we can construct such a term from the primitive terms we have available. And there seems to be no reason to use a language in which this is possible.

Although the principle of indifference cannot be applied in the present system to properties, it can be, and is, applied to the individuals in a reference class. But this amounts to no more than counting each individual in a reference class once. This form of the principle of indifference is of course also found in frequency interpretations of probability, to which we now turn.

2. The main insight of the frequentist interpretation of probability is that we cannot claim that the probability that an A is a B is p, unless we have grounds for claiming that the frequency with which A's are B's is p. As we saw in Chapter Ten (MT-10.24), if we interpret 'an A' as 'something about which we know no more than that it is an A', and if A is an acceptable reference term, it follows under ordinary circumstances that the probability that it is a B is p, if and only if, we know that the proportion of A's that are B's is p. As we also saw (MT-10.1), if a is a random member of A with respect to belonging to B, the probability that it will belong to B will be p if and only if we know that the frequency of B's among A's is p.

Now the two main locutions in which 'probability' figures are the abstract one, in which we talk about the probability that *an* A is a B; and the practical one in which we talk about the probability that some specific object – the next toss, the weather tomorrow, Mr. Jones – will have a certain property. The former locution, as we have construed it, leads by MT-10.24 to the same results as the frequency interpretation. In the latter locution it is generally (but not always) the case that the individual is a random member of a certain class (which class is made obvious by the context) and MT-10.1 leads to the same result as the frequency interpretation. Thus in the vast majority of occasions in which abstract reference is made to probability, there is agreement between the interpretation offered here and the frequency interpretation.

There are other contexts, those, for example, in which we talk about the *unknown* probability that an A is a B, or in which we speak hypothetically: 'Suppose that the probability that an A is a B were p'; 'Suppose that the probability that Mr. Jones will die within the year is q'. These contexts too can be accommodated in the present framework by the device, already mentioned, of considering *hypothetical* bodies of knowledge. The unknown probability that an A is a B, is the probability, relative to a body of knowledge that contains a true and precise statistical statement about the frequency of B's among A's, that an object about which this hypothetical rational corpus knows no more than that it is an A will also be a B. Similarly, when we 'suppose' that the probability that Mr. Jones will die within the year is q, we are supposing that the frequency with which death occurs in a certain class (made clear by the context) is known in a certain hypothetical rational corpus to be q.

There are differences, of course. For one thing, the frequency interpretation provides (in one way or another) for the applicability of the standard probability calculus, which the interpretation offered here does not. But MT-10.18 shows that there are circumstances under which the orthodox probability calculus applies to our probabilities. This is often so in the case of ideal or hypothetical rational corpora.

Under a strict frequency interpretation of probability, none of our probabilities are literally acceptable, because we have assigned probabilities only to statements, to particular cases, and the frequency interpretation, construed strictly, will only allow probabilities to be assigned to classes, or to statement forms containing free variables. (Thus we may speak of the probability of B's among A's, where A and B are classes, or we may speak of the probability of Px given Qx where this will be the measure of $\{x : Px \wedge Qx\}$ in $\{x : Qx\}$.) But this difference in applicability between our theory and the frequency theory seems to be a matter of frequency theorists being pushed, by accidental characteristics of their own theories, into positions which are quite clearly unacceptable. What is good and plausible in the frequency theory can easily enough be accommodated in the present theory; and we are not stuck with those consequences of frequency interpretations of probability that frequentists themselves accept only reluctantly.

3. What is characteristic of the logical (logical range) interpretations of probability is the determined effort to save those probabilities that frequentists have reluctantly ruled illegitimate. These logical probabilities are essentially the same as the classical number-of-favorable-cases probabilities. The manner of counting cases is more sophisticated, and the 'cases' may even receive unequal weights (on some good logical ground), but the basic idea is that cases are counted for the whole language (rather than for the small part of a language representing a particular kind of experiment, such as rolling a die). There is no reflection of any such probability in our system. There is no way, except in the case of patently pathological languages, in which a set of 'possible cases' can be construed as a reference class, according to our definitions MD-7.39 and MD-7.40.

Nevertheless the intent and use of probability construed as it has been construed here is precisely that of logical probability. Probability statements are logical truths: Given a rational corpus (for classical logical probability, read 'evidence statement'), and given a statement, there is exactly one logically determined value of the probability of that statement, relative to that rational corpus. This probability, on the view presented here, is an interval, rather than a point; if we relax the traditional stipulation that a rational corpus be logically closed and consistent, the probability may sometimes not exist – namely, when it is not the same for every strand of the rational corpus. But it is still intended to be legislative for rational belief, and it is still intended to be logically true. Of course the statements *mentioned* in our probability statements will not generally be logically true; they are 'accepted' (if we are talking about someone's real rational corpus), or 'acceptable', or 'true' (if we are talking about an ideal rational corpus) or 'supposed', (if we are talking about a hypothetical rational corpus), and so on.

To say that probability is legislative for rational belief (something that has been said since Keynes) is to say something that requires explanation. The classical explanation (since Ramsey) is that a person's degree of belief in a statement is represented behavioristically by the least odds at which he is willing to bet on the truth of that statement. In Chapter Five I have already alluded to certain obscurities in this explanation: the criterion should be the odds at which the person is willing to *make book* on the statement (rather than the least odds at which he would bet in its favor); and in real life there might well be a range of odds at any of which the person would be willing to make book on the statement. There seems to be no reason why this whole range of odds shouldn't be regarded as characterizing the person's belief in the statement. It is unnatural, perhaps, to call an interval a 'degree', but the notion of 'degree' of belief is somewhat metaphorical to begin with. In terms of this expanded notion of 'degree of belief', the probabilities we have defined are to be legislative for degree of belief in the following sense:

> If a person's rational corpus is, so far as he knows, represented by the set of statements S, and if the probability of the statement T, relative to S (in the sense *PROB*), exists, and is the interval $[p, q]$, then if the person is rational, he will *not* make book on T at odds of r:1, unless $p \leqslant r/(r+1) \leqslant q$.

The phrase 'so far as he knows' is required, since, as I have construed rational corpora, they contain not only an infinity of statements, but theorems that have never been proved. Yet since, so far as I know, the axioms for set theory that I have adopted are consistent, I think that all statements that follow from these axioms are part of my rational corpus. They are part of what I have committed myself to, and I know that I have committed myself to them, even though I do not know which statements those are.

The reason for the negative form of the consequent, is that it does not seem at all unreasonable for someone to refuse to make book on a statement. What does seem positively irrational, is for a person to be willing to make book, and to be willing to do so at odds that fall outside his legislated probability interval.

Subject to these refinements, our probabilities perform the same function that logical probabilities are designed to perform. They are designed to serve not only as a guide to bookmaking and to betting, but as a guide to life in general – that is, a guide to rational expectation. As we shall see, statistical generalizations themselves, on this view, can properly be assigned probabilities. This was a goal of logical interpretations of probability, though a goal often abandoned.

5. The subjectivistic interpretation of probability is closely related to the logical interpretation. All that is given up is the attempt to derive probabilities from a sophisticated counting of possible cases. Probabilities

are rationalized (coherent or strictly coherent) betting ratios. They are the odds that a person could reasonably post for betting on a whole set of statements. They must be coherent (or strictly coherent) since if they were not a canny bettor could put the bookmaker in a position where he is bound to lose (or where he cannot win, and might possibly lose).

The proponents of the subjectivistic interpretation of probability make much of the fact that if an individual employs betting odds that correspond to his degrees of belief, and if these degrees of belief satisfy the axioms of the probability calculus, the individual will never have a book made against him. As has already been remarked, there are a number of serious lacunae in this argument; in particular, the probability calculus provides the individual with nothing, beyond algorithms for computation, that straight-forward deductive logic does not provide him with.

Nevertheless, if we suppose that the individual can be put into the position of *having* to make book on a finite number of statements – i.e., of posting odds on each of the statements, and then accepting any bets regarding those statements at those odds that anyone cares to offer him – then it is of interest to ask whether or not the individual can post odds that satisfy both the constraints of coherence (or strict coherence) and the constraints imposed by the above mentioned principle of rationality. Now there is no doubt that a rational agent, going into the bookmaking business seriously, should post odds on a finite set of statements that are coherent, and if he is going to post odds on an infinite set of statements, they should be strictly coherent. This is true, and has nothing at all to do with probability, subjective or otherwise. Taken in conjunction with the principle stated above relating betting odds to rational degress of belief, there is a connection: the odds posted by the bookmaker should conform to his probability intervals.

Now it seems possible that for certain sets of statements there should be no way of posting coherent odds that conform to these probability intervals. This would mean no more than that we can conceive of certain situations in which rationality is unattainable. I would not regard it as a serious defect in the interpretation that has been offered for probability, and the principle relating it to 'belief', if such situations should be conceivable. They are certainly not common in ordinary life. But it is an extremely interesting fact that under typical constraints, there is a set of betting odds, for any finite number of statements, that conform to the probability intervals. More precisely, if w is a normal and probabilistically terminating rational corpus, then if S is any finite set of statements, there exists a function from S to R, which is a coherent probability function, and whose value, for any statement in S, lies in the probability interval of S, relative to w.

MT-12.1 $Normal_L\ w \wedge w$ *is statistically closed* $\wedge \bigvee_z z$ *is basis*$_L$ *for* $w \wedge$

$$\wedge\, CPC(w) \wedge w \text{ is } P\text{-}Term \wedge s \in St^n_L \wedge \bigwedge_{i<n} p\text{-}prob_L(s_i, w) =$$

$$= [p_i, q_i] \rightarrow \bigvee_P (P \text{ is } fun \wedge \mathscr{R}s \subset \mathscr{D}P \wedge \mathscr{R}P \subset [0, 1] \wedge$$

$$\wedge \bigwedge_{x, y \in \mathscr{D}P} (P(\ulcorner x \vee y \urcorner) = P(x) + P(y) - P(\ulcorner x \wedge y \urcorner) \wedge$$

$$\wedge\, P(x) + P(\ulcorner \sim x \urcorner) = 1) \wedge \bigwedge_{i<n} p_i \leqslant P(s_i) \leqslant q_i)$$

Proof:

(1) Assume antecedent

(2) $[\bigvee_{x, y, z, b}]\, b$ *is n-sequence* $\wedge \bigwedge_{i<n} (RAN_L(x, y, z, b_i, w) \wedge$
$\wedge \ulcorner z\langle x \rangle \in b_i \urcorner\, B_{w, L}\, s_i)$

By the construction used in MT-10.16, and MT-9.8

(3) $\bigwedge_{i<n} PS_{w, L}(\ulcorner S\langle y, z, b_i, p_i, q_i \rangle \urcorner, y, z, b_i)$ By construction

(4) $\sim \{s\colon \bigvee_{i<n} s = \ulcorner S\langle y, z, b_i, p_i, q_i \urcorner \rangle\} \vdash_L \ulcorner 0 \neq 0 \urcorner$

(5) $\{s\colon \bigvee_{i<n} s = \ulcorner S\langle y, z, b_i, p_i, q_i \rangle \urcorner\} \vdash_L \ulcorner \bigvee_r (r$ *is n-sequence* $\wedge$
$\wedge \bigwedge_{i<n} S\langle y, z, b_i, r_i, r_i \rangle) \urcorner$ (1), (4)

(6) $[\bigvee_{d, m}] \vdash_L \mathscr{D}d = \mathbb{R}^m \urcorner \wedge \sim \{\ulcorner D\langle y, z, \{d\} \rangle \urcorner\} \cup$
$\cup \{s\colon \bigvee_{i<n} \ulcorner s = S\langle y, z, b_i, p_i, q_i \rangle \urcorner\} \vdash_L \ulcorner 0 \neq 0 \urcorner$

Let r be a sequence of the sort mentioned in (5); select at most 2^n points c to represent every non-empty intersection of the b_i and their complements. Assign weights $w(c)$ satisfying (1) $\sum$ w$(c) = 1$; (2) $\sum_{c \in b_i} = r_i$. This is possible since the statements $\ulcorner S\langle y, z, b_i, r_i, r_i\rangle\urcorner$ are consistent. Let d be the corresponding distribution function.

(7) Let $P = \imath f\Big(f \text{ is fun} \wedge \mathscr{D}f = \{t : \bigvee_b (\vdash_L b \in \mathbb{B}^m \wedge$
$\wedge \ulcorner z\langle x\rangle \in b\urcorner\, B_{w,L}\, t\} \wedge \bigwedge_{t\in\mathscr{D}f} f\langle t\rangle = \int_b \mathrm{d}d\Big)$

(8) $\mathscr{R}s \subset \mathscr{D}P \wedge \mathscr{R}P \subset [0, 1]$ (2), (7)

(9) sh~~o~~w $\bigwedge_{x,y\in\mathscr{D}P} (P(\ulcorner x \vee y\urcorner) = P(x) + P(y) - P(\ulcorner x \wedge y\urcorner) \wedge$
$\wedge P(x) + P(\ulcorner \sim x\urcorner) = 1)$

(10) $x', y' \in \mathscr{D}P$ ass.

(11) $[\bigvee_{b,b'}] \vdash_L b,\ b' \in \mathbb{B}^m \wedge x' B_{w,L} \ulcorner z\langle x\rangle \in b\urcorner \wedge y' B_{w,L} \ulcorner z\langle x\rangle \in b'\urcorner$
$\wedge \ulcorner x' \vee y'\urcorner\, B_{w,L} \ulcorner z\langle x\rangle \in b \cup b'\urcorner \wedge$
$\wedge \ulcorner x' \wedge y'\urcorner\, B_{w,L} \ulcorner z\langle x\rangle \in b \cap b'\urcorner \wedge$
$\wedge \ulcorner \sim x'\urcorner\, B_{w,L} \ulcorner z\langle x\rangle \in \mathbb{R}^m - b\urcorner$
(7), (10), MT-8.11, 8.12

(12) $P(x' \vee y') = \int_b \mathrm{d}d + \int_{b'} \mathrm{d}d - \int_{b\cap b'} \mathrm{d}d =$ (7), (12)
$= P(x') + P(y') - P(\ulcorner x' \wedge y'\urcorner)$

(13) $P(x') + P(\ulcorner \sim x'\urcorner) = \int_b \mathrm{d}d + \int_{\mathbb{R}^m - b} \mathrm{d}d = 1$ (7), (12)

(14) sh~~o~~w $\bigwedge_{i<n} p_i \leqslant P(s_i) \leqslant q_i$

(15) $i < n$ ass.

(16) $\sim p_i \leqslant P(s_i) \leqslant q_i$ ass.

(17) $\sim p_i \leqslant \int_{b_i} \mathrm{d}d \leqslant q_i$ (7), (2), (16)

impossible, by (6).

On the other hand, it turns out that the constraint of coherence on conditional bets is not satisfied. We might have anticipated this, since coherence demands the symmetry of independence, and independence is not symmetrical in the present system. Consider again the experiment described in Chapter Eleven: an experiment consists of tossing a coin and drawing a ball from an urn. Our body of knowledge consists of the deductive consequences of: half the tosses of the coin land heads; 10% of the balls in the urn are green; 20% of the lead balls in the urn are green; if the second toss lands heads, the seventh ball is lead. Let s be the statement that the second toss lands heads, and let t be the statement that the seventh ball is green. Knowing nothing about the proportion of balls that are lead, the probability of s is surely a half. The probability of t might be determined by the measure of green balls in the set of all trials of the experiment, less the set of trials in which heads occurs on the coin toss and a non-lead ball is drawn from the urn. We may compute that the frequency of green balls in this set lies *between* 0.1 and 0.2, and this determines the appropriate probability for t. The conditional probability for t given s is clearly 0.20. The conditional probability of s given t, on the other hand, is still $\frac{1}{2}$. But for coherence, the conditional probability of s given t should be $P(s) \cdot P(t \mid s)/P(t) = 0.5 \times 0.2/0.1 = 1$; alternatively, $P(t) = P(s) \cdot P(t \mid s) + P(\bar{s})\ P(t\ \bar{s}) = 0.15$; so that $P(s \mid t) = \frac{2}{3}$.

It might be argued that the probability of s and the probability of s given t should be modified in the light of our presumed knowledge of 's only if the seventh ball is lead'. But how should it be modified? We have not supposed ourselves to know anything about the frequency of lead balls – except that not more than half can be lead. Let e be the sequence of outcomes of the trials, so that $e_i(1)$ is the outcome of the draw from the urn on the ith trial, and $e_j(0)$ is the result of the toss of the coin on the jth trial: s is then $e_2(0) \in \text{Heads}$; t is $e_7(1) \in \text{Green}$. Let r be $e_7(1) \in \text{Lead}$. The biggest plausible reference class is:

$$\{\langle u, v\rangle : \bigvee_{x,y} (\langle u, x\rangle, \langle y, v\rangle \in \mathscr{R}e)\} = \mathrm{A}$$

But knowing that if the second toss landed heads, the seventh draw yeilded a lead ball, there is a part of this reference set that we know that $\langle e_2(0), e_7(1)\rangle$ does not belong to, namely:

$$\{\langle u, v\rangle : u \in \text{Heads} \wedge \sim v \in \text{Lead}\} = B$$

The proportion of Heads in $A - B$ varies from 0 to 1, as the proportion of lead balls varies from 0 to $\frac{1}{2}$; thus our knowledge that the item in question belongs to $A-B$ doesn't yield any knowledge conflicting with our knowledge that half the A's are Heads. The probability of s is a half, as advertised.

The proportion of green balls among the elements of $A-B$ may be anything from 0.1 to 0.2. Since there is no conflict between this knowledge, and our knowledge of the proportion of green balls in A as a whole, and the latter is more precise, we take the probability of t to be 0.1. (If we suppose that we know that the proportion of lead balls lies between 0.1 and 0.4, then the proportion of green balls in $A-B$ is known to lie between 0.111 and 0.167, which is then the appropriate probability.)

If we add s to our body of knowledge, then the appropriate reference set for t is:

$$\{\langle u, v\rangle : \langle u, v\rangle \in A \wedge v \in \text{Lead}\}$$

and the probability of t turns out to be 0.20, as it should.

If we add t to our body of knowledge we are left with an interval that includes $\frac{1}{2}$. (If the proportion of lead balls is between $\frac{1}{10}$ and $\frac{4}{10}$, the proportion of heads in $A-B$ is between $\frac{1}{9}$ and $\frac{2}{3}$, and since $A-B$ is a cross product, this will also be the proportion in $(A-B) \cap \{\langle u, v\rangle : v \in G\}$), and therefore the stronger knowledge about A will yield that the probability of s, even given that we know t, is a half.

There is no coherent belief function which satisfies these constraints for *conditional* probabilities. This means that if one is going to make book on s and t and accept conditional bets, one is forced to abandon rationality as it is determined by the underlying frequencies. This is not to say that one will have to compute one's bets in accordance with frequencies that one knows do *not* hold: it means that one must compute one's odds in accordance with frequencies which *might*, for all one knows, hold, but which one has no adequate evidence for. To be *forced* to make book is to be forced to act as if one knew more about frequencies than in fact one knows or has grounds for knowing. This explains the extra sense of discomfort one feels on being forced not only to make book, but to make book on conditional bets.

The interpretation of probability presented above is thus compatible with the subjectivistic interpretation, in the sense that given a finite set of statements it is possible to assign coherent subjective betting quotients to them in such a way that these numbers fall in our interval probabilities, but not in the sense that our conditional probabilities conform to the subjective probabilities obtained from conditionalization.

6. Probability, as Bishop Butler said long ago, is the very guide of life. Interpretations of probability have associated with them certain approaches to decision making. Subjectivistic and logical interpretations of probability are wedded to a strict Bayesian principle of decision making: act so as to maximize your expected utility. Frequentistic interpretations of probability usually employ also another principle – a principle of minimizing maximum loss, or maximum regret, or the like. It was pointed out in Chapter Six that even the Subjectivist and Logical interpretations can be in conflict, despite the fact that they employ the same general principle; and of course both can be in conflict with the various non-Bayesian principles. Although there is a form inter-reducibility, none of the classical approaches provides for a genuine reconciliation. I claim that the present approach does provide for a genuine (and natural) reconciliation.

Formulations of subjective probability are often simultaneously formulations of subjective utility. Like probabilities, utilities turn out to be real-valued. Some economists find this unrealistic, and would prefer to see utilities construed as interval valued. On the interpretation of probability that we are expliring here, *expected* utilities will generally turn out to be intervals, regardless of the nature of basic utilities. I shall assume that the utilities of various 'worlds' or 'acts' or 'consequences' or 'states of affairs' are simply given (by economists or by philosophers or as a result of a dialectic interaction of both groups), and, for the sake of simplicity, that they are real valued. Since expected utilities, in terms of which we must make our decisions, are interval valued in any case, only the most minor changes would be required if we chose to regard the basic utility function to be interval valued.

To be more precise, we consider utility to be a real valued random quantity. In the sense in which utility is of interest to decision theory, we shall regard it as a function of an act and the value of some other random quantities. For example, the utility of the act of buying a certain truckload of corn for a certain price depends on the total weight of the truckload, the quality of the corn, the degree to which the kernels have fallen off the ears, etc. It also depends, of course, on my 'circumstances', or rather, what

I reasonably believe my circumstance to be. But for present purposes we assume that the relevant utilities given for the problem at hand.

Let us look at the simplest possible example. Suppose that we have a choice between two acts, a_1 and a_2, and the value to be associated with each act only depends on the truth or falsity of a certain statement *s*. Let X_s be the characteristic function of *s* – i.e., that random quantity having the value 1 if *s* is true and the value 0 if *s* is false. Our utilities may then be represented:

$$U(a_1, 1) = u_1$$
$$U(a_2, 0) = u_2$$
$$U(a_1, 0) = u_3$$
$$U(a_2, 1) = u_4$$

We suppose our body of knowledge to be *w*, and that $PROB_L(s, w) = [p, q]$.

What is the expected utility of a_1? It will be an interval, whose lower bound is the infimum of $\{u_1 r + u_3(1-r) : p \leqslant r \leqslant q\}$, and whose upper bound is the supremum of the same set. Let us suppose that $u_1 \leqslant u_3$ and $u_2 \leqslant u_4$. Then the expected utility of a_1 is the interval $[(u_1 q + u_3(1-q)), (u_1 p + u_3(1-p))]$ and that of a_2 the interval $[(u_4 p + u_2(1-p)), (u_4 q + u_2(1-q))]$.

If the probability interval is degenerate – i.e., if $p = q$, then the utility-intervals will also be degenerate, and the classical principle of maximizing expected utility will apply directly. We should choose act *a* if and only if $u_1 p + u_3(1-p) > u_4 p + u_2(1-p)$.

In the general case it may still be true that the minimum of the expected utility of a_1, $u_1 q + u_3(1-q)$, is greater than the maximum of the expected utility for a_2, $u_4 q + u_2(1-q)$: then surely the arguments for the classical principle apply, and we should choose a_1. Similarly, if $u_4 p + u_2(1-p) > u_1 p + u_3(1-p)$, a_2 is the correct choice.

One possible case remains: the intervals overlap. The expected utility of a_1 may be [3, 5] and the expected utility of a_2 may be [4, 6], or one interval may include the other. In either case, the principle of maximizing expected utility doesn't yield a directive. There are Humean arguments against choosing either action: we should not choose a_1 since *so far as we know* a_2 may be superior. This does not mean that in point of fact a_2 may be the best act, as it would be if we knew that *s* were true – that is trivial, and irrelevant in our human situation. It means that someone who differed from us *only in having true and precise statistical knowledge where ours is vague*, so that the principle of maximizing expected utility could be applied, would find that the expected utility of a_2 was higher than that of a_1. And of course there would be a parallel argument against a_2.

There a number of alternatives one might consider. One could look at the average of the endpoints of the intervals, and choose the action for which this average was highest. One could adopt a minimax strategy, and look at the lower endpoints of the intervals, and choose that strategy whose minimum expectation was the greatest. There are other possibilities as well. It does not seem unreasonable to suppose that what strategy should be regarded as appropriate should depend on the context. It is here, at any rate, where we have run out of the guidance offered by the classical principle of maximizing expected utility, that the fancy principles developed in modern decision theory come to play an important role.

In a later chapter, we shall consider in detail the manner in which these relatively recent principles operate. In every case, however, their operation will be bounded and constrained by the pervasive principle of rational action: insofar as it is possible, maximize expected utility.

CHAPTER THIRTEEN

BAYESIAN INFERENCE

1. Following Carnap – part way – we shall construe the outcome of a statistical inference as the assignment of a probability to a statement. Sometimes this is done wholesale – we arrive at a distribution yielding probabilities for a whole family of statements, as when we infer that the length of a table is three feet one inch, with a standard deviation of half an inch. This does not mean, of course, that we suppose that the length of the table has a standard deviation of half an inch, nor, indeed, that the length of the table is somehow indeterminate. (On an atomic level, it may well be that the length of the table is indeterminate in some sense, or that this is one way of interpreting the formalism of atomic theory. This has nothing to do with our measurement of the table. If we restricted ourselves to length predicates corresponding to even multiples of one millimeter, our statistical statement about the length of the table would still – almost – hold.) Taken in conjunction with (a locution that should be taken quite literally) the conventional theory of measurement, the statement about the length of the table provides us with a shorthand for an infinite number of probability statements about the length of the table: the probability is 0.63 that it lies between $3'\ 1\frac{1}{2}''$ and $3'\ \frac{1}{2}''$, and so on.

Note that in making a statistical inference (assigning a probability to a statement, relative to a body of knowledge), we do not necessarily come to accept the statement said to be probable. Do we have here a contrast with deductive inference? Not really. We say that the deductive inference of *S* form the set of statements *w* is valid whether or not *w* represents our body of knowledge, and whether or not, if it does represent our body of knowledge, we actually take the psychological leap of accepting *S*. To say that *S* may be deductively inferred from *w* is to observe that a certain formal relationship holds between the set of statements *w* and the statement *S*. If *w* does in fact represent someone's body of knowledge, and *w* is indeed a rational corpus, and this relation does in fact obtain between *S* and *w*, then we might feel that *S* ought also to belong to *w*. (Or we might not: if the person whose body of knowledge *w* represents doesn't know that w implies S, we might hold him guiltless for failing to incroporate S into w.)

Similarly, to say that the probability of S, relative to w, is p, is simply to observe that a certain formal relationship obtains between w and S. This relation holds whether or not w represents anybody's body of knowledge, or is a rational corpus, or whether or not anyone knows that it holds. If, in fact, w is a rational corpus, and the body of knowledge of a certain person, then we might feel that that person ought to have a degree of belief in S falling in the interval p. (Or we might not: if the person whose body of knowledge w represents doesn't know that the probability of S given w is p, we might hold him guiltless for failing to have the appropriate degree of belief in S.)

Under certain circumstances (which we shall explore in a later chapter) we might even propose that the person should accept S, provided that the probability of S, relative to w, was high enough, and provided certain other conditions are satisfied. There is this difference, though. Under no circumstances will a high probability for S lead to the inclusion of S in w. This is easy to see: Suppose that $PROB_L(S, w) = [p, q]$, and that $1 - \epsilon < < p < 1$. If on this ground w is taken to include S, we have $PROB_L(S, w) = = [1, 1]$, contrary to our assumption. Thus S may be a member of a new rational corpus w^*, but it will not be a member of w.

2. There are four kinds of statements that are of particular concern in connection with statistical inference.

(1) We may be concerned with the probability that the value of a certain random quantity for a certain object will fall into a certain Borel set. Thus we may be concerned with the probability that the next toss of a certain coin will yield heads – i.e., that $X_H(t_n) \in \{1\}$, where t_n is the next toss, and X_H is the characteristic function of heads. Or we may be concerned with the probability that between $\frac{45}{100}$ and $\frac{55}{100}$ of the next ten thousand tosses yield heads, i.e., that $Fr_H(T_{10\,000n}) \in [0.45, 0.55]$, where Fr_H is the function whose value is the proportion of heads, and $T_{10\,000n}$ is the next ten thousand tosses. Or we may be concerned with the probability that the diameter of a particular ball bearing is 'satisfactory', i.e., falls into a certain interval $[d_l, d_u]$, or we may be concerned with the probability that a certain fraction of a shipment of such ball bearings have diameters that fall into the acceptable interval.

We are concerned here in each case with the probability of a particular statement. The classical Bayesian analysis requires that we know an underlying distribution statement, in each case. In the case of a frequentistic Bayesian it is required that we have in our body of knowledge an appropriate statistical hypothesis; in the case of a subjectivistic Bayesian, it is merely required that we have examined our dispositions to behave, and have thus isolated the distribution which we are committed to accepting by our tendencies to behave in certain ways. In the simplest case, (a) we may have in our body of knowledge a statistical statement describing the behavior of coins, or the coin at hand, or a certain family of ball bearings. From this distribution statement we may derive distribution statements concerning sequences of 10000 tosses, or shipments of ball bearings. In a more complicated case, (b) we may have in our body of knowledge a joint distribution statement concerning (say) a population of coins, and the flipping behavior of each sort of coin in the population. For example, we may suppose that the coin in question is chosen from a set of coins, $\frac{1}{3}$ of which are double headed, and the rest of which are normal. If the coin and toss are random ones in the appropriate sense, the probability of heads on the next toss is $\frac{2}{3}$; the probability of between 0.45 and 0.55 heads among the next ten thousand tosses is $\frac{1}{3} \times 0$ plus $\frac{2}{3}$ of the appropriate probability for a normal coin. We may know of the ball bearings that they are produced by one of k machines, each of which produces ball bearings whose diameters are roughly normally distributed with mean m_k and standard deviation s. In any event, we have to deal with a single distribution statement in our body of knowledge (though I have hedged slightly in the case of the ball bearings by saying 'roughly' normally distributed). A somewhat generalized Bayesian approach, such as is possible in the framework presented here, would allow that there be a family of distributions represented in our body of knowledge, and would thus allow for interval probabilities. Thus we might know of the coin that its tosses were binomial with respect to yielding heads, with a parameter p of which we know that it lies between 0.4 and 0.6. Similarly, if we are considering a coin drawn from a population of coins, and then tossed we may be dealing with a family of distributions represented by saying, for example, that the proportion of coins of style k in the population is between p_k and q_k, and the long run frequency of heads for coins of style k is between r_k and s_k.

It is furthermore clear that in addition to knowledge of certain distributions, we require that the object in question be a random member, relative to what we know, of the reference class in question. Assuming that we have observed the last toss of the coin, it is not a random member of the set of tosses with respect to yielding heads. Assuming that we have measured the ball bearing, it is not a random member of the set of ball bearings with respect to having a satisfactory diameter.

(2) Sometimes we are concerned, not with the probability of a single statement, but with the probabilities of a whole family of statements. Thus we may be concerned with the probability, for arbitrary *m*, that the next *n* tosses of a certain coin will yield less than *m* heads. Given knowledge of a distribution, we can represent this probability as a function of *m*, provided that 'the next *n* tosses' is a random member of the set of all sets of *n* tosses with respect to yielding *m* heads, for arbitrary *m*. As was pointed out in Chapter X, this is often the case. There is a formal similarity between asserting generally (for arbitrary *numerical expression m*)

$$PROB_L(\text{'}Fr_H(T_{10\,000n})\leqslant m\text{'}, w)=$$
$$=\left[\sum_{j<m+1}\binom{10000}{j}\left(\frac{1}{2}\right)^{10\,000},\ \sum_{j<m+1}\binom{10000}{j}\left(\frac{1}{2}\right)^{10\,000}\right]$$

and the assertion (represented by a statement in our body of knowledge), that the distribution of heads in the increasing sequence of sets of tosses of the coin is binomial with parameter $\frac{1}{2}$ – or, more specifically, that in the non-descreasing sequence of sets of ten thousand tosses, the proportion with the property of exhibiting less than *m* or less heads in the long run approaches $\sum_{j<m+1}\binom{10000}{j}\left(\frac{1}{2}\right)^{10\,000}$. Nevertheless the similarity is only formal. The probability statement is a metalinguistic statement, *about* another statement (and a set of statements representing a body of knowledge); the statistical distribution statement is about a reference class, and not about the object $T_{10\,000n}$ at all. Since there are among the 10000 members of $T_{10\,000n}$ exactly so many heads – whatever it will be or has been – there is no question of this number having a distribution. It is what it is. What the probability statement gives us is guidance regarding the *statements* that the number is less than *m*, for arbitrary integer expression '*m*'. Since the distribution statement says nothing at all about

$T_{10\,000n}$, it cannot by itself, give us any information about that particular object. To be sure, the distribution statement, combined with a general metalinguistic assertion about randomness, can give us the probability statement – but that is another thing.

This is an important distinction to be clear about, for what statistical inference gives us, according to subjective or Bayesian views, is often a *posterior* distribution of probability, quite on a par with the prior distribution. According to the present view, neither of these is properly speaking a distribution; they concern particular statements, and distributions do not. They are rather shorthand ways of specifying the probabilities of whole classes of particular statements. It would be correct to speak of prior and posterior distributions in a body of knowledge: the former would be a joint distribution, and the latter a conditioned distribution obtained from the former by adding a certain statement (the statement of the observation) to the body of knowledge w. These are straight-forward, honest-to-goodness, statistical distributions. They lead to probabilities just when certain requirements concerning randomness are met.

(3) The phrase 'statistical inference' is often used in a somewhat more narrow sense than I have used it in the title to this chapter. It is used specifically when the statement to which a probability is assigned is itself a statistical statement; thus when we infer, with the help of some evidence, that the probability is p that the binomial parameter characterizing a certain coin experiment is in the interval [0.4, 0.6], we are doing statistical inference in the narrow sense. In the strict Bayesian case, we begin with a prior distribution for q, the binomial parameter, together with a distribution for a set of outcomes dependent on q; we perform an experiment (make observations) which enter into our body of knowledge to yield a posterior distribution in that new body of knowledge, relative to which the probability of heads on the next toss has a new value; or relative to which the conditional distribution of q is different from the prior distribution of q. Thus as a typical result, we shall have 'the probability is p that the q of this coin lies in the interval [0.4, 0.6]'.

Similarly, we may obtain, by making measurements on a set of ball bearings, 'The probability is r that the distribution of diameters in the set of ball bearings produced by this machine is Normal (m_k, s)'.

(4) In the same way as in the style of statement mentioned in (2), we

may obtain whole bundles of probability statements at once. Thus we may, for arbitrary numerical expressions m, have

$$PROB_L(\text{`}\imath k(S\langle R, Z, b, k, k\rangle)\leqslant m\text{'}, w)=[f(m), f(m)], \quad \text{or}$$
$$PROB_L(\text{`}\imath k(D\langle R, Z, \{g\langle k\rangle\}\rangle\leqslant m\text{'}, w)=[f_1(m), f_2(m)].$$

As before, we observe that such statements are *not* distribution statements. They are metalinguistic statements, about certain expressions of L or sets of expressions of L. It is customary (among subjectivists) to refer to similar statements as posterior distribution statements. But this is surely inappropriate: if what we have is an expression giving the probability that *the* binomial parameter k characterizing the reference set R and the subset Z is less than m, this is not a *distribution* for k. As frequentists were once wont to point out, k is what it is, and it makes no sense to speak of its having a certain distribution. It would make sense to speak of our *belief* about k having a certain distribution; but if this is what has a certain distribution, it is hard to know what to make of the likelihoods: the distribution of the number of heads on n trials, for example, conditioned – not on a given value for k, but – on our belief about k having a certain value. We can make subjectivistic sense of these locutions, of course, but not without a certain sense of strain.

3. There are two main differences between Bayesian Inference as discussed in this chapter, and Bayesian Inference as commonly construed by most contemporary writers. As a result of the development of the subjectivistic interpretation of probability, Bayesian Inference has come to be associated with a form of inference in which prior probability distributions are taken to represent prior opinions, whether founded or unfounded, whether based on evidence or on whimsey, whether rational or insane. In the present system, all probabilities are based (in part) upon statistical distributions that are known or reasonably believed to hold, or, in hypothetical considerations, assumed to hold. The other difference is that all Bayesian Inference, however probability is construed, as assumed to start from one single prior distribution. In the case of subjectivistic Bayesians, it is a matter of doctrine that there can be only one prior distribution, for a given person. He has the prior distribution of beliefs he has, and that is that. (It is also true, however, that the exact form of his prior distribution may have little influence on the outcome

of his inference. This is taken account of, and indeed is extremely important to, subjective Bayesians.) In the case of frequentists, the existence of a single known prior distribution is a necessary condition of Bayesian inference; without such a distribution, Bayes' theorem is simply inapplicable. In the present system there will often (as in real life almost always) be a *family* of prior distributions that will serve as the statistical basis for the inference. This quite clearly and naturally accounts for the subjectivist observation that in the face of evidence, quite a wide variety of prior distributions may lead to quite a narrow group of posterior distributions; and equally clearly and naturally it allows for a large expansion in the scope of applicability of Bayes' theorem over the scope assigned to it from the traditional frequentist view.

4. In this section we shall look in greater detail at examples such as those alluded to in section 2. We shall consider first cases in which there is a single prior distribution and we are seeking to make an appropriate probability statement about the next object, or the next n objects. We then generalize, by allowing our prior knowledge to be vague. We shall at the same time consider statistical inference proper (where we are assigning probabilities to statistical hypotheses or distribution statements) in the case where we have a single prior distribution, and also without this severe constraint on our prior knowledge.

Case I: Suppose that we have three urns full of coins. In the first urn, the coins have a long run propensity to land heads (when tossed in our standardized apparatus) $\frac{1}{4}$ of the time; in the second, this propensity is $\frac{1}{2}$; and in the third it is $\frac{3}{4}$. We suppose that there is a method M by which we choose an urn and a coin, and that it is such that a coin is chosen by M from each urn with equal frequency. A *test* consists of choosing a coin by M, and then tossing it n times and observing the number n_h of heads. We also suppose that we know the tosses of a coin are independent, so that we know ourselves to be observing a Bernoulli process. Let U be the random quantity whose value $U(t)$ is 0, 1, or 2, according as the first, second, or third urn is chosen on trial t. Let H be the random quantity whose value $H(t)$ is m/n just in case m of the n test tosses result in heads, and let H_k be the random quantity whose value $H_k(s)$ is j/k just in case s is a subsequent set of k tosses of the same coin, and j of them result in heads.

Let T_i be the finite set of trials of this experiment performed before the ith date, where we consider a denumerable sequence of dates and the i plus first date is later than the ith date. Thus T is an omega sequence. We further assume that the union of the range of T is provably denoted by a member of the set of reference terms RT_L of our language. We assume that our body of knowledge is normal, probabilistically terminating, and deductively closed.

It follows that we know the joint distribution of U, H, and H_k in the sequence T. Specifically, let F be a function whose domain is $\mathbb{R}^3$, and whose value for (r_1, r_2, r_3) *is*

$$F(r_1, r_2, r_3) = \sum_{\substack{x_r \leqslant r_1 \\ x \in \{0,1,2\}}} \sum_{\substack{y_r \leqslant r_2 \\ y \in n+1}} \sum_{\substack{z_r \leqslant r_3 \\ z \in k+1}} \frac{x+1}{3} \times$$
$$\times \binom{n}{y} \left(\frac{x+1}{4}\right)^y \left(\frac{3-x}{4}\right)^{n-y} \binom{k}{z} \left(\frac{x+1}{4}\right)^z \left(\frac{3-x}{4}\right)^{k-z},$$

if all of r_1, r_2, r_3 are non-negative, and 0 otherwise.

Then, if w is our body of knowledge, we are assuming that

$$\text{`}D \langle \imath f\, \mathscr{D} f = T \wedge \bigwedge_{x \in T} f\langle x \rangle = \langle U\langle x \rangle, H\langle x \rangle, H_k\langle x \rangle \rangle),$$
$$T, \{F\} \rangle\text{'} \in w.$$

Given this rather complete knowledge of the situation, we can compute all of the marginal and conditional distributions that interest us. For example,

$$\text{`}S\langle T, U, \{0\}, \tfrac{1}{3}, \tfrac{1}{3} \rangle\text{'} \in w$$

$$\text{`}S\langle T, H, \{(m/n)_r\}, \frac{1}{3}\binom{n}{m}\left(\frac{3^m + 3^{n-m}}{4^n} + \left(\frac{1}{2}\right)^n\right),$$
$$\frac{1}{3}\binom{n}{m}\left(\frac{3^m + 3^{n-m}}{4^n} + \left(\frac{1}{2}\right)^n\right)\rangle\text{'} \in w$$

$$\text{`}S\langle T, H_k, \{(j/k)_r\}, \frac{1}{3}\binom{k}{j}\left(\frac{3^j + 3^{k-j}}{4^n} + \left(\frac{1}{2}\right)^k\right),$$
$$\frac{1}{3}\binom{k}{j}\left(\frac{3^j + 3^{k-j}}{4^n} + \left(\frac{1}{2}\right)^k\right)\rangle\text{'} \in w$$

$$\text{`}S\langle\ T'(T' \textit{ is fun} \wedge \mathscr{D} T' = \omega \wedge \bigwedge_{i \in \omega} T'\langle i \rangle =$$
$$= T\langle i \rangle \cap \{x : U\langle x \rangle = 0_r\}), H,$$

$$\{(m/n)_r\}, \binom{n}{m}\left(\frac{1}{4}\right)^m\left(\frac{3}{4}\right)^{n-m}, \binom{n}{m}\left(\frac{1}{4}\right)^m\left(\frac{3}{4}\right)^{n-m}\rangle\text{'}\in w$$

And so on.

According to Theorem 10.1, under these circumstances we will have, for a specified trial t^*.

$$PROB_L(\text{'}H_k\langle t^*\rangle = j/k\text{'}, w) =$$
$$= \left[\frac{1}{3}\binom{k}{j}\left(\frac{3^j+3^{k-j}}{4^k}+\left(\frac{1}{2}\right)^k\right), \frac{1}{3}\binom{k}{j}\left(\frac{3^j+3^{k-j}}{4^k}+\left(\frac{1}{2}\right)^k\right)\right]$$
if and only if $RAN_L(\text{'}t^*\text{'}, \text{'}T\text{'}, \text{'}H_k\text{'}, \text{'}\{(j/k)_r\}\text{'}, w)$.

Recall that the assertion regarding randomness will be true just in case we don't know anything in w that will *prevent* it from being true, i.e., just in case we don't know anything in w which will render the union of the range of T an inappropriate reference class. In particular, we suppose that we do not know anything which places t^* in a subset of the union of the range of T in which the frequency of the occurrence of j heads on the k trials differes from that in T itself.

$$\left[\frac{1}{3}\binom{k}{j}\left(\frac{3^j+3^{k-j}}{4^k}+\left(\frac{1}{2}\right)^k\right), \frac{1}{3}\binom{k}{j}\left(\frac{3^j+3^{k-j}}{4^k}+\left(\frac{1}{2}\right)^k\right)\right]$$

is thus the prior probability that j out of the k secondary tosses will hield heads.

Now let w' be just like w, but for containing the logical consequences of the sentence '$H(t^*) = m/n$'. We take w' to be probabilistically terminating, normal, and statistically closed. By Theorem 11.1, we will have

$$PROB_L(\text{'}H_k\langle t^*\rangle = j/k\text{'}, w') =$$
$$= \left[\binom{k}{j}\frac{3^{n-m+k-j}+2^{n+k}+3^{m+j}}{4^k(3^m+2^n+3^{n-m})}, \binom{k}{j}\frac{3^{n-m+k-j}+2^{n+k}+3^{m+j}}{4^k(3^m+2^n+3^{n-m})}\right]$$
if and only if $RAN_L(\text{'}t^*\text{'}, \text{'}\imath T'(T' \text{ is } fun \wedge \mathscr{D}T' =$
$= \omega \wedge \bigwedge_{i\in\omega} T'\langle i\rangle = T\langle i\rangle \cap \{x : H\langle x\rangle = m/n\})\text{'}, \text{'}H_k\text{'}, \text{'}\{j/k\}\text{'}, w')$.

Thus

$$\left[\binom{k}{j}\frac{3^{n-m+k-j}+2^{k+n}+3^{j+m}}{4^k(3^m+2^n+3^{n-m})}, \binom{k}{j}\frac{3^{n-m+k-j}+2^{k+n}+3^{j+m}}{4^k(3^m+2^n+3^{n-m})}\right]$$

will be the posterior probability that j of k tosses in the second sequence will land heads, provided that the randomness conditions are met.

The same classical analysis may be applied to obtain the prior and posterior probability that one or another or the urns has been drawn from on t^* – i.e., that the long run frequency of heads in tosses of the coin selected is $\frac{1}{4}$ or $\frac{1}{2}$ or $\frac{3}{4}$. We will have:

$$PROB_L(\text{`}U\langle t^*\rangle = 0\text{'}, w) = [\tfrac{1}{3}, \tfrac{1}{3}] \quad \text{if and only if}$$
$$RAN_L(\text{`}t^*\text{'}, \text{`}T\text{'}, \text{`}U\text{'}, \text{`}\{0\}\text{'}, w).$$

Since we know that the long run frequency of heads is $\frac{1}{4}$ if and only if $U(t^*) = 0$, the probability that the long run frequency of heads for the coin we have selected on trial t^* is $\frac{1}{4}$ is also $[\frac{1}{3}, \frac{1}{3}]$. Relative to w', t^* is not a random member of T with respect to having a U-value in $\{0\}$ but rather (presumably) of $\imath T'(T' \textit{ is fun} \wedge \mathscr{D}T' = \omega \wedge \bigwedge_{i \in \omega} T'\langle i\rangle = T\langle i\rangle \cap \cap \{x : H\langle x\rangle = m/n\})$; and we have

$$\left[\frac{3^{n-m}}{3^m + 2^n + 3^{n-m}}, \frac{3^{n-m}}{3^{n-m} + 2^n + 3^m}\right] = PROB_L(\text{`}U\langle t^*\rangle = 0\text{'}, w')$$
$$\text{if and only if} \quad RAN_L(\text{`}t^*\text{'}, \text{`}\imath T'(T' \textit{ is fun} \wedge \mathscr{D}T' =$$
$$= \omega \wedge \bigwedge_{i \in \omega} T'\langle i\rangle = T\langle i\rangle \cap \{x : H\langle x\rangle = m/n\})\text{'}, \text{`}U\text{'}, \text{`}\{0\}\text{'}, w')$$

Thus, given the satisfaction of the randomness conditions, the posterior probability that $U\langle t^*\rangle = 0$ is

$$\left[\frac{3^{n-m}}{3^m + 2^n + 3^{n-m}}, \frac{3^{n-m}}{3^m + 2^n + 3^{n-m}}\right],$$

and the posterior probability that the coin we have selected would yield heads in the long run $\frac{1}{4}$ of the time is also

$$\left[\frac{3^{n-m}}{3^m + 2^n + 3^{n-m}}, \frac{3^{n-m}}{3^m + 2^n + 3^{n-m}}\right].$$

This is a highly simplified and artificial example. Its analysis along classical lines and its analysis along subjectivistic lines would coincide, and coincide with the treatment given here. A more interesting example is provided by the subjectivistic treatment of Bernoulli processes with prior beta distributions. The attractiveness of the beta distribution for the subjectivist is that it can have a wide variety of shapes, according to the values assigned to its two parameters. Its density function is

$$f_\beta(p, r, n) = \frac{1}{B(r, s)} p^{r-1}(1-p)^{s-1} \quad \begin{array}{l} 0 \leqslant p \leqslant 1 \\ s = n - r \\ 0 < r, 0 < s \end{array}$$

where

$$B(r, s) = \int_0^1 t^{r-1}(1-t)^{s-1}\, dt.$$

If we suppose ourselves to be dealing with a Bernoulli process with a parameter p, whose distribution is given by a beta density with parameters r, n; and if we observe r' successes out of n' trials, then the posterior distribution of p is a beta distribution with parameters $r + r'$, $n + n'$. This would apply, for example, if we supposed ourselves to be dealing with a set of coins, among which the characteristic long run frequency of heads, p, was known to have a specific beta distribution. The treatment would be perfectly analogous to that of the preceding example, with the discrete prior distribution represented by the urns replaced by the continuous distribution given by the beta function.

It is hard to take such a case too seriously, however, from our point of view, since there are so few coins in the world. It is also a little difficult to swallow from a subjectivistic point of view, for it implies that our distribution of beliefs regarding p is absolutely continuous. This awkwardness (and inconvenience) has not been lost on statisticians of subjectivistic persuasion. It is all very well to demand that "The decision maker must either resolve his decision problem by using direct unaided intuition or else decide exactly what prior distribution he is willing to take as a basis for action."[1] But in point of fact 'the decision maker' (aside from feeling quite properly resentful at having such impossible demands imposed on him) may be *willing* to take as a basis for action quite a variety of 'prior distributions'. On the analysis given here, that willingness is eminently rational. Furthermore, as the subjectivists go on to show, it is often quite irrelevant which particular distribution of a certain wide class of distributions is chosen as 'the' prior distribution. We do know something about coins – on the basis of our own experiences, on the basis of the reports of others. It may be difficult to quantify precisely what we know about them simply because it is difficult to integrate our various sources of knowledge. But it would not be unreasonable to suppose that the distribution of p-values among coins is known to be a beta distribution bounded by beta distributions with parameters lying between n_l, n_u, r_l, r_u. Put more precisely, this would be to say that for every distribution F in the set z, where z is the minimal set of distributions, of which it is known that one characterizes the set of coins, and for every real number r, there is a distribution F_{b_1} and a distribution F_{b_2}, both of which are beta distributions, both of which belong to z, and which are such that $dF_{b_1}(r) \leqslant dF(r) \leqslant dF_{b_2}(r)$. (Obviously, except in a degenerate case, different functions F_{b_1} and F_{b_2} are required for different values of r.) If this is the case, then the conditional distribution of p given the observation of n tosses, r of which are heads, will be one of a set z_c, where for every function F in z_c and every real number r, there are beta distributions F_{b_1} and F_{b_2} belonging to z, such that the conditional density of F at r lies between the conditional densities of F_{b_1} and F_{b_2} at r.

For example, let us suppose that we have good physical reasons for believing that the distribution of p among the set of modern coins is symmetrical with respect to heads and tails. We consider (on the basis of our general experience with coins, our knowledge of physical processes – that is, on good inductive grounds) that the set $z_b = \{F_\beta(p, r, s): \frac{1}{4} \leqslant r \leqslant 10 \wedge \wedge \frac{1}{4} \leqslant s \leqslant 10\}$ includes all the beta distributions possible for p. Then the conditional density of p, given that 100 trials have been performed, of which 53 have landed heads, is bounded by the densities of beta distributions with parameters lying in the intervals (53.25, 63) and (47.25, 57).

The posterior distribution of p is thus almost entirely concentrated between 0.4 and 0.65. For short sequences of tosses, we might as well use a p of 0.5. But for long sequences of tosses (or for high stakes) we would want to take account of the vagueness of the posterior distribution of p.

It is interesting to generalize the conditions of this problem. Suppose that we do not know that p has a beta distribution, but merely that whatever distribution it has, its density at any point r is between the densities of a couple of beta distributions that also belong to z. Under what conditions will the conditional density for the same point r lie between the conditional densities obtained from these same two beta distributions? We shall consider the general problem.

Let z be a set of n-dimensional distribution functions. (In the particular case at hand, these will be 2-dimensional joint distribution functions for p and the number of heads among n tosses.) Let a sequence of n_1 indices s be given, and a special subclass w of z, such that for every x in $\mathbb{R}^{n_1}$ and every F in z, the value of the s-marginal density of F evaluated at x lies between two s-marginal densities of members of w evaluated at x. That is: there will be two functions G^l and G^u such that G^l, G^u belong to w and $f_{{}_sG^l}x \leqslant f_{{}_sF}x \leqslant f_{{}_sG^u}x$, in the notation of the last section of the Appendix. (For our example, w is the set of joint β and binomial distributions, and the marginal ${}_sG^l$ and ${}_sG^u$ are the bounding β-distributions F_{b_1} and F_{b_2}.) Let t be the appropriate sequence of the remaining n_2 indices, so that $A(F, n, n_1, n_2, s, t)$ holds for every F in z. We require that for every F in z, the s-marginal distributions be all absolutely continuous (as they are in our example) or all discrete; and we require that the t-marginal distributions be all discrete (as they are in our example) or all absolutely continuous. We assume, finally, that the conditional distribution for the t-variables, *given* a value for the s-variables, is the same for every F in z. In the notation of the appendix: for every F and G in z, ${}_tF_x = {}_tG_x$. (This holds for our example, since, given p, the distribution of the number of heads on n trials is assumed to be binomial with parameters n and p. More generally, in Bayesian inference, the likelihoods – that is, the distribution of the observations given the parameters – are supposed to be given and uncontroversial.) Thus, in particular,

(i) $$f_{{}_tG^l_x}y = f_{{}_tG^u_x}y = f_{{}_tF_x}y.$$

From what we have assumed so far, it follows that

(ii) $$\bigwedge_x \bigvee_{G^l G^u} \bigwedge_y f_{G^l}O(y, x, t, s) \leqslant f_F O(y, x, t, s) \leqslant f_{G^u}O(y, x, t, s)$$

(since $f_F O(y, x, t, s) = f_{{}_sF}x \cdot f_{{}_tF_x}y$, by T 0.52.)

We may now apply theorem 0.53 to obtain, for every x in $\mathbb{R}^{n_1}$,

$$f_{{}_sF_y}x = \frac{f_F O(y, x, t, s)}{f_{{}_tF}y} = \frac{f_{{}_tF_x}y \cdot f_{{}_sF}x}{f_{{}_tF}y}$$

We would like to know when the functions G^l and G^u in w that are picked out according to our antecedent assumption as satisfying

$$f_{{}_sG^l}x \leqslant f_{{}_sF}x \leqslant f_{{}_sG^u}x$$

for all F in z, also satisfy

$$f_{sG_y^l}x \leqslant f_{sF_y}x \leqslant f_{sG_y^u}x$$

for all F in z. A necessary and sufficient condition is that

$$f_{tF}y \cdot f_{sG^l}x \leqslant f_{tF}y \cdot f_{sG^u}x.$$

5. Let us look once more at the question of stable estimation briefly alluded to in Chapter Five. It was observed there that posterior Bayesian probabilities are often only mildly affected by variations in prior probabilities. For example, my prior distribution of belief concerning the p of a certain coin may be quite different from yours; but our posterior distributions of belief, after having observed a few hundred tosses together, will in the ordinary course of events be very nearly the same. This phenomenon, the coming together of posterior distributions generated by diverse prior distributions and given evidence, often enables the subjectivist statistician to deal with two problems simultaneously. He has, as already remarked, the difficult problem of determining the prior joint distribution of his client (or himself); to do this in detail seems an extremely difficult job, and handbooks for budding subjectivist statisticians usually suggest that knowledge of the deciles, or even the quartiles, of the prior distribution is as far as it is worth trying to go, unless there are special features in the distribution. Furthermore, the statistician is often concerned with a number of people whose prior distributions of belief are different; he would like to be able to say something to all of them, and indeed sometimes to force them into near (or near enough agreement. If the exact shape of the prior distribution function doesn't matter much, then we do not have to spend much effort finding out about that exact shape; and the posterior distribution for a number of people whose prior distributions vary will turn out to be about the same.

But what is it for a prior distribution to be 'gentle'? What does it mean to say that its 'exact' character doesn't matter 'very much'? Without rather precise answers to these questions, we are not in a position to advise anybody about anything. The theory of stable estimation is the Bayesian theory that provides answers to these questions. It has an extremely useful counterpart in the present theory. My exposition of the Bayesian approach (and my selection of a particular example) will be modeled on that of Edwards, Lindeman, and Savage.[2] After the exposition, we shall see how the theory can be thought of more simply and more naturally in our epistemic framework.

Suppose that we are concerned about the distribution of a random quantity X in a certain population Z. We suppose that we know the form of the distribution of X (a typical Bayesian assumption, though hard to swallow in Bayesian terms), and that it depends on a single parameter θ. Now we must look on θ in turn as coming from some population of populations; in Bayesian terms and more simply, we merely suppose that θ in turn has a prior distribution. Let R be a certain evidential result – say the result of measuring X for n members of Z. The quantity R will be a random quantity (with domain Z^n), whose distribution will also depend on the single parameter θ. We have, then, to be concerned with the following densities (or frequency functions):

$f_X(x/\theta)$	the density function for X, given θ
$f_R(r/\theta)$	the density function for R, given θ
$f_T(\theta)$	the prior density of θ
$f_R(r)$	the prior density of R
$f_T(\theta/r)$	the density of θ given r

It follows perfectly straightforwardly from Bayes' theorem that:

$$(1) \qquad f_T(\theta/r)=\frac{f_T(\theta)\,f_R(r/\theta)}{f_R(r)}=\frac{f_R(r/\theta)\,f_T(\theta)}{\int f_R(r/\theta)\,f_T(\theta)\,d\theta}$$

Under certain circumstances $f_T(\theta/r)$ may be approximated by

$$(2) \qquad f_T^*(\theta/r)=\frac{f_R(r/\theta)}{\int f_R(r/\theta)\,d\theta}$$

One obvious such set of circumstances is that in which the frequency function $f_T(\theta)$ is constant over some interval $[\theta_1, \theta_2]$, and zero elsewhere. It is easy enough to imagine unnatural ways in which these circumstances can arise. For example, we might imagine a machine that turns out cookie cutters in such a way that the diameters of the cookie cutters are uniformly distributed between d_1 and d_2. (It would be more natural to suppose that the distribution is roughly normal, but it is conceivable that it could be rectangular.) Then we wish to make some inference concerning the diameter of a particular cookie cutter by measuring the cookies it produces, where we know that the diameters of the cookies produced by a cookie cutter of diameter d are distributed normally with mean d and fixed standard deviation s. Let m be the result of measuring n cookies produced by cutter c. (We ignore the additional uncertainties introduced by measurement itself.) Then we make assumptions about randomness: we assume (1) that c is a random member of the population of cookie cutters, with respect to having a D-value in any given interval relative to our prior body of knowledge; and we assume (2) that relative to our body of knowledge *after* our observation of m, c is a random member of the set of cookie cutters yielding the observation m, with respect to having a D-value in any given interval. (And hence, of course, with respect to c having a D-value in any given Borel set.) It now follows that the probability that the diameter of c lies in any Borel set b is given by the uniform density of D between d_1 and d_2 before our observation, and that the diameter of c lies in any Borel set b is given by the conditional density

$$f^*_D(d/m) = \frac{f_M(m/d)}{\int f_M(m/d)\, dd}$$

The assumption we made about the prior distribution of D was a strong one: that it is uniformly distributed between d_1 and d_2. The argument will go through, however, with a vastly weaker set of assumptions. Edwards, Lindeman, and Savage[3] list three assumptions under which the posterior density of θ given r is satisfactorily approximated by Equation (2), i.e., under which $f_T(\theta/r)$ is close to $f^*_T(\theta/r)$.

There is a set B of possible values of θ such that:

(3) It is practically certain that $\theta \in B$, given r, on the assumption that $f_T(\theta)$ has a fixed value in B and is zero elsewhere; i.e., for small α:

$$\int_{\bar{B}} f^*_T(\theta/r)\, d\theta \leqslant \alpha \int_B f^*_T(\theta/r)\, d\theta$$

(4) The prior frequency function $f_T(\theta)$ must not change much in B – i.e., for some number k and some small number β, we have

$$k \leqslant f_T(\theta) \leqslant (1+\beta)\, k$$

(5) The prior frequency function $f_T(\theta)$ must be bounded by a multiple η (which may be as large as 100 or 1000) of k, i.e., $f_T(\theta)$ is nowhere very much larger than its relatively uniform values in B. Given these three assumptions, we may prove a number of theorems relating f_T^* to f_T. An alternative form of assumption (5), that is easier to work with, but which Edwards, Lindeman and Savage find harder to intuit, is

$$\int_{\bar{B}} f_T(\theta/r)\, d\theta \leqslant \gamma \int_B f_T(\theta/r)\, d\theta \quad \text{for a given small number } \gamma.$$

The relation between (5) and (6) is given by $\eta \leqslant \gamma/\alpha$.

Given these assumptions, we may prove a number of theorems relating f_T^* to f_T:

(7) $$\int f_R(r/\theta)\, f_T(\theta)\, d\theta \geqslant \frac{k}{1+\alpha} \int f_R(r/\theta)\, d\theta$$

(8) $$\int f_R(r/\theta)\, f_T(\theta)\, d\theta \leqslant (1+\gamma)\,(1+\beta)\, k \int f_R(r/\theta)\, d\theta$$

(9) Let δ and ε be defined by: $1-\delta = \dfrac{1}{(1+\beta)\,(1+\gamma)}$;

$1+\varepsilon = (1+\beta)\,(1+\delta)$

Then $(1-\delta) \leqslant \dfrac{f_T(\theta/r)}{f_T^*(\theta/r)} \leqslant (1+\varepsilon)$

(10) $$1-\gamma \leqslant \int_B f_T(\theta/r)\, d\theta$$

(11) If $C \subset B$, then $$1-\delta \leqslant \frac{\int_C f_T(\theta/r)\, d\theta}{\int_C f_T^*(\theta/r)\, d\theta}$$

(12) For all C, $$\left| \int_C f_T(\theta/r)\, d\theta - \int_C f_T^*(\theta/r)\, d\theta \right| \leqslant \max(\delta, \varepsilon) + \gamma + \alpha$$

The example used by Edwards, Lindeman, and Savage to illustrate stable estimation is this: you feel somewhat ill, and propose to take your temperature with a clinical thermometer yielding readings with a standard deviation of 0.05° and a negligible systematic error. You get a reading of 101.0°. Remember that you had some prior distribution $f_T(\theta)$ for temperature before you took the reading; we are looking now for the conditional distribution of temperature, *given* the thermometer reading. Let B be the region extending from 100.8° to 101.2° – four standard deviations on each side of the observation. Then α is less than 10^{-4}. The number k in (4) is essentially irrelevant; what matters is the number β, which represents the increase in $f_T(\theta)$ over its minimum value in the interval 100.8° to 101.2°. Edwards, Lindeman, and Savage argue for 0.05 as a reasonable value for β. Finally let us consider η. It may seem reasonable to suppose that there is no hundredth of a degree outside of B which is more than twice as likely as the least probable 100th of a degree inside B; but just to be on the safe side let us take η to be 100. We find, then, that $\gamma = \alpha\eta = 10^{-4} \times 10^2 = 0.01$; $\delta = 1 - [(1+\beta)(1+\gamma)]^{-1} < 0.06$; and $\varepsilon = (1+\beta)(1+\alpha) - 1 < 0.051$. By (11), then,

$$1 - \delta \leqslant \frac{\int_C f_T(\theta/r)\, d\theta}{\int_C f_T^*(\theta/r)\, d\theta}$$

for any C in B, so that for computing the conditional probability that θ is in C, we will not be off by more than 6%. By (12), for any C, whether in B or not, the difference between the measure of C computed on the supposition of uniform prior density in B and the true conditional measure of C will be less than 0.068.

Now let us abandon the unrealistic and demanding assumption that we have a precisely defined prior distribution for our temperature. Are we worse off? Not at all. While we may not have accumulated a great store of knowledge concerning the general relationship between the way we feel and the temperature we actually have, we do have some knowledge about it – enough to narrow down the possible distributions to a set S, each member of which satisfies assumptions (4) and (5). We may then go through exactly the same argument as the subjectivist goes through, to conclude that whatever member of S be true, the frequency with which θ will lie in C will differ from the number calculated on the basis of a uniform prior by less than 0.068; that for any C included in B the difference between the artificial and the true measure of C will be less than 6%; and so on. These measures bound the measures which yield the probabilities we are interested in, subject to conditions regarding randomness analogous to those mentioned earlier. That is, if my temperature today is a random member of the set of my temperatures on days when I feel as I do today, and observe a thermometer reading of 101.0, then the *probability* – the regular, epistemic probability – that my temperature falls in a certain interval C falls within the intervals calculated from (11) and (12) above.

Thus the theory of stable estimation, though developed to aid the subjectivist statistician confused about his prior distribution, is splendidly suited to provide bounds on the epistemological probability interval, just so long as the conditions on randomness are satisfied. Indeed, in many respects, the theory of stable estimation seems to fit the present approach to probability even better than it fits the subjectivistic approach.

6. There is really nothing more to Bayesian inference (in principle) from our point of view than we have discussed in these few examples. The major points of difference between our form of Bayesian inference and the subjectivistic form are (1) we require that the prior distributions be based on statistical knowledge, rather than mere opinion; and (2) we do not require that a single prior distribution be employed, but rather allow that our prior knowledge may be represented by a whole family of distributions. Where the standard subjectivist analysis claims that the posterior distribution of a parameter is largely determined by the likelihood function when the prior distribution is 'gentle', our analysis would claim that the family of posterior distributions in our body of knowledge will be narrow and determined largely by the likelihood function when the family of prior distributions includes none that is significantly 'ungentle'. Furthermore, for the posterior distribution to determine probabilities requires that certain conditions regarding randomness be met. To make a probability statement about the population parameter, given knowledge of the outcome of a sampling test, we require that the population be a random member of the set of populations whose sampled subsets are such as that we have observed, with respect to having a population parameter in a specified Borel set. For example: a selected coin subjected to a test consisting of n trials, r of which results in heads, is to

be a random member of the set of all coins subjected to a test consisting of n trials, of which r produce heads, with respect to having a long run frequency of heads in (say) the interval (0.5, 0.6). Alternatively, we may require the complex object consisting of the selection of a coin and the performance of n tosses of the coin, to be a random member of the cross product of coin selections and the set of n-fold sets of tosses yielding r heads. To make a probability statement about the likelihood that a future set of m trials will result in s heads, we may require that the selection of the coin, the test consisting of n trials, of which r result in heads, and the performance of m subsequent trials, be a random member of the set of all selections of a coin, followed by the performance of n tirials, r of which result in heads, followed by m subsequent trials, with respect to having s heads turn up on the subsequent trials. These conditions regarding randomness are complex, but they lie at the very heart of the application of probability as a guide to the future. In the subjectivistic account they are concealed by subjective judgments; in the account offered here, they are objective in that they depend explicitly and formally on the body of knowledge, real or hypothetical, that is being considered.

NOTES

[1] Pratt, Raiffa, and Schlaifer, *Introduction to Statistical Decision Theory*, New York, 1965, Chapter 11, page 11.

[2] Edwards, Lindeman, and Savage, 'Bayesian Statistical Inference for Psychological Research', *Psychological Review* **70** (1963), 193–242.

[3] *Ibid.* Alternatively, see Kyburg, *Probability Theory*, Englewood Cliffs, 1969, pp. 243–252.

CHAPTER FOURTEEN

THE FIDUCIAL ARGUMENT

1. The fiducial argument has been of considerable interest for a long period of time, despite the fact that no one (except possibly R. A. Fisher himself) was altogether clear as to what a fiducial argument is. When it applies it has a great deal of appeal, and there are relatively uncontroversial circumstances under which it applies. In the present section, we shall investigate such an uncontroversial example within the framework provided by the preceding chapters.

Suppose that we know that z has a normal distribution in a certain reference sequence, and that that is all we know about z. Let the reference sequence be y. This is just to say that our statistical knowledge about y and z is represented by the following distribution statement:

$$(1) \qquad D\langle y, z, \{N(\mu, \sigma): \mu \in \mathbb{R} \wedge \sigma \in \mathbb{R}^{+}\}\rangle,$$

where $N(\mu, \sigma)$ represents the normal distribution function with parameters μ and σ.

Now consider the sequence of n-sequences drawn from the union of the range of y; let

$$(2) \qquad y^{n*} = \imath f\,(f \text{ is } \omega\text{-sequence} \wedge \bigwedge_{i \in \omega} (f\langle i\rangle = \\ = \{s: s \text{ is } n\text{-sequence} \wedge \bigcup \mathscr{R}s \subset y\langle i\rangle\}))$$

We define the following three random quantities on y^{n*}:

$$(3) \qquad \bar{z}_n = \imath f\,(f \text{ is } fun \wedge \mathscr{D}f = y^{n*} \wedge \bigwedge_{t \in \mathscr{D}f} f\langle t\rangle = \frac{1}{n} \sum_{i<n} z\langle t\langle i\rangle\rangle)$$

The quantity $\bar{z}_n$ is just the average of the z-values of the items in a sample of n; it is otherwise known as the sample mean.

$$(4) \qquad z_s = \imath f\,(f \text{ is } fun \wedge \mathscr{D}f = y^{n*} \wedge \bigwedge_{t \in \mathscr{D}f} f\langle t\rangle = \\ = \sqrt{\sum_{i<n} (\bar{z}_n\langle t\rangle - z\langle t\langle i\rangle\rangle)^2})$$

This is otherwise know as the sample standard deviation.

(5) $$z^* = \imath f \left(f \text{ is } fun \wedge \mathscr{D} f = y^{n^*} \wedge \bigwedge_{t \in \mathscr{D} f} f\langle t \rangle = = \sqrt{n-1}\, \frac{\bar{z}_n \langle t \rangle - \mu}{z_s \langle t \rangle} \right)$$

The quantity z^* is the pivotal quantity required for fiducial inference; we shall have considerably more to say about it shortly.

From (1) and the definitions (2)–(5) it is provable that the distribution of z^* in y^{n^*} is given by the student t distribution with $n-1$ degrees of freedom; we denote this distribution by t_{n-1}. Thus we have as an item of statistical knowledge:

(6) $$D\langle y^{n^*}, z^*, \{t_{n-1}\}\rangle$$

Let us now suppose that a is a member of y^{n^*} – i.e., an n-sequence of members of $\bigcup \mathscr{R} y$. In particular let us suppose that, relative to our body of knowledge, a is a random member of $\bigcup \mathscr{R} y^{n^*}$ with respect to having a z^*-value in b, where b is any purely mathematical term denoting a Borel set. That is, we suppose that

(7) $$RAN_L(a, y^{n^*}, z^*, b, w)$$

where w represents out body of knowledge. It then follows (from MT-10.1) that the probability that $z^* \langle a \rangle \in b$ is exactly $\left[\int_b dt_{n-1}, \int_b dt_{n-1}\right]$:

(8) $$PROB_L(\ulcorner z^* \langle a \rangle \in b \urcorner, w) = \left[\int_b dt_{n-1}, \int_b dt_{n-1}\right]$$

For example, suppose that $n=25$, and that we choose b to be $(-2.787, 2.787)$. That means that (due to (6) and (7)) the probability that $z^* \langle a \rangle \in$ $\in (-2.787, 2.787)$ is $[0.99, 0.99]$.

But (again referring to definitions (2)–(5))

(9) $$\vdash_L \ulcorner z^* \langle a \rangle \in (-2.787, 2.787) \leftrightarrow \bar{z}_n \langle a \rangle - -2.787 (z_s \langle a \rangle / \sqrt{n-1}) \leqslant \mu \leqslant \bar{z}_n \langle a \rangle + 2.787 (z_s \langle a \rangle / \sqrt{n-1}) \urcorner$$

This biconditional is thus in our body of knowledge, and, again subject to (7), the probability is [0.99, 0.99] that the parameter μ lies within the indicated limits.

Nothing has been said so far about the time of sampling. The term a, for example, may be 'the next n items from the reference sequence y we happen to look at'. But it may also be 'the last n items from the reference sequence y we looked at'. In the former case, we have

$$(10) \qquad RAN_L(a, y^{n*}, \bar{z}_n, b, w)$$

and therefore

$$(11) \qquad PROB_L(\ulcorner \bar{z}_n\langle a\rangle \in b \urcorner, w) = [0, 1]$$

We also have

$$(12) \qquad RAN_L(a, y^{n*}, z_s, b, w)$$

and therefore

$$(13) \qquad PROB_L(\ulcorner z_s\langle a\rangle \in b \urcorner, w) = [0, 1]$$

as well as the randomness and probability statements already mentioned.

In the latter case, we suppose ourselves to know the values of $\bar{z}_n\langle a\rangle$ and $z_s\langle a\rangle$, and thus $PROB_L(\ulcorner z_s\langle a\rangle \in b\urcorner, w) = [1, 1]$ or $[0, 0]$ according as $z_s\langle a\rangle$ is in b or not; similarly for $\bar{z}_n\langle a\rangle$. We do not have either (10) or (12). But under the circumstances described, we do still have (7) and therefore (8) and therefore also

$$(14) \qquad PROB_L(\ulcorner z_n\langle a\rangle - 2.787(z_s\langle a\rangle/\sqrt{n-1}) \leqslant \mu \leqslant \bar{z}_n\langle a\rangle + \\ + 2.787(z_s\langle a\rangle/\sqrt{n-1})\urcorner, w) = [0.99, 0.99]$$

All of this holds (barring peculiar contents in w) for any borel term b, and thus we have a general posterior 'distribution' for μ induced by the distribution we have for z^*. There are, of course, any number of Borel sets whose measure under the t_{n-1} distribution is 0.99 – it is our desire to get the most information out of our evidence that leads us to choose the symmetrical interval $(-2.787, 2.787)$. Furthermore, note that μ itself is not a random quantity distributed in any reference sequence. Thus the insights of Neyman and others who refuse to regard fiducial probabilities as 'real' probabilities are to a certain extent vindicated. Fisher's counter-argument [1] is beside the point. Nevertheless, although μ has no distribution in this instance on the view of probability presented here, we may obtain probability statements about μ that are just as good – and just as soundly based in statistical knowledge – as probability statements about z^*. It is this indirect statistical basis that provides for the possibility of sampling indefinitely 'to demonstrate the correct frequency'.

2. A number of conditions must be met before the fiducial inference will go through. Some of these conditions are reflected in a natural and simple way in our epistemological system. One such condition concerns 'recognizable subsets'. The random quantity z^* defined in (5) has as its domain the set of n-sequences of items from the domain of z. Since z^* has the same distribution for every set of n-sequences from every normal population, the derived distribution "is applicable to the enlarged set of all pairs of values [(z^*, μ)] obtained from all values of [μ]. The particular pair of values of [μ] and [z^*] appropriate to a particular experimenter certainly belongs to this enlarged set, and within this large set the proportion of cases satisfying the inequality [9] is certainly equal to the chosen probability P. It might, however, have been true... that in some recognizable subset to which this case belongs, the proportion of cases in which the inequality was satisfied should have some value other than P. It is the stipulated absence of knowledge *a priori* of the distribution of [μ], together with the exhaustive character of the statistic [z^*], that makes the recognition of any such subset impossible, and so guarantees that in this particular case ... the general probability is applicable."[2]

There are a number of comments that may be made on this passage. First, while it is true that the distribution is applicable to the enlarged set obtained by considering a set of populations, with a normally distributed random quantity defined on each of them, and, corresponding to each population, a set of n-sequences constituting samples from that population, it is not *necessary* to consider any population other than the domain of z. The particular n-sequence we sample from the population y^{n*} may well be a random member of y^{n*} with respect to having a z^* value less than 2.787, relative to our body of knowledge; if so, the probability is 0.99 that its value is less than this, and thus the probability is 0.99 that μ lies within the limits mentioned in (9).

We may *also* consider the enlarged set referred to by Fisher. In this enlarged set, comprising a set of populations, in each of which a certain quantity is, we know, normally distributed, we have several random quantities to consider. The mean of the normally distributed quantity in each population is now a perfectly good random quantity, whose domain is the set of populations under consideration. The random quantity z^* is also a perfectly good random quantity – though its meaning changes from population to population, since each population is the reference set of a different random quantity z – and in each population it has a student's distribution with $n-1$ degrees of freedom. What we require, in order that the argument go through from our point of view, is that the n-sequence sampled be a random one with respect to having an absolute z^*-value less than 2.787. We must, of course know which population we are sampling (else we would not know how to calculate $\bar{z}$ and z_s and knowledge about z^* would not give us knowledge about the appropriate μ). If we suppose that the random quantity is the same for each population (though it may have a different mean and variance in each population) we may, in effect, draw a population and a sample, compute the value of $\bar{z}$ and z_s, and then *if* the pair (population, sample) is a *random member* of the set of pairs of populations and n-samples from them, we may convert a distribution for z^* into a 'distribution' for μ in the particular population sampled. This condition of randomness will be met just in case we know of no set of terms (k, b, c, d) such that we know $c(k) \in d$ if and only if $|z^*(a)| \leqslant 2.787$; k is known to belong to b; and the frequency with which objects in b have c-values in d is different from 0.99; and there is no proper subset of b in which a corresponding frequency *is* 0.99; and there is no product set of which b is a projection, in which a corresponding frequency is 0.99. Now this clearly includes Fisher's stipulation that there be 'no recognizable subset leading to some value other than P', but it is equally clearly rather more broad. It is furthermore quite inadequate simply to stipulate that the inference will be valid just in case we have no knowledge of the distribution of μ among the populations. (We may, for example, have it on good authority that the z-value of the first item in our sample will lie within 0.01 of μ, with probability 0.999. If this conflicts with the knowledge we obtain from $z^*(a)$, it is the former which should prevail, since it is based on a narrower reference class.)

Fisher goes on to say, "... had there been knowledge *a priori*, the argument of Bayes could have been developed, which would have utilized all the data, and which would in general have led to a distribution *a posteriori* different from that to which the fiducial argument leads. Bayes' method in fact calculates the distribution of μ in a particular subset of pairs of values $[(z^*, \mu)]$ defined by $[z^*]$, and to which therefore the observation belongs."

It is not generally the case, however, that we are either in total ignorance regarding the distribution of μ in a population of populations, or we have knowledge of the exact distribution of μ, in order to apply Bayes' theorem in a straightforward way. And Fisher's suggestions do not give us any way of dealing with the common instances in which we have some, but by no means complete, knowledge of the distribution of μ. Our own suggestions with regard to this question will be mentioned in the following chapter.

Within a classical framework, Robert Buehler takes the question of relevant subsets seriously, and provides explicit definitions of 'relevant' and 'semirelevant' subsets. The considerations he adduces are presented in a classical frequentist framework, and therefore it is difficult to make the connection to bodies of knowledge. Nevertheless he correctly sees that what is at issue is a "generally accepted notion of an 'appropriate reference set' for inferences."[3]

In the present system, knowledge concerning the distribution of μ in the population of populations will interfere with the randomness of a in y^{n*} with respect to having a z^*-value in b, just in case that knowledge leads to differing measures for some equivalent statement. Although there is no simple way in which to give a general characterization of the kind of knowledge which will have this effect, the general principle does not seem excessively difficult to follow in practice.

3. Another problem arising for fiducial inference concerns the existence of a variety of 'exhaustive estimates' and a corresponding variety of 'pivotal quantities'. In the case of a single continuous parameter, the requirements of sufficiency and monotonicity are apparently adequate to make the fiducial distributions unique.[4] Where several parameters are involved, however, the situation is strikingly different.[5] One response of Fisher was to insist that the joint fiducial distribution of several parameters should be built up 'rigorously by a step by step process'.[6] But Demptster has now exhibited a step by step approach that leads to "alternative inconsistent answers even in the basic two parameter situation of sampling

from the normal distribution."[7] Barnard apparently feels that at least in the constructions offered by Stein, Creasy, and Fieller, differences of underlying logical structure can be detected, so that in fact these are not alternative fiducial distributions for exactly the same situation; the underlying logical structures are different. As Barnard writes: "It is evident that a fiducial distribution, especially one involving several parameters, is a frail thing, easily upset by additional information. In this respect, however, it does not differ very much from more familiar probability distributions. If we are playing poker against an invisible opponent, or one with a poker face, the hand he holds has a perfectly definite (though complicated) distribution randomly chosen from the set of cards other than those we hold; but if he has not a poker face, and we notice a twinkle in his eye, our probability distribution is upset, in a way it is difficult to specify precisely."[8]

In a similar vein, J. S. Williams,[9] by distinguishing "between one event conditional on another, and one event given the occurrence of another," in a manner much like that in which we distinguished between stochastic and epistemological independence, and between stochastic and epistemological conditionality, argues that there is a formal and explicit way to unpack the logical differences encountered in different formulations of fiducial arguments, and that in the case of those fiducial arguments that are epistemologically valid (my phrase) there exist restrictions whose satisfaction is sufficient to warrant that validity.

Whether or not Dempster's alternative constructions depend on variation in the underlying logical framework is hard to tell; in any event even Fisher produced a fiducial distribution for the means of a bivariate normal distribution[10] which is inconsistent with the marginal distribution derived from his general fiducial distribution for the means, variances and correlation coefficient for the bivariate normal distribution.[11]

The Behrens-Fisher problem, concerning differences of two means of normally distributed quanties with unknown variances, has been the most common cource of difficulties and doubts (perhaps of inconsistencies) concerning the fiducial argument. Nevertheless it is clear that the difficulties have a much deeper and more general domain than is indicated by this one problem, pervasive though the problem is. It may be that the considerations developed in our chapters concerning randomness and probability will yield an explication of the logical differences underlying allegedly inconsistent fiducial analyses. On the other hand, it may be that just as the set of reference terms had to be made dependent on a language, and could not be construed as the whole set of terms of a language (cf. Chapter 8), so also not every expression (functor) in a language that denotes a function having as its domain a reference class and having its range included in the set of *n*-sequences of reals, will turn out to be a plausible random quantity. It may be, that is, that just as in selecting a language in which to perform a bit of science or decision making we implicitly choose a set of classes that may function as reference classes, so in selecting a language in which to perform a bit of science or decision making we implictly choose a set of functions that may function as random quantities. If this is so, one would suspect that the difficulties would arise mainly in multidimensional random quantities, in which the components were not logically independent. This, indeed is the direction Williams' argument takes.

There is a third alternative that lies between these two. The fiducial argument, according to Fisher, will only work when we have continuous parameters and continuous observables. Yet according to the foregoing framework, continuous distributions must be regarded as the limiting cases of what are, strictly speaking, discrete distributions. Now whenever one begins with something discrete and approaches something continuous it can make all the difference, as we all know, exactly how that limit is approached. This, too could be a factor in the 'paradoxes' concerning multiparameter distributions. If this were a factor, it might (or might not) suggest an underlying difference in the logic of what is 'taken for granted', and thus might lead us back to the first alternative: namely, that there is a solution to these difficulties implicit in our discussions of randomness.

NOTES

1 Fisher, R. A., *Statistical Methods and Scientific Inference*, New York, 1956, p. 59.

2 *Ibid.*, pp. 54–55.

3 *Ibid.*, p. 55.

4 Buehler, R. J., 'Some Validity Criteria for Statistical Inferences', *Annals of Mathematical Statistics* **30** (1959), 845–863, p. 860.

5 Brillinger, D. R., 'Examples Bearing on the Definition of Fiducial Probability with a Bibliography', *Annals of Mathematical Statistics* **33** (1962), 1349–1355.

6 Brillinger, *op. cit.*; Mauldon, J. G., 'Pivotal Quantities for Wishart's and Related Distributions, and a Paradox in Fiducial Theory', *J. Royal Statistical Society, Series B* **17**

(1955), 79–85; Stein, C., 'An Example of Wide Discrepancy between Fiducial and Confidence Intervals', *Annals of Mathematical Statistics* **30** (1959), 877–880; Creasy, M. A., 'Limits for the Ratio of Means', *J. Royal Statistical Society, Series B* **16** (1954), 186–194; Fieller, E. C., 'Some Problems in Interval Estimation', *J. Royal Statistical Society, Series B* **16** (1954), 175–185.

[7] Fisher, *op. cit.*, p. 172.

[8] Dempster, A. P., 'Further Examples of Inconsistencies in the Fiducial Argument', *Annals of Mathematical Statistics* **34** (1963). 884–891, p. 884.

[9] Barnard, G. A., 'Logical Aspects of the Fiducial Argument', *Bulletin of the International Statistical Institute*, Ottawa, 1963, pp. 870–883, p. 880.

[10] Williams, J. S., 'The Role of Probability in Fiducial Inference', *Sankhyā, Series A* **28** (1966), 271–296.

[11] Fisher, R. A., 'Discussion of the Symposium on Interval Estimation', *J. Royal Statistical Society, Series B* **66** (1954), 212–213.

[12] Fisher, R. H., *Statistical Methods and Scientific Inference*, New York, 1956.

CHAPTER FIFTEEN

CONFIDENCE METHODS

1. The requirements of the fiducial inference, as laid out by Fisher, entail that the observable quantity as well as the parameter about which the inference is being made be distributed continuously. A simple, pervasive, and important instance in which this condition is not met, is that of estimating the parameter p of a binomial distribution. From a philosophical point of view, in which our chief concern is with the problem of inductive logic, this may well be the most important problem of all: it is the problem, basically, if analyzing the inference that the proportion of A's that are B's is likely to be close to the proportion of *observed* A's that are B's.

The official doctrine of confidence intervals was described briefly in Chapter Two. We shall recapitulate it here, as it applies to this special case, first in order to see what the conditions are under which the theory is applicable in the epistemological framework; and second in order to reveal the profound differences in the analysis of the underlying mathematics on the classical and epistemological views.

Official Doctrine: We have a population P, of which an unknown proportion p have the property Q. We can design an experiment of the following sort: by a randomizing method – one which draws each member of the population with the same long run frequency, and draws the members independently so that each sample of n is drawn with the same long run frequency – we shall draw a sample of n members of the population. The frequency of Q's in such a sample of n, say f_n, is a perfectly good random quantity. Given the value of p, we can compute the probability that f_n will have a certain value, by means of Bernoulli's theorem, taking account of the fact that our method of drawing the sample is 'random' in the appropriate frequency sense. Let us choose a nice large number, such as 0.95; we may then compute, for each value of p, limits $f_{n*}(p)$ and $f_n^*(p)$ such that the probability is 0.95 that f_n will lie between $f_{n*}(p)$ and $f_n^*(p)$ and the difference between $f_{n*}(p)$ and $f_n^*(p)$ is a minimum. (That seems a desirable property.) Thus if p is 0.2, the probability that r or less of the n items in our sample will be Q's is

$$\sum_{x=0}^{x=r} \frac{n!}{x!(n-x)!} p^x (1-p)^{n-x}$$

We find that value of r for which this is less than 0.025 (equal tailed intervals being shortest); and then we find that value of r for which the upper tail has a measure of at most 0.025. If we do this for a lot of values of p, we get a graph like Figure 9.

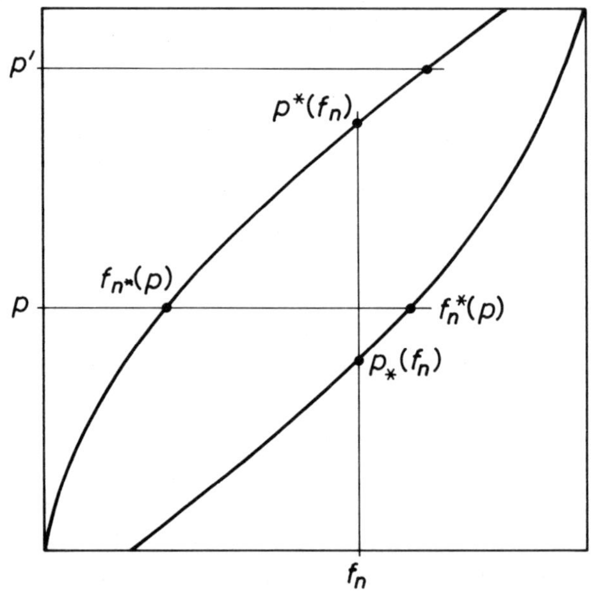

Fig. 9.

Let f_n^* and f_{n*} correspond to the upper and lower curves. Thus for *every* value of p the probability that the interval (f_{n*}, f_n^*) will cover f_n is 0.95. Now look at the curves as functions (not single valued, to be sure) of f_n: $p^*(f_n)$ and $p_*(f_n)$. Can we speak of the probability that the true value p will fall in the interval $(p_*(f_n), p^*(f_n))$? No, since p is not a random quantity. But the interval (p_*, p^*) is a random quantity (being a function of random quantities), and it makes good sense to ask what the probability is that that interval will cover p. Consider a fixed value of p. What is the probability that p^* will be less than p? That is a perfectly good question (in classical terms), and the answer is that the probability is less than 0.025, for, for p^* to be less than p entails that f_n is less than $f_{n^*}(p)$, which happens less than 0.025 of the time. Similarly the probability that p_* is more than p is at most 0.025. Thus the probability is at least 0.95 that the random interval (p_*, p^*) covers p. Now, at long last, we perform our experiment. We observe f_n. As prearranged, we assert that p is in the interval $(p_*(f_n), p^*(f_n))$. We do so, the argument goes, because we will be right 95% of the time in the long run, if we make a habit of making such assertions. We do not and cannot say that the probability that p is in that interval is 0.95. Nor can we say that that *particular* interval includes p with probability 0.95. As announced earlier, p is not a random quantity, and is what it is – either in or out of that interval.

Four things about the official doctrine should be noted: First, that the method of sampling P must be such that each sample of n has the same probability (frequency) of being sampled.

Second, we must decide in advance how large the sample is to be, and what our 'confidence coefficient' is to be.

Third, the outcome of the confidence method is not a probability statement concerning p, but a confidence statement concerning p.

Fourth, there must be no way in which we can use a Bayesian argument: that is, P must not be regarded as being in turn drawn from a population P^*, in which the distribution of subpopulations with various fractions of Q in them is 'known'.

2. The conditions for the application of the confidence method in the epistemological framework are quite different. First, there is no need for our sample to be drawn by a method which draws each equinumerous sample equally often in the long run. (To have such knowledge about a method of sampling would be already to know a lot about the world, which would be inappropriate for a basic inductive method; and, as we shall see, such knowledge is irrelevant to the cogency of the argument anyway.) We do require, instead, that the sample be a random member of the set of n-sequences of items from P (distinct, or not, according as our sampling is without or with replacement) with respect to a certain random quantity (to be specified) falling in a certain Borel set (to be specified). What is the random quantity? To keep things simple, let it be

that quantity z, whose domain is the subsets of P, and whose value for a given subset of P is 1, if that subset bears a frequency of Q's f_n such that f_n lies between $f_{n*}(p)$ and $f_n^*(p)$. The Borel set in question is, of course $\{1_r\}$. When will it be the case when this assertion regarding randomness is false? Precisely when we *know*, or have adequate *reason* to believe, that this particular sample S comes from a subset of the sample space in which the frequency with which z-values fall into the Borel set $\{1_r\}$ differs from that frequency in the sample space as a whole. Or, of course, if in our body of knowledge there are terms a, b, c, d, such that $\ulcorner b(a) \in c \urcorner$ is connected in our body of knowledge to $\ulcorner z(S) \in \{1_r\} \urcorner$, and our knowledge about a, b, c, and d, to which a is known to belong, *prevents* S from being a random member of the sample space with respect to having a z-value in $\{1_r\}$. In typical cases this latter alternative is not one that needs to be taken very seriously, although there do arise special instances of confidence interval arguments in which just this sort of thing undermines the argument. What is more typical, more to the point, and more often the subject of some degree of controversy, is the possibility that S is known to belong to a subset of the sample space in which the frequency with which z-values fall in $\{1_r\}$ is different from the frequency in the whole sample space. It is this, in fact, that inspires the classical requirement that the sampling method be statistically random. It is worth observing, then, that the statistical randomness requirement is neither necessary nor sufficient to insure randomness in the epistemological sense required for the cogency of the inference.

Suppose that the sampling is done by a method we have every reason to believe is statistically random – i.e., which is such that in the long run, every possible sample of n will be chosen equally often, or which will, in the long run, select every point of the sample space equally often. Let T be the particular sample we have obtained. It is perfectly possible (indeed, given the assumption of randomness, it is statistically inevitable!) that T should consist entirely of items with the property R, and that we know (a) that the frequency of Q among items with R is markedly different from the frequency of Q among items lacking R; and (b) that only a small proportion of the items in P have the property R. In this state of knowledge, although T is a perfectly adequate 'statistically random' sample, it is not epistemologically random. We know that it belongs to a special subset of the sample space (that in which all the items have the property R), and that the frequency with which such items have z values in $\{1_r\}$ is very low. If the frequency of Q in T is f_t, we cannot assert that the probability is $1-\varepsilon$ that p lies between $p_*(f_t)$ and $p^*(f_t)$.

The expected comment of the confidence limiter, is that there is a confidence limit argument in favor of stratified sampling, and that that is what is appropriate here. But although there is an argument in favor of stratified sampling, it is false that there is an argument against unstratified sampling. The conditions stipulated for the application of the confidence limit method are met: the experiment is such (in virtue of the statistical randomness assumption) that the computed interval will, in the long run, cover the true value of the parameter at least $1-\varepsilon$ of the time.

Neither is statistical randomness necessary for the epistemological randomness of the sample. Suppose that there is some natural spatial or temporal ordering of the items in P, and that we take as our sample V, obtained by choosing every 10th item of P (in the natural ordering) until we have a sample of n. There is no way in which this sample can be construed as havfng been obtained by a statistically random method (unless you want to go to such an extreme as to consider God the Divine Dice Shaker). And yet, in relation to perfectly plausible and natural bodies of knowledge, there is no reason that V should not be a random member of the set of sample points with respect to having a z-value in $\{1_r\}$. Indeed, a body of knowledge in which V would not be a random member of P^n would be quite unusual: we would have to know that there was a connection between the presence or absence of Q, and location in the ordering. And knowing this, we would be interested in the details of that relation rather than the bare proportion of Q's among P's.

The second requirement from the classical point of view is that the experiment must be designed in advance. We must decide what our confidence coefficient is to be, how large the sample is to be. We may incorporate a randomized procedure in order to obtain exactly the right confidence coefficient. We must definitely not interrupt our experiment part way through, just because we have enough information to settle the question we had in mind to start with. Optional stopping is out. And the likelihood principle – the principle that says that the evidential import of the evidence is completely embodied in the likelihood function – is false.

From the present point of view, the pragmatics of the situation are irrelevant. It couldn't matter in the least whether we decided before, or after, or never what our confidence coefficient is to be: The evidence is what it is, and has the epistemological import that it has, regardless of our hopes, fears, expectations, intentions or what have you, but not, of course regardless of what we *know*. On the other hand there are special cases in which these attitudes may have a bearing on the characterization of an experiment. Let us consider, as a case in point, optional stopping. Suppose that we do not choose n in advance, but rather collect items of P until f_n is a half. We cannot 'choose' to do that, since it may well be that it will never be the case that f_n is a half. But it is conceivable that in the process of collecting items of P, we will get a sequence of frequencies of Q that generally tend to lie around a quarter, but, due to a long run of Q's at a certain point, the frequency goes up to a half. If we want to show that the frequency of Q is a half, the argument goes, and we are free to stop collecting evidence whenever we want, we'll stop then, and our results will be misleading. The counterargument is that the likelihood function for that experiment is just the same as the likelihood function for an experiment which happened to choose in advance a sample of exactly that size, and that therefore the evidential content of the two experiments is the same. From the present point of view, what counts is simply whether or not the sample is a random member of the set of subsets of its size, relative to what we know. It is possible that it is not: in which case the argument is undermined. If so, it is not undermined simply by the fact that the experiment was terminated arbitrarily, but by the fact that the set of equinumerous experiments, *terminated in that way*, is known to exhibit an abnormal frequency of instances in which the z-values belong to $\{1_r\}$. For example, consider a coin tossing experiment with the following 'termination rule': Begin tossing the coin; at the first occurrence of tails, terminate the experiment temporarily; continue to toss, and reinitiate the experiment with the first occurrence of heads; again terminate the experiment with the first occurrence of tails, but keep tossing; and so on. A possible sequence might be: (bracketed outcomes are not part of the experiment):

THTHHT [T] HT [TTT] HHT [TTT] HHT [T]
HTHHHT [T] HT [TT] HTHTHHT [T] HHHHT...

In this case we know perfectly well that the proportion of z values in $\{1_r\}$ is less than it is in general. Of course this experiment doesn't have the same likelihood function as a normal one either; but it is the identifiability of a subset in which the frequency of z-values in $\{1_r\}$ is abnormal that is what counts epistemologically.

On the issue of randomization to obtain the right confidence coefficient, we may observe the following: In ordinary cases, the experimenter knows the outcome of the auxilliary experiment, and hence knows to which subset of the possible results the 'observed' result belongs. Suppose, in order to achieve exactly the right confidence coefficient, we design the following experiment: we draw a sample of n, and observe f_n; we then draw from a deck of k cards, of which j are red, and if we get a red card, we assert that p lies in $(p_{r*}(f_n), p_r^*(f_n))$; if we get a black card, we assert that p lies in $(p_{b*}(f_n), p_b^*(f_n))$. We know that p lies in the former interval with frequency $1-\varepsilon_1$ and in the latter interval with frequency $1-\varepsilon_2$; and $j/k(1-\varepsilon_1)+(1-j/k)(1-\varepsilon_2)$ is just exactly our desired confidence coefficient. (To be sure, this is the wrong sort of example for that sort of thing; for a more plausible example, see Chapter VI.) In ordinary cases, the experimenter knows the outcome of the auxiliary experiment, and hence knows to which subset of the possible results the 'observed' result belongs, and hence knows whether the probability is $1-\varepsilon_1$ or $1-\varepsilon_2$ that p lies in the computed interval. The probability is determined by what he knows. Now it is also possible to consider cases in which the overall frequency is indeed the probability. The manager, who has an assistant who actually does the testing, may well not know in a particular case whether a red card or a black card was drawn, and for him, the probability may be the canonical $(j/k)(1-\varepsilon_1)+(1-j/k)(1-\varepsilon_2)$.

According to Pratt (as mentioned in Chapter VI) randomization in this sense is not followed by practical statisticians (though perhaps it should be, according to the classical confidence method theory). As Birnbaum says, "The usual practise in applied statistics,... is to restrict consideration to nonrandomized confidence region estimators." In point of fact it seems that there are cases where randomization should be employed (quality control is one), as well as cases where it should not (most informative inference), and the epistemological approach differentiates clearly and plausibly between them.

Another remark concerning the classical interpretation of the confidence method was that it did not issue in probability statements. It is obvious that under the epistemological interpretation of probability, the result of a confidence interval inference *is* a probability, and a probability in the same way and same manner as a probability issuing from a Bayesian inference or a fiducial inference.

The fourth point mentioned was that in order for the confidence interval techniques to be applicable, there must be no way in which we can use a Bayesian argument: the population P must not be regarded as being drawn from a population of populations, P^*, in which the distribution of populations with various fractions of Q in them is known. Here again there is a region of vagueness and confusion, on which the present approach throws some light. Everything can in some sense be regarded as being drawn from a population of populations. What counts is not whether there is any way of construing the population P as being 'drawn' from a superpopulation P^*, but rather whether there is any way of doing so which will yield information about the frequency of Q in P, given the observed sample. We are interested in informative inference. We thus want to find the shortest interval such that the probability is $1-\varepsilon$ that p lies in that interval. In principle the problem is no different from those we have considered before. Consider the shortest $1-\varepsilon$ confidence interval obtainable from our observations: $(p_*(f_n), p^*(f_n))$. If the set S is a random member of P^n with respect to having a z-value in $\{1_r\}$, then the probability that p lies in this interval is (at least) $1-\varepsilon$. If S is not a random member of P^n with respect to having a z-value in $\{1_r\}$, in virtue of our prior knowledge about the distribution of p in the set of proportions P^*, of which P is a member, it is because we are led to a different probability by that knowledge. Let f_g be that function whose domain is the union of the power sets of the members of P^*, and whose value for any argument is just the proportion of Q's in that subset of a member of P^*. Suppose that in the sample S that we actually observe the proportion is $f_g(S)=f_n$. Let the set of distributions of p, given the observation f_n, be D_r; and let

$$p_i=\inf\left\{\int_{p_*(f_n)}^{p^*(f_n)} d\,w : w\in D_r\right\},\quad p_s=sup\left\{\int_{p_*(f_n)}^{p^*(f_n)} d\,w : w\in D_r\right\}.$$

Several cases may arise. If $1-\varepsilon$ lies between p_i and p_s, the probability that p lies between $p_*(f_n)$ and $p^*(f_n)$ is given by the confidence coefficient $1-\varepsilon$ – since that represents stronger statistical knowledge. If $1-\varepsilon$ is less than p_i, or if $1-\varepsilon$ is greater than p_s, the Bayesian knowledge of the distribution of p in P^*, together with the observation f_n conflicts with the statistical knowledge represented by the confidence coefficient. To resolve this conflict (without going to some altogether new set of terms) we must find a subset of one of these sets which agrees appropriately with the other, or we must show that one of these sets is a marginalization of a product set which has a subset agreeing appropriately with the other. (See the definitions of PND_L and SND_L in Chapter Nine.) Our intuitions suggest that the Bayesian inference should be given priority here – but how? The reference set (for simplicity we here replace reference to reference *sequences* by reference to reference *sets*) for the Bayesian conditional probability is $R=\{\langle a,b\rangle : a\in P^* \wedge b\in a^n \wedge f_g(b)=f_n\}$; an appropriate random quantity is z, whose value for the pair $\langle u,v\rangle$ is $\langle 1_r\rangle$ if $p_*(f_n)\leqslant p\leqslant p^*(f_n)$, and $\langle 0_r\rangle$ otherwise. As remarked, we have $\ulcorner S\langle R, z, \{\langle 1_r\rangle\}, p_i, p_s\rangle\urcorner$ in our body of knowledge.

How do we deal with the question of randomness in this complex situation? Let us list the various terms involved explicitly. For the original confidence interval analysis, we have:

$x'=S$, the sample actually chosen from P.

$y'=$an increasing sequence of n-membered subsets of P

$z'=$the function whose value is $\langle 1_r\rangle$ for a member u of y' if $p_*(f_g(u))\leqslant P_p\leqslant p^*(f_g(u))$, and $\langle 0_r\rangle$ otherwise, where P_p is the proportion of Q's in P.

$b'=\ulcorner\{\langle 1_r\rangle\}\urcorner$

For the Bayesian analysis, we have:

$x=\ulcorner\langle S, P\rangle\urcorner$, where S is the sample chosen, and P the population chosen y is a sequence such that

$\bigcup \mathscr{R}y=R=\{\langle u,v\rangle : v\in P^* \wedge u\in v^n \wedge f_g(u)=f_n\}$

$z=$the function whose value is $\langle 1_r\rangle$ for $\ulcorner\langle u,v\rangle\urcorner$ if $p_*(f_n)\leqslant P_u\leqslant p^*(f_n)$, and $\langle 0_r\rangle$ otherwise.

$b=\ulcorner\{\langle 1_r\rangle\}\urcorner$

We now show that even if our knowledge about y' and about y differs, y' does not prevent $RAN_L(x, y, z, b, w)$ by difference.

Let

$$y''' = \text{an increasing sequence of subsets of } P^*$$
$$y'' = \ulcorner \imath f\,(f \textit{ is } \omega\textit{-sequence} \wedge \bigwedge_{i \in \omega} f\langle i\rangle = \{w^n : w \in y'''\langle i\rangle\})\urcorner$$
$$y'''' = \ulcorner \imath f(f \text{ is } \omega\textit{-sequence} \wedge \bigwedge_{i \in \omega} f\langle i\rangle = y'\langle i\rangle \cap \{w : f_g(w) = f_n\})\urcorner$$
$$y^* = y$$
$$y^{*\prime} \text{ is sequence such that } \bigcup \mathscr{R} y^{*\prime} = \{\langle u, v\rangle : v \in P^* \wedge u \in v^n \wedge \wedge f_g\langle u\rangle = f_n\} = \bigcup \mathscr{R} y$$
$$x''' = P$$
$$b'' = b$$
$$z'' = z$$

Then,

$$\ulcorner \bigcup \mathscr{R} y^* \; Perm \bigcup \mathscr{R} y \urcorner \in w$$
$$\ulcorner \bigcup \mathscr{R} y^* \urcorner \; Incl_{w,L} \ulcorner \bigcup \mathscr{R} y'' \times \bigcup \mathscr{R} y''' \urcorner$$
$$\ulcorner \bigcup \mathscr{R} y^{*\prime} \urcorner \; Incl_{w,L} \ulcorner \bigcup \mathscr{R} y'''' \times \bigcup \mathscr{R} y''' \urcorner$$
$$\ulcorner \bigcup \mathscr{R} y'''' \urcorner \; Incl_{w,L} \ulcorner \bigcup \mathscr{R} y' \urcorner$$
$$\ulcorner \langle x', x'''\rangle \urcorner \; M_{w,L} \ulcorner \bigcup \mathscr{R} y^{*\prime} \urcorner$$
$$\ulcorner z''\langle\langle x', x'''\rangle\rangle \in b'' \urcorner \; B_{w,L} \ulcorner z\langle x\rangle \in b \urcorner$$
$$\ulcorner x' \prec x \urcorner \in w$$
$$PS_{w,L}(\ulcorner S\langle y, z, b, p_i, p_s\rangle \urcorner, y, z, b)$$
$$PS_{w,L}(\ulcorner S\langle y^{*\prime}, z'', b'', p_i, p_s\rangle \urcorner, y^{*\prime}, z'', b'')$$
$$\ulcorner S\langle y, z, b, p_i, p_s\rangle \urcorner \; E\text{-}STR_{w,L} \ulcorner S\langle y^{*\prime}, z'', p_i, p_s\rangle \urcorner$$

The conditions of MD-9.2 are thus satisfied, so that

$$(x', y', z', b') \; PND_L(x, y, z, b, w),$$

and therefore,

$$\sim y' \; D_L(x, y, z, b, w), \text{ by MD-9.3.}$$

Thus if we have *enough* prior information so that the Bayesian posterior interval actually *differs* from the confidence coefficient, then it is the case that the Bayesian form of the inference takes precedence. But we do not require (what is impossible or vague) that there be *no way* of assigning the population about which we are making an inference to a set of populations.

In general one may expect that confidence sets will be the appropriate yield of statistical inference in the absence of much knowledge, and that, as knowledge of a subject matter increases, conditional Bayesian intervals will become more often relevant. Randomized methods will only be relevant when we are in that rare circumstance of (a) have a repeated process of testing going on, and simultaneously (b) we have not obtained enough information from the repetitions to narrow down the background distributions to the point where Bayesian techniques are appropriate. The question of when Bayesian intervals based on prior known distributions supercede confidence intervals is a question which cannot even be raised in any of the alternative approaches to statistical inference: Bayesians and those who offer logical interpretations of probability will find themselves obliged to say 'always'; frequentists will essentially find themselves obliged to say 'never', since they say that only when prior distributions are *known* are the Bayesian techniques appropriate;[2] but it is never the case that a prior distribution is 'known' with perfect accuracy. The practising statistician will of course use that most precious commodity, judgment; but it is surely advantageous to replace judgment by reasoned argument when that is possible.

3. Another procedure of informative inference, and one which in number of applications in the sciences is probably greater than any other, is hypothesis testing. In psychology, in economics, in sociology, in medicine, the dominant mode of statistical inference is the test of a hypothesis, and the result of a piece of scientific inquiry is most often that a certain (null) hypothesis is rejected at the 0.05 (or 0.01) level. One does not generally obtain from a test of a hypothesis, a probability that the hypothesis is true or false, although one may in special cases. For example, one may test the hypothesis that the mean of a certain population does not lie in the interval (m_1, m_0), and use a confidence interval (or fiducial) argument to obtain rejection at the 0.05 level; this amounts precisely to asserting that the probability is 0.95 that the mean of the population *does* lie in the mentioned interval.

A number of the comments made concerning confidence methods apply equally to hypothesis testing. In particular, randomized tests are no help: we know the result of our card drawing as well as we know the result of the test, and if we decide to 'reject the hypothesis', given certain evidence, when a red card is drawn, and we get that evidence and draw a red card, the characteristics of that particular inference are just those that apply to the particular type experiment in which a red card is drawn, and not those of the whole experiment. For example, if, in order to get a test of size 0.05 we require to reject the hypothesis when E is observed a third of the time, and we accomplish this end by drawing a card from a deck of twice as many black cards as red, then in case we observe E, and draw a red card, the experiment is not one of rejecting the hypothesis at the 0.05 level: it is one of rejecting the hypothesis at the level corresponding to what we do when we observe a red card. What should determine our epistemological attitude toward a hypothesis is what we know at a given time, not what we knew some other time, or in general.

Furthermore, the same comments on statistically random samples are to the point: that a sample from a population be drawn in a statistically random way (so that each equinumerous subset of the population has the same statistical chance of being drawn) is neither a sufficient nor a necessary condition of the kind of randomness appropriate to an epistemological context, in which we are trying to learn something. Epistemological randomness, which in this context means randomness with respect to some property such as 'leading to the false rejection of the hypothesis H', depends on what we know (or reasonably believe) and not at all on the long run properties of the method by which we select our sample.

But there is no general and natural way of turning statistical tests of hypotheses into arguments which yield probability statements concerning the hypotheses except when they admit of a confidence interval or fiducial interpretation. It is thus not altogether clear that tests of statistical hypothesis are appropriate as a vehicle of scientific inference. This is not to say that they may not be useful guides to scientific investigation: a sensible hypothesis that fails to be rejected at the 0.01 level may be worthier of investigation than an analogous sensible hypothesis that only fails to be rejected at the 0.05 level. But most of the import of the 'level of significance', it seems to me, depends on the particular nature of the hypothesis, and the context in which it is offered.

There is one form of context, concerned with relatively small significance levels, in which an inductive argument does go through. We shall, however, consider this in another chapter, when we are concerned with the general problem of induction.

NOTES

[1] Birnbaum, Allan, 'Concepts of Statistical Evidence', in Morgenbesser *et al.* (eds.), *Philosophy, Science, and Method*, New York, 1969, pp. 112–143; p. 125.
[2] Lehman, E. L., *Testing Statistical Hypotheses*, New York, 1959, p. 13: "This assumption is usually not warranted in applications."

CHAPTER SIXTEEN

EPISTEMOLOGICAL CONSIDERATIONS

1. There is a sense of 'inference' we have not begun to deal with yet. We have followed Carnap in regarding the outcome of a statistical inference as the assignment of a probability (or something like a probability, such as a confidence coefficient) to a simple or complex statistical hypothesis, or to a simple or complex singular prediction. From the point of view of an impossibly pure Bayesianism, this might be all we should ever want or need to do. But purity in induction is no less difficult of attainment than other kinds of purity, and may equally inhibit fruitfulness.[1]

In this work we have already, from the beginning, committed ourselves to an approach according to which certain empirical generalizations – statistical distribution statements – can come to be members of our bodies of knowledge. Of course some statistical distribution statements, being logically true, can come to be accepted in our bodies of knowledge simply on grounds of that logical truth. But to limit our acceptance of statistical distribution statements to these would very seriously cramp our statistical style.

What we require, and have not yet touched upon, is a principle which will allow one to *accept* statistical hypotheses, or, as one might also say, detach statistical hypotheses from their supporting evidence. There has been a certain amount of controversy concerning both the necessity and the desirability of an inductive acceptance rule.[2] There has also been some controversy about its possibility. The resolution of the former controversy is not part of our program – but it should be observed that there is no way to resolve it without having established whether or not inductive rules are possible. A number of inductive rules have been proposed[3] that are associated with varying degrees of consistency and cogency in the resulting body of knowledge. In most instances, however, the languages for which these rules have been designed to work, if characterized explicitly at all, are far weaker than the language about which we have been talking.

Let us see, then, if we can devise a rule of acceptance that can lead to bodies of knowledge of the sort we have been concerned with.

2. The first thing we must observe is that one cannot 'accept' into a body of knowledge something that is only probable relative to that body of knowledge. Such a rule would have the effect that if S has probability p, relative to K, then S should be included in K, i.e., S should have probability 1, relative to K. We therefore distinguish between the body of knowledge relative to which the probability of a statement is evaluated, and the body of knowledge of which that statement may become a member on the grounds of its probability (perhaps among other things).

Following, in part, the development in my *Probability and the Logic of Rational Belief* [4], we shall suppose that every body of knowledge is characterized by a number, which we may call its degree of practical certainty. The appropriate body of knowledge to be concerned about depends on the context: in some situations, a probability (relative, clearly, to a body of knowledge of higher level) of 0.90 might count as a sure thing, as practical certainty; in other contexts a probability $1–10^{-4}$ might barely count as practical certainty. There is a trade-off here: the more demanding our notion of practical certainty, in general, the less we can claim to know in our body of knowledge.

Some philosophers have suggested that, at least in some contexts, a probability of over a half should count as a level of acceptance. That it should be over a half seems reasonable, since otherwise, both S and its negation might qualify for acceptance. But observe that similar objections could be brought to bear on any level of acceptance. Suppose that (as in the lottery paradox [5]) a probability of $1-1/k$ is taken as an acceptance level. Let P_0, $P_1, \ldots$, P_k be a set of $k+1$ statements that are exhaustive, exclusive, and of equal probability. Each $\ulcorner \sim P_i \urcorner$ will have a high enough probability to be accepted; but since the statements are exhaustive the denial of the conjunction of the denials of all the statements will be acceptable, leading to a body of knowledge that is strictly inconsistent. There are a number of reactions one might have: one might say that any number of the $k+2$ statements in question might be accepted, but not all of them, thus preserving strict consistency for a body of knowledge at the expense of the uniqueness of the body of knowledge built on given evidence. In a way, this defeats the purpose of having a logic of rational belief, for a large part of that purpose is to provide a framework within which people can come to reasonable agreement concerning what, given certain evidence, is worthy of rational belief.

Another reaction is to say that in such circumstances one should suspend belief in each of $\ulcorner \sim P_i \urcorner$ precisely on the grounds that in this case symmetry considerations, combined with probabilistic acceptance would lead to a strictly inconsistent body of knowledge. We should observe at this point that this reaction seems to throw a monkey-wrench into the machinery of statistical inference. Let us suppose that we are concerned about a binomial parameter p, and (for simplicity) that we have a continuous prior distribution for this parameter. Given a body of evidence, we have a posterior distribution for p. The posterior distribution will also be continuous, and we will therefore be able to construct k plus 1 intervals, such that the posterior probability that p is any given one of these intervals is exactly $1/(k+1)$. It seems to follow that we should suspend judgment about the location of p, at least with respect to this set of alternatives. But again we seem to have landed in a region of arbitrariness, since what we accept depends on what alternatives we are considering; and surely one should be free to 'consider' any alternatives. Yet another reaction is to devise a rule of inductive acceptance that is not purely probabilistic.[6]

Our own approach will be to suppose that the body of knowledge built on the basis of a given body of evidence (all in a given language) is unique; but that it need not be strictly consistent. The techniques developed in Chapter Eight and those following give us a way of handling rational corpora that are not strictly consistent. On the other hand, for some statements our inductive acceptance rule will not be purely probabilistic.

An alternative philosophical position takes the degree of certainty of anything with a right to membership in anything worthy of the title 'body of knowledge' to be 1. This would be the body of knowledge of a pretty thorough skeptic. We shall have occasion to refer to such a body of knowledge, but it is surely inappropriate to the purposes of our analysis to deal only with such a body of knowledge.

In real life what counts as 'practical certainty' depends on a number of factors – for example, what is at stake, perhaps personal tendencies toward optimism or pessimism or a general attitude toward risk taking, and the like. We do not assert that there are specific degrees of certainty appropriate to specific contexts that all men should adopt. We shall speak generally of a body of knowledge of level *r*, or of level of practical certainty *r*, or, more simply, of level *r*.

Our rule of acceptance must have two parts, one of which will stipulate the acceptability of certain simple statements on the basis of 'direct observation' or 'direct experience'. We can tell by looking at him that crow number seven is black, and thus accept, without further ado, the statement 'crow number seven is black'. Now to be sure, there are difficulties in spelling out the details of this rule – crows might change color, we might be mistaken, etc., and to begin to examine it seriously would clearly lead to a morass of epistemological perplexities that it would be inappropriate to go into here. The standard epistemological procedure (since Descartes) is to suppose that there are some items in our body of knowledge that can be certified (rendered certain) by our experience, and which in turn will serve as evidence for other items. One might move toward common-sense realism, or toward some brand of phenomenalism, to put flesh on these bones. One might conjecture that the bodies of knowledge that one ended up with, moving in either of these directions, would (at levels less than one) contain just about the same statements. But to prove that would require a lot of epistemological argument. On the other hand, one might also suppose that the best way of construing those basic evidence statements is to regard them as statements rendered directly acceptable in a body of knowledge *of a certain level* by experience or observation. Perhaps the way to look at the observation that crow number seven is black is this: The experience is such that 'crow number seven is black' is directly acceptable into any body of knowledge of level less than 0.95. This is to be contrasted with: 'There is a squirrel at the top of the tree' which experience might warrant as directly acceptable only into bodies of knowledge of level less than 0.75.

Nothing in what follows depends on the particular epistemological treatment of these primitive evidential statements. We suppose merely that there is a class of statements, some of which may become members of the body of knowledge of level *r*, directly on the basis of experience and observation. They may of course also become members of the body of knowledge of level *r* in virtue of their probability; thus if I have a good squirrel dog who is right 75% of the time, I may use that knowledge to infer (in my body of knowledge of level 0.75) 'there is a squirrel at the top of the tree', even when I have not seen the squirrel at all, but only observed the behavior of my dog.

Let $O_L(s, i, r)$ – where s is a statement of L, i is an index of the bearer of a rational corpus, and r its level – mean that s has been warranted by the observational experience of the bearer i of a rational corpus of level r. The 'bearer' may be construed as a person or as a society of people, such as a scientific community. We could now index rational corpora according to their bearers, but alternatively, and perhaps more clearly, we can propose a relation B_L between a bearer (person or society) and the set of statements of L to which he is committed, at a level r of practical certainty. Thus $B_L(i, x, r)$ asserts that x is the set of statements to which the individual i is committed at that level. A natural condition of rationality in general is that the set of statements to which i regards himself as committed should be (at least) a rational corpus:

(i) $B_L(i, x, r) \rightarrow RC_L\, x$

A reasonable part of an acceptance rule is then the requirement,

(ii) $O_L(s, i, r) \wedge B_L(i, x, r) \rightarrow s \in x$

On some epistemological views, once a statement s has gotten into i's rational corpus x by occurring in the relation $O_L(i, s, r)$, it is there forever. There is no advantage, obviously, in dating the experience that resulted in the inclusion of s in the rational corpus x born by i – i.e., there is no advantage in putting s in the rational corpus x just in case there is a timer prior to the present *at which* the bearer i of s observed s to hold. There is no advantage, because the problem is that of *getting rid* of statements that *were* 'observed to hold'. There are good reasons (*now*) for rejecting 'There was a pink elephant in the tavern last night'. The date of the observation has no bearing on this problem.

The most straight-forward way of dealing with the problem of the rejection of statements that were once accepted on observational grounds is that suggested earlier – namely, to regard 'acceptance' as acceptance into a rational corpus of given level. (If this done, that the statement will appear in lower level rational corpora follows on probabilistic grounds.) We need merely stipulate that s is acceptable in a rational corpus of given level just in case (a) it is observed at that level to be the case, and (b) it is consistent with what is derived as acceptable – is rendered acceptable on probabilistic grounds – relative to rational corporal of higher levels. We shall see shortly that there is no problem about getting rid of statements that are accepted on probabilistic grounds.

3. The classical problem of induction is the problem of how one can come to accept general statements – even statistical hypotheses – on an observational basis. It has already been observed that statistical statements (statistical hypotheses) can be rendered probable by the evidence provided by instances. But this does not make things simple. We shall proceed by steps.

The first step in developing a probabilistic acceptance rule is that in which statements that are 'probable enough' are accepted. Thus in the million ticket lottery it seems reasonable to accept 'ticket 300 132 will not win', because there is only one chance in a million of being wrong. One would accept it in the rational corpus of level 0.999 999, on the grounds that, relative to a higher level rational corpus, its probability was at least that much.

But we cannot simply say that s should belong to the rational corpus of level r born by i just in case there is a higher level rational corpus born by i relative to which the minimum probability of s is at least r. We must reject this attractive proposal for two reasons:

(1) Let $r_0 < r_1 < r_2$. It is perfectly conceivable that s will be more probable relative to a rational corpus of level r_2 than relative to a rational corpus of level r_1, and thus that the minimum probability relative to the body of knowledge of level r_2 will exceed r_0, while its minimum probability relative to a body of knowledge (of the same individual) of level r_1 does not exceed r_0. For example, relative to no general information at all, the probability is very high indeed that a very large sample will be very nearly representative of the proportion of P's that are Q's. This is so even though the sample should reveal a secular trend – e.g., if the Q's occur more and more frequently as the sampling procedure progresses. Relative to this highest level rational corpus, then, it is very probable that the proportion of Q's in the population is close to the proportion observed in the sample. But it will also be probable, relative to that highest level rational corpus, that the frequency of Q is increasing, so that this hypothesis (in some form) will occur in bodies of knowledge of level lower than the highest. This knowledge in the rational corpus

of level r_1, for example, prevents the sample from being a random member of the set of n-membered samples from P, with respect to having a proportion of Q's close to that in P as a whole, relative to the body of knowledge of level r_1; and hence the probability will not be so high that that proportion is close to the sample ratio. In this situation a hypothesis that is very probable relative to the highest level rational corpus is not so probable relative to a lower level of rational corpus.

(2) We would like our bodies of knowledge to be statistically closed (MD-8.8) in order to be assured that strongest statistical statements exist. But if p is a binomial parameter, say, having a continuous posterior distribution f_p, according to a body of knowledge of level r_1, then there are any number of Borel sets b such that $\int_b df_p$ is at least r_0. In particular there will be two, b_1 and b_2 such that

$$\int_{b_1} df_p > r_0, \int_{b_2} df_p > r_0 \quad \text{and} \quad \int_{b_1 \cap b_2} df_p < r_0 .$$

If we were simply accepting into the rational corpus of level r_0 whatever had at least that probability we would have statistical statements corresponding to the first two integrals in the rational corpus of level r_0, but not that corresponding to the third integral, even though that statement is a consequence of the other two. We cannot simply *impose* deductive closure for statistical statements on pain of contradiction: divide [0, 1] into subintervals $b_0, b_1, \ldots, b_{n-1}$ such that for every $i < n, \int_{b_i} df_p < 1 - r_0$. Then $\int_{[0,1]-b_i} df_p > r_0$, for every i, hence is acceptable. But the deductive closure of these statements is clearly inconsistent! On the other hand, if we accept all the probable statements, but do not impose deductive closure, then the body of knowledge is not statistically closed, and we no longer have the strongest statistical statements we require in the definition of probability.

So much for a purely probabilistic acceptance rule.

The way to handle the first problem seems simple enough: we have already let the context determine a level of practical certainty r_0; we let it also determine (as ordinary language seems to) a level of *moral certainty* r_1. A statement will be morally certain for an individual or institution i, just in case its probability, relative to the logical consequences of the observations of i more certain than r_1, is greater than r_1. We then state that s is practically certain for i, when (a) the lower bound of the probability of s, relative to the body of moral certainties of level r_1 is greater than r_0, and (b) there is no level of moral certainty r_1^* between r_1 and r_0 such that the lower bound of the probability of s, relative to the body of moral certainties of level r_1^* is less than r_0.

The second problem is far more difficult to handle. Like many of the philosophical problems associated with induction it is the sort of problem that undergoes serious qualitative changes as the quantitative complexity of the problem increases by small increments. (For example, it is a very different thing to define logical range probabilities in predicate calculi without identity, and to define them in predicate calculi with identity and operation expressions.)

In discussing confidence intervals, we concerned ourselves with those of minimum length. This provides a clue to the principle we wish to operate here. Let $B(p, b)$ be the family of two-point distributions with parameter p belonging to the Borel set b. Given certain evidence there will be a (large!) number of Borel sets b such that the minimum probability of $\ulcorner D\langle y, z, B(p, b)\rangle\urcorner$ is at least r. We accept that statement $\ulcorner D\langle y, z, B(p, b)\rangle\urcorner$ which is such that $\sup b - \inf b$ is a minimum, if there is one. In general, b will be an interval, and in the case of confidence interval inference it will be precisely the 'shortest' confidence interval for a given confidence level.

There may exist instances of background information in which there exists no 'shortest' confidence interval even for this case. In general however, we may obtain a principle governing the acceptance of statistical statements of the form $\ulcorner S\langle y, z, b, p, q\rangle\urcorner$ from that clue: we should accept into our body of knowledge of level r that statistical statement about y, z, and b, whose minimum probability exceeds r relative to the rational corpus representing moral certainty and for which $q-p$ is a minimum, or if there is more than one exhibiting the same minimum value $q-p$, we accept the statement $\ulcorner S\langle y, z, b, p, q\rangle\urcorner$ where $[p, q]$ is the union of minimal intervals. This last clause is arbitrary, but it yields a unique statistical statement ⌜$S\langle y, z, b, p, q\rangle$⌝ as acceptable at level r, on the basis of given statistical evidence.

To obtain a general principle of inductive acceptance we need to generalize this principle in two ways: the first generalization is to distributions that depend on more than one parameter; the second is to statements in general.

Let F now be a family of distributions characterized by n-parameters $p_0, \ldots, p_{n-1}$, and let b an n-dimensional Borel set, so that $F\langle b\rangle$ is the set of distributions of the family F with the n-tuple of parameters $\langle p_0, p_1, \ldots$ $\ldots, p_{n-1}\rangle \in b$. For an n-dimensional Borel set b, let

$$_ib = \{x : \bigvee_{y, z} (y \in \mathbb{R}^{i-1} \wedge z \in \mathbb{R}^{n-1} \wedge \widehat{yxz} \in b\}.$$

$D\langle y, z, F\langle b\rangle\rangle$ of those satisfying the probability requirement, that in some sense minimizes the set b. There are a number of ways in which we might measure the 'extent' of the Borel set b:

(i) $\underset{i<n}{Pr}(sup_i b - inf_i b)$ would give a measure of the volume of the minimal enclosing 'rectangle'.

(ii) $\sum_{i<n} (sup_i b - inf_i b)$ would give a measure of the 'perimeter' of the enclosing rectangle.

(iii) $\sum_{i<n} (sup_i b - inf_i b)^2$ would give a measure of the (square of) the diagonal of the enclosing rectangle.

We have no guarantee that minimizing any of these measures will pick out a unique Borel set b satisfying our other conditions.[7]

The third measure – at least in other contexts – turns out to be mathematically pleasanter than the others, and so we adopt that as our standard. To take care of instances in which that does not pick out a unique family of distributions $F(b)$, we stipulate that the accepted distribution statement is to be $D\langle y, z, F\langle b^*\rangle\rangle$, where b^* is the union of the minimal Borel sets b picked out by (iii).

It is clear that this proposal only works when a specific family of distributions of the form $\{f\widehat{xa} : a \in b\}$ is being considered, where x represents the n-dimensions of the random quantity z, and b is a Borel set included in the m-dimensional parameter space.

How do we treat the more general case? How do we deal with distributions of m parameter form, that differ only trivially from those of the form $\{f\widehat{xa}:a\in b\}$ and hence are equally well supported? If the differences are trivial, then the problem does not seem to be a very serious one. Let us suppose, for example, that $F=\{f\widehat{xa}:a\in b\}$, and that G is a more general set of distributions such that G is as well supported by the evidence as F. If it is the case that there is a small number ε, such that for every distribution function g in G there is a distribution function f in F (i.e., there is an a in b) such that

$$\bigwedge_{c\subset\mathbb{R}}\left|\int_c dg-\int_c df\right|<\varepsilon,$$

then it seems that serious errors will not be introduced by our restricting our considerations to distributions in F. (This is not true, it should be noted, if we are to be concerned with conditional distributions: it may well be that

$$\left|\frac{\int_{b\cap c} dg}{\int_c dg}-\frac{\int_{b\cap c} df}{\int_c df}\right|\gg\varepsilon$$

holds for some g in G and f in F.)

As a practical matter, then, we can often restrict our considerations to parametricizable families of distributions. It would nevertheless be interesting to have criteria that could be applied quite generally. It is not altogether clear how to go about providing such criteria, but the following might work:

Suppose that we are talking about the acceptability of statement s at level r, given that we have a body of knowledge of level r^*. Let $\mathfrak{F}$ be the set of all those sets F of distribution functions such that, relative to the body of knowledge of level r^* (and relative to the body of knowledge of any level between r^* and r), the minimum probability is at least r that $D(y, z, F)$. How do we pick out a unique member of $\mathfrak{F}$? If every member of every member of $\mathfrak{F}$ can be expressed in the form $\{f\widehat{xa}:a\in b\}$, where $b\in\mathbb{B}^n$, then we accept the logical consequences of $\ulcorner D(y, z, F^*)\urcorner$, where $F^*=\{f\widehat{xa}:a\in b^*\}$ is the unique member of $\mathfrak{F}$ picked out by our previous rule. If the union of $\mathfrak{F}$ contains distributions of a number of different forms, we apply the previous rule to the distributions of each form, and take F^* to be the union of the sets thus picked out. For example, let $\mathfrak{F}=\{\{\{f_1\widehat{xa_1}:a_1\in b\}:b\in B_1\}\cup\cup\{\{f_2\widehat{xa_2}:a_2\in b\}:b\in B_2\}$. Let our previous principle (of minimizing the diagonal and taking our parameter set to be the union of those parameter sets that minimize the diagonal) be applied to each of the two forms of distributions; suppose the results are $\{f_1\widehat{xa_1}:a_1\in b_1^*\}$ and $\{f_2\widehat{xa_2}:a_2\in b_2^*\}$. Then F^* will be the union of these two sets.

Let us call this rather informally stated rule R, and let us construe it as a selection function that picks out the appropriate distribution statements from a set of distribution statements of adequately high probability.

How about statements of forms other than $\ulcorner D\langle y, z, f\rangle\urcorner$ and $\ulcorner S\langle y, z, b, p, q\rangle\urcorner$? We may allow them to become incorporated into our rational corpus of level r, just when their probability relative to the set of moral certainties of level r' exceeds r, provided there is no set of moral certainties of level between r and r' that leads to a lower minimum probability than r, and provided that the statement does not entail any statements of the first two forms not warranted as members of that body of knowledge in their own right. To be sure, this means that we may have in our body of knowledge statements of the form $\ulcorner s\rightarrow S\langle y, z, b, p, q\rangle\urcorner$ and statements of the form s, where the statistical statement is not entailed by the strongest statistical statement about y, z, and b we can accept according to the first criterion. But this is all right: since we do not have deductive closure, $\ulcorner S\langle y, z, b, p, q\rangle\urcorner$ will *not* be a member of that body of knowledge, although it will be a member of its logical consequences.

In order to obtain statistical closure, we must stipulate that distribution and statistical statements that are implied by those we propose to accept also be accepted. We would like to be sure that the result of applying our proposed rule to distribution statements will yield just those statistical statements that we first considered.

Let $K(z, b)$ be that function whose domain is the same as that of z, and whose value is 1 for a given object just in case the z-value of that object belongs to the Borel set b; and whose value is 0 otherwise. Recall that $B(p, b)$ is the family of two-point distributions with parameter p belonging to b. Then we have as a logical equivalence:

$$\ulcorner S\langle y, z, b, p, q\rangle \leftrightarrow D\langle y, k(z, b), B(r, (p, q))\rangle\urcorner$$

Since p is the infimum of the parameters r such that the two point distribution with parameter r belongs to $B(r, (p, q))$, and q is the in supremum of same set, the rule we are talking about will pick out the statistical statement just in case it picks out the distribution statement.

We shall now formalize this. Let $MC_L(i, r)$ be the set of moral certainties (of level r) for the individual or institution i in the language L. This will consist of just those statements whose minimum probability, relative to the set of logical consequences of the observation statements more certain than r for i, is greater than r, provided that all those logical consequences of statements in $MC_L(i, r)$ that are of the form $\ulcorner D\langle y, z, F\rangle\urcorner$ or of the form $\ulcorner S\langle y, z, b, p, q\rangle\urcorner$ are also logical consequences of the unique distribution statement regarding y and z picked out by the rule R.

MD-16.1 $$MC_L(i, r) = \{x: \bigvee_{p,q} (PROB_L(x, L\text{-}Cn\ (\{w: \bigvee_{r'>r} O_L(w, i, r')\} \cap \cap St_L) = [p, q] \wedge r \leqslant p) \wedge \bigwedge_{y, z, F, p', q', b} ((\ulcorner D\langle y, z, F\rangle\urcorner \in \in L\text{-}Cn(\{x\}) \rightarrow \ulcorner D\langle y, z, F\rangle\urcorner \in L\text{-}Cn(\{R\{y': \bigvee_{F', p, q} (y' = = \ulcorner D\langle y, z, F'\rangle\urcorner \wedge PROB_L(y', L\text{-}Cn(\{w: \bigvee_{r'>r} O_L(w, i, r')\} \cap \cap St_L) = [p, q] \wedge r \leqslant p)\}\})) \wedge (\ulcorner S\langle y, z, b, p', q'\rangle\urcorner \in L\text{-}Cn(\{x\}) \rightarrow \rightarrow \ulcorner S\langle y, z, b, p', q'\rangle\urcorner \in L\text{-}Cn(\{R\{y': \bigvee_{F', p, q} (\ulcorner D\langle y, z, F'\rangle\urcorner = = y' \wedge PROB_L(y', L\text{-}Cn(\{w: \bigvee_{r'>r} O_L(w, i, r')\} \cap St_L) = = [p, q] \wedge r \leqslant p)\}\}))))\}$$

The definition of the set of practical certainties of level r, relative to moral certainties of level r^* is parallel; we simply replace reference to the body of observational certainties by reference to the set of moral certainties:

MD-16.2 $$PC_L(i, r^*, r) = \{x: \bigwedge_{r'} (r < r' \leqslant r^* \rightarrow \bigvee_{p,q} (PROB_L(x, MC_L(i, r')) = = [p, q] \wedge r \leqslant p)) \wedge \bigwedge_{y, z, F, p', q', b} ((\ulcorner D\langle y, z, F\rangle\urcorner \in L\text{-}Cn(\{x\}) \rightarrow \rightarrow \ulcorner D\langle y, z, F\rangle\urcorner \in L\text{-}Cn(\{R\{y': \bigvee_{F'} (y' = \ulcorner D\langle y, z, F'\rangle\urcorner \wedge$$

$$\wedge \bigwedge_{r'} (r < r' \leqslant r^* \rightarrow \bigvee_{p,q} (PROB_L(y', MC_L(i, r')) =$$
$$= [p, q] \wedge r \leqslant p)))\}\})) \wedge (\ulcorner S\langle y, z, b, p', q'\rangle\urcorner \in L\text{-}Cn(\{x\}) \rightarrow$$
$$\rightarrow \ulcorner S\langle y, z, b, p'\, q'\rangle\urcorner \in L\text{-}Cn(\{R\{y' : \bigvee_{F'} (y' = \ulcorner D\langle y, z, F'\rangle\urcorner \wedge$$
$$\wedge \bigwedge_{r'} (r < r' \leqslant r^* \rightarrow \bigvee_{p,q} (PROB_L(y', MC_L(i, r')) =$$
$$= [p, q] \wedge r \leqslant p)))\}\}))))\}$$

We would now like to show that these sets of statements are rational corpora. For this we need a premise, namely that the set of statements s such that $O_L(s, i, r)$ is consistent, and indeed strictly consistent. This, though a common enough assumption, is not altogether plausible when applied to our whole body of knowledge, and provides yet another reason for preferring a probabilistic perception principle which would allow for the direct inclusion of observation statements in the moral certainties of a given level less than the highest. We also need a premise that is not at all implausible, namely, that the set of sentences s such that $O_L(s, i, r)$ is finite. It is hard to imagine intelligible grounds for wanting to reject this. We also need to suppose that our observational sentences are relatively weak. We can observe that a certain object (say a crow) belongs to a certain class (that of black things, or of things such that there is a bigger one that is a crow, and a smaller one that is a crow). We surely cannot observe the truth of most sentences that entail an infinite number of logically independent sentences – so we cannot 'observe' the truth of 'All crows are balck'. We can observe the truth of some such sentences – e.g., 'There is no crow in the urn'. A statistical statement ('half the crows are black') doesn't entail very many sentences ('There are at least two crows', 'There is at least one black crow', 'There is at least one non-black crow') but sometimes, at least, it is not the sort of statement that can be entailed by a finite number of observations. ('Half the balls in the Urn are black' is a statistical statement that *may* be entailed by a finite number of observation sentences.) Some writers have suggested that there is a special class of basic sentences that can be certified observationally. But it is hard to pick out such a class, particularly in the face of the fact that one can accept such general statements as 'There are exactly ten balls in the Urn' on observational grounds. The difficulty of characterizing in an abstract and general way those sentences we may come to accept on observational grounds precludes the possibility of formal proofs. Nevertheless, informal arguments are possible, since we all have a reasonably clear idea of the sorts of sentences t that may occur in the first place of $O_L(t, i, r)$, and they may be of interest since they will support, if only informally, the general applicability of the ideas developed in the earlier chapters.

MT-16.1 $\sim\ulcorner 0 \neq 0 \urcorner \in L\text{-}Cn(\{s: O_L(s, i, r)\}) \wedge \{s: O_L(s, i, r)\} \prec \omega \wedge$
$\wedge \frac{1}{2} < r \rightarrow MC_L(i, r)$ is RC_L

domain of $PROB_L$; by MD-10.4, then, $x \in St_L$. Suppose that x belongs to $MC_L(i, r)$ and that $y \in St_L \cap L\text{-}Cn(x)$; then $\ulcorner x \rightarrow y \urcorner \in St_L$ and $\ulcorner x \rightarrow y \urcorner \in$ $\in L\text{-}Cn(\{w: \bigvee_{r'>r} O_L(w, i, r')\})$. Since $L\text{-}Cn(\{w: \bigvee_{r'>r} O_L(w, i, r')\}) \cap St_L$ is deductively closed with respect to statements, it is statistically closed; and it satisfies the cross product condition (MD-10.2).

We may suppose that this set of statements is $Normal_L$, since it contains only those statistical statements that are derivable from the set of observation statements, and none of those can involve true but unprovable statements of arithmetic (MD-8.9). Finally $L\text{-}Cn(\{w: \bigvee_{r'>r} O_L(w, i, r')\} \cap St_L$ has a basis, namely, the finite conjunction of the members of $\{s: O_L(s, i, r')\}$. It follows from MT-10.20 and MD-10.3 and 10.4 that $L\text{-}Cn(\{w: \bigvee_{r'>r} O_L(w, i, r')\} \cap St_L$ is its only strand, and thus that $p\text{-}prob_L$ and $PROB_L$ coincide for this special case. The set $L\text{-}Cn(\{w: \bigvee_{r'>r} O_L(w, i, r')\})$ is both probabilistically terminating and statistically terminating, because it is deductively closed and has as a finite basis the weak observation statements. In particular, if there are only a finite number of logically independent consequences of the observation statements, the infinite strings of statements required for the failure of either probabilistic or statistical termination are not forthcoming. It now follows that $PROB_L(y, L\text{-}Cn(\{w: \bigvee_{r'>r} O_L(w, i, r')\} \cap St_L)$ is different from the empty set, and thus by

MT-10.17 that its minimum probability is greater than the minimum probability of x, and thus greater than r. Therefore if x belongs to $MC_L(i, r)$, so does y. Suppose y is of the form $\ulcorner D\langle a, b, c\rangle\urcorner$ or $\ulcorner S\langle a, b, c, d, e\rangle\urcorner$. Then by MD-16.1 y must be a consequence of a statement the form $\ulcorner D\langle a, b, f\rangle\urcorner$ picked out by the rule R; and therefore y too will be probable enough to be accepted and y too will be such that its logical consequences follow from $\ulcorner D\langle a, b, f\rangle\urcorner$. Finally, $MC_L(i, r)$ is non-empty, by MT-10.6, together with observations already made. Then, by MD-10.1, $MC_L(i, r)$ is a rational corpus.

MT-16.2 $\quad \sim\ulcorner 0\neq 0\urcorner\in L\text{-}Cn(\{w: \bigvee_{r'>r} O_L(w, i, r')\})\wedge\{w: \bigvee_{r'>r} O_L(w, i, r')\}\prec$
$\prec\omega\wedge\frac{1}{2}<r\rightarrow\bigwedge_{w}(w \text{ is } Strand_L \text{ of } MC_L(i, r)\rightarrow w \text{ is statistically}$
$\text{closed}\wedge Normal_L\ w\wedge S\text{-}Term_L\ w\wedge P\text{-}Term_L\ w\wedge$
$\wedge\bigvee_{z} z \text{ is } Basis_L \text{ for } w)$

Argument:

That $MC_L(i, r)$ is statistically closed is obvious from MD-16.1. That it is $Normal_L$ follows from an argument similar to the foregoing one: Godelian complications can't be introduced in the process of making empirical inferences. That there is a basis (perhaps no longer finite) follows from the fact that if there is an infinite sequence of statements, each logically implied by its successor, and which is such that for any finite n, the nth statement has a minimum probability greater than r, then there exists a 'limiting' statement, whose probability is also greater than r or equal to r. (If at some point the probability of the nth statement is equal to r, the probabilities of all the succeeding statements must also be r, and the same is true of the limiting statement.) $MC_L(i, r)$ is not deductively closed, but for it to fail to be probabilistically terminating, requires the existence of an infinite sequence of ever stronger statistical statements, each of which is strongest about its subject matter. Only a finite number can be obtained from the finite empirical data that is available, so that an infinite number must be logical truths; but then we may proceed to a limiting statement which is a logical truth and is stronger than all the others. It follows that each strand of $MC_L(i, r)$ is probabilistically terminating. We do not know of any reference term that it denotes an infinite set; therefore no inconsistency can be introduced by supposing that the term t has a finite denotation. But if w is a strand of $MC_L(i, r)$, it is consistent; and so the addition of $\ulcorner t<\omega \urcorner$ should preserve consistency; and so the deductive closure of that set should also be consistent. But if the strand w is not statistically terminating, the deductive closure of $w \cup \{\ulcorner t<\omega \urcorner\}$ is not consistent.

Let us now look at sets of practically certain statements. We first observe that a set of practically certain statements (subject to the same assumptions about our observational knowledge as before) is a rational corpus; and then we observe that each strand of a set of practically certain statements has the nice properties that we have just observed the strands of sets of morally certain statements to have.

MT-16.3 $\sim \ulcorner 0 \neq 0 \urcorner \in L\text{-}Cn(\{w: \bigvee_{r'>r} O_L(w, i, r')\}) \wedge$

$\wedge \{w: \bigvee_{r'>r} O_L(w, i, r')\} \prec \omega \wedge \frac{1}{2} < r < r^* \wedge$

$\wedge\, CPC(MC_L(i, r^*)) \rightarrow PC_L(i, r^*, r)$ is RC_L

Argument:

As before, it is clear that everything that belongs to $PC_L(i, r^*, r)$ is a statement of L, and that $PC_L(i, r^*, r)$ is a set. As before, suppose that x belongs to $PC_L(i, r^*, r)$ and that y is a statement implied by x. Then by MT-16-2 and the assumption that $MC_L(i, r^*)$ satisfieds the cross product condition, the conditions for the application of MT-10.17 are satisfied, so that the probability of y, relative to the set of moral certainties of level r^*, is at least r. Then y will be included as a member of $PC_L(i, r^*, r)$ unless there is some intermediate level of moral certainty according to which it has a lower probability than r; but if that is the case, then x too will have a lower probability than r relative to that body of knowledge. On the other hand, if y has the form $\ulcorner D\langle a, b, c\rangle \urcorner$ or $\ulcorner S\langle a, b, c, d, e, f\rangle \urcorner$, it will belong to $PC_L(i, r^*, r)$ just in case it follows from a unique distribution statement picked out by R. If it does not follow from that distribution statement then, by MD-16.2, x will not belong to $PC_L(i, r^*, r)$, contrary to our assumption.

MT-16.4 $\sim \ulcorner 0 \neq 0 \urcorner \in L\text{-}Cn(\{w: \bigvee_{r'>r} O_L(w, i, r')\}) \wedge$

$\wedge \{w: \bigvee_{r'>r} O_L(w, i, r')\} \prec \omega \wedge \frac{1}{2} < r < r^* \rightarrow$

$\rightarrow \bigwedge_w$ (w *is Strand*$_L$ *of* $PC_L(i, r^*, r) \rightarrow w$ *is statistically closed*

$\wedge$ *Normal*$_L$ $w \wedge$ *S-Term*$_L$ $w \wedge$ *P-Term*$_L$ $w \wedge \bigvee_z$ z *is Basis*$_L$ *for* w)

Argument:
That $PC_L(i, r^*, r)$ is statistically closed is obvious from MD-16.2, and thus w is statistically closed. That w is $Normal_L$ follows from the same arguments used before. That w has a basis, that it is statistically terminating, and that it is probabilistically terminating all are evidenced by arguments similar to those given for MT-16.2.

What does all this mean? It means that the bodies of knowledge (both sets of practical certainties and sets of moral certainties) that we define (in part) by means of probability relations – in fact, which we define by means of an almost completely probabilistic acceptance rule – are such as to support probability statements. We may apply metatheorem 10.5, for example, to show that partial probabilities of statements of L exist relative to any strand of a set of practical certainties or of a set of moral certainties. In turn, this means that complete probabilities ($PROB_L$) exist for any statement of L, relative to any body of moral certainties or any body of practical certainties, subject only to the condition that there not be two stands relative to which that statement has different partial probabilities.

4. Statements which come to be regarded as moral or practical certainties, at one point in the career of i, may perfectly well come to be rejected at a later point. There is obviously no difficulty in supposing that relative to one set of observation statements $\{s: O_L(s, i, r)\}$ a certain statement may have a probability greater than r, and thus be a moral certainty of level r, while relative to an expanded set of observation statements it no longer has this high a probability, and thus no longer is a member of the set of moral certainties of level r. The same, equally obviously, is true of sets of practical certainties.[8]

Nothing has been said about the acceptance of (non-statistical) laws and theories. The treatment of these matters lies beyond the scope of these investigations, and clearly requires more detailed consideration of acceptance rules, together with a detailed consideration of the nature of scientific laws and thoeries. It is likely that such an analysis would lead to the conclusion that in many cases, at least, what is involved in the acceptance of a theory is in part the acceptance of a language; and we have said nothing about what might lead us to choose one language over another. A natural kind of consideration is that the practical certainties that emerge from the use of one language L rather than another L' might be more useful, in the sense that they would embody more statistical statements with parameters close to 0 or close to 1.

5. Back in Chapter Seven we introduced the two predicates S and D; we provided explicit set theoretical definitions for them, according to which they are automatically given perfectly straight-forward extensional interpretations. If the reference sets (the union of the range of y in $\ulcorner D\langle y, z, f\rangle\urcorner$ or $\ulcorner S\langle y, z, b, p, q\rangle\urcorner$) are thought of as composed of possible entities, e.g., possible results of measuring a table, then these explicit definitions are plausible enough. But then we are faced with the classical problem of relating 'possible' entities to actual ones. Suppose we construe the 'true value' of the length of a table as the mean of the set of all possible measurements of the table. These possible measurements we know to be normally distributed with mean equal to the true value of the length of the table, and with variance s^2. Now, however, we may in fact *know* that we shall measure the table exactly five times. Then we *know* that the distribution of *actual* measurements of the table is not at all normal, but is discrete, and in fact contains at least one point with measure $\frac{1}{5}$. There are two things we can do to justify the use of the normal analysis. First, we may stipulate that all those statistical statements relative to which we evaluate probabilities are to concern sets of entities not known to be finite. We have already seen that we must not know these sets of entities to be infinite. We must therefore be in limbo regarding their cardinality. We then build this condition into our definition of probability. There are several awkwardnesses involved in this approach. We must make our analysis of the result of measuring the table relative to a hypothetical rational corpus in which it is *not* known that there are only five measurements – we can't use the body of knowledge we *really* have. And we can never use our explicit and complete knowledge of an actual finite class to yield probabilities: we can never argue that the probability of drawing a black ball from an urn containing five black and five white balls is a half; nor that the holder of a ticket in a five-thousand ticket lottery has a probability of winning of 0.0002.

Second, we may reject the extensionality of the predicates S and D. A better approach would be to introduce two new non-extensional predicate S^* and D^*. We then interpret these new predicates in the following way. We introduce the notion of a model of the language Lin the usual way.

(Due to our assumption that there exists a strongly inaccessible cardinal, we know that there are models of our language.) Consider a set of sentences of L that we may regard as embodying our physical theories of the world. (True, we have regarded the problem of picking out such a set of sentences as beyond the scope of our endeavors here; but we assume the problem not impossible to solve.) Let this set of sentences be T. We restrict our considerations to models in which the sentences of T are true. One model m is an extension of another, just in case the universe of m includes the universe of m', and, further, the assignments of m, restricted to the universe of m', are just the assignments of m'. Let t be a reference term of L. We say that m is a t-maximal extension of the model m^*, just in case m is a model of our set, and either what m assigns to t is already infinite, or there is no extension m' of m such that what m assigns to t is properly included in what m' assigns to t. We say that $\ulcorner D^* \langle x, y, f \rangle \urcorner$ holds in a model m^*, just in case $\ulcorner D \langle x, y, f \rangle \urcorner$ holds in (almost) every t-maximal extension of m^*, where $t = \bigcup \mathscr{R}x$.

This approach has the following attractive features. Suppose that t is a term with a cardinality restriction built in, e.g., 'the set of k balls in Urn i'. Then in every model t must have as its extension just these k balls. Furthermore, if we know $\ulcorner D \langle x, y, f \rangle \urcorner$, we know $\ulcorner D^* \langle x, y, f \rangle \urcorner$, since the union of the range of x must consist of those same k balls in every extension of the model in which we know that $\ulcorner D \langle x, y, f \rangle \urcorner$ is true. Suppose that t is the set of tosses of a certain coin made yesterday, and that we know that a certain proportion landed heads. Again cardinality is built into t, and that same proportion will characterize every extension of the model in which that bit of knowledge holds. On the other hand, if we consider the set of measurements of the table, or the set of balls drawn from the urn, or the set of throws of the coin, there is no cardinality restriction built into the reference term, and in certain models that reference term may have an arbitrarily large cardinality. $\ulcorner D^* \langle x, y, f \rangle \urcorner$ will be true in our world just in case in almost all $\ulcorner \bigcup \mathscr{R}x$-maximal extensions of our world it is true. Of course we can have perfectly good grounds for thinking that this is the case: the sample on which we based our acceptance of $\ulcorner D \langle x, y, f \rangle \urcorner$ is usually a perfectly good sample on which to base the acceptance of $\ulcorner D^* \langle x, y, f \rangle \urcorner$. We then base all probabilities on non-extensional statements of the form $\ulcorner D^* \langle x, y, f \rangle \urcorner$.

A problem which has attracted the attention of some philosophers is the following: consider an Urn U, that we know to contain five white and five black balls. There is a sense in which we want to say that the probability is a half that the next ball to be drawn from that urn will be white. But there is also a sense in which, given a long series of draws (with replacement), of which a third have resulted in a white ball, we want to say that the probability of *drawing* a white ball from that urn is a third. This has suggested to some philosophers that there must be two distinct notions of probability operating here. Our analysis provides an explication of both senses. Let x be the monotone increasing sequence of sets each of which is the set of balls in the urn, so that the union of the range of x is simply the set of balls in the urn. We know that half of the objects in this set are white. Since we have a built in cardinality restriction, we even have the intensional distribution statement in our body of knowledge. 'The next ball to be drawn from the urn' is a perfectly good definite description of a member of the union of the range of x, i.e., of the set of balls in the urn. Now if, relative to our body of knowledge, 'the next ball to be drawn from the urn' is a *random* member of this set with respect to being white, then the probability is a half that it will be white, This explains the first intuition. Now suppose that we have drawn a lot of balls from the urn, and we consider the set x', which is a monotonically increasing set of the form x_i', where x_i' is the set consisting of the first i drawings of balls from the urn. Suppose that we have examined the results of k draws, and we want the probability of that the $k+1$'st draw will result in a white ball. Even if we know that that is the last ball we shall draw, there are extensions of our model in which arbitrarily many balls are drawn from the urn. We suppose that, relative to our observations of k balls, we are practically certain that very close to a third of the union of the range of x' yields white balls. 'The next draw of a ball' is known to be a member of the union of the range of x', just as the next ball to be drawn is known to be a member of x. If it is a random member of the former set, with respect to yielding a white ball, then the probability that it is white is $\frac{1}{3}$. To see that the set of draws of balls from the urn, rather than the set of balls in the urn, is the correct reference set, we must examine more closely what a draw of a ball from the urn is. A natural way to construe it is as an ordered pair, consisting of a ball, together with an ordinal representing the particular drawing on which it is drawn. (It cannot be simply the ball, since the same ball can be drawn on distinct drawings.) The reference set of draws consists of the set of all pairs consisting of a ball, and an ordinal representing the occasion of its being drawn. For example the drawing *d* might consist of the drawing of ball *A* on the 75th draw. Consider the cross product of the set of balls and the set of ordinals ω. Now consider the subset of this set in which each ball occurs only in pairs with the ordinal numbers of the occasions on which it is drawn. This is the reference set, $\bigcup \mathscr{R}x'$. Now we observe that if the proportion of white ball-drawings (limit of relative frequency) in a maximal extension of $\bigcup \mathscr{R}x$ is *different* from the proportion of white balls in $\bigcup \mathscr{R}x$, then the former represents the appropriate reference set, since it is (known to be) a subset of the cross product of $\bigcup \mathscr{R}x$ with something else (cf. MD-9.2). The situation is parallel to that in which we accept a fiducial inference, unless there is a Bayesian inference which yields a different probability. The existence of a product set matching the latter saves the randomness of the pair $\langle$population, sample$\rangle$ in the Bayesian reference class; the same principle is operating here.

We thus have a perfectly explicit and straight-forward criterion for when the accumulating observations of frequencies of draws that yield white balls takes over from the knowledge that half the balls in the urn are white. It takes over precisely when, in our body of practical certainties we have a statistical statement about the frequency of draws of white balls which *differs* from our knowledge of the frequency of white balls in the urn. This does not occur quickly (ordinarily), but it does occur, and the more quickly, the more the apparent frequency of white draws differs from the known ratio of white balls in the urn.

There are a large number of problems that remain to be dealt with. I can claim only that the framework of the foregoing chapters does help to solve some of the problems of statistical inference. Even in this regard, there are a large number of technical and mathematical difficulties to be dealt with before the resulting system can be applied with ease and confidence. There are a large number of epistemological problems underlying the metalinguistic predicate *O* that I so casually introduced in this chapter. There are serious problems concerning the acceptance of laws and theories – and even concerning what laws and theories are – that must be dealt with before we have a clear picture of scientific inference. There is the very basic question of the grounds on which we choose one language over another. I hope however, that the framework we have developed here is one in which the quest for answers to all these problems will not be so frustrating as it has been in the alternatives that we have hitherto had to choose among.

NOTES

[1] I have elsewhere – 'Two World Views', *Nous* **4** (1970), 337–348 – attempted to characterize two approaches to scientific knowledge, a pure one, in which inference is Bayesian and Theoretical entities are mythical; and an impure one, in which one accepts hypotheses, even when the hypotheses commit one to theoretical entities.

[2] Lakatos, Imre (ed.), *The Problem of Inductive Logic*, Amsterdam, 1968, contains a good sample of that controversy.

[3] Hintikka, Jaakko and Hilpinen, Risto, 'Knowledge, Acceptance, and Inductive Logic', in Hintikka and Suppes (eds.), *Aspects of Inductive Logic*, Amsterdam, 1966; pp. 1–20; Lehrer, Keith, 'Justification, Explanation, and Induction', in Swain (ed.), *Induction, Acceptance, and Rational Belief*, Dordrecht, 1969; Hilpinen, Risto, 'Rules of Acceptance and Inductive Logic', *Acta Philosophica Fennica* **21**, Amsterdam, 1968.

[4] Middletown, Conn., 1961.

[5] First described in *Probability and the Logic of Rational Belief*, Middletown, 1961.

[6] See K. Lehrer, 'Justification, Explanation, and Induction', in *Induction, Acceptance and Rational Belief*, Swain (ed.), Dordrecht, 1969; and I. Levi, *Gambling with the Truth*, New York, 1967.

[7] So far as I know!

[8] Note that observation statements themselves may come to be rejected – namely, when their negations come to be probable enough, relative to a higher level body of knowledge, to be members of a body of knowledge of the appropriate level.

THE MATHEMATICAL BACKGROUND

1. The mathematical fragments to be mentioned in the chapter serve a dual purpose. They provide the mathematical underpinning for the discussion of the first six chapters. Although most of that discussion is informal, some mention is made of mathematical facts and entities; while a vague or partial understanding of these mathematical matters should suffice for the points being made, some readers will feel better if there is available an explicit (if abbreviated) exposition of the mathematical theory and definitions involved. The second purpose of these fragments is to provide an explicit characterization of the metalanguage we will employ in stating definitions and results in chapters after the first six. While the main points (I hope) are intelligible without the details of the background provided here, claims of theoremhood and the adequancy of definitions can only be substantiated against such a background. The mathematics to follow (perfectly conventional and ordinary, by the way) will thus serve as part of the object language in the informal discussion of the first six chapters, and as part of the formal object language as well as the metalanguage to be employed in Chapter Seven and its successors.

2. The logical notation we shall use is based on that of Kalish and Montague; disjunction will be represented by '$\vee$', conjunction by '$\wedge$', the conditional if-then by '$\rightarrow$', the biconditional by '$\leftrightarrow$', and negation by '$\sim$' membership by '$\in$'. Universal and existential quantification will be represented by '$\bigwedge$' and '$\bigvee$', respectively. '$\bigwedge_{a \in b} \phi$' is an abreviation for '$\bigwedge a(a \in b \rightarrow \phi)$';'$\bigvee_{a \in b} \phi$' is an abbreviation for '$\bigvee a(a \in b \wedge \phi)$'; '$\overset{1}{\bigvee}_{a} ..a.$' is short for '$\bigvee a(..a. \wedge \bigwedge b(...b. \rightarrow b = a))$', which means that there is exactly one a such that.... The definite description, 'the unique a such that...' will be written '$\imath a(...)$'; the description will always be understood to denote; if it is improper, that is, if there is no a such that..., or if there is more than one a such that..., '$\imath a(...)$' will be taken to denote the empty set, 0.

3. We begin with a set of axioms for Zemelo-Fraenkel set theory with individuals. The primitive predicates are '$\in$' for membership, and '$\sum$' for 'is a set'.

A-1: *Zemelo's Axiom*

$$\bigwedge x \bigvee y(\sum y \wedge \bigwedge z(z \in y \leftrightarrow Fz \wedge z \in x)),$$

where y is not free in F.

This is an axiom schema. Suitable inference rules permit us to infer instances of A-1 obtained by proper substitution on the schematic predicate letter 'F'.

A-2: *Montague Closure Schema*

$$\bigwedge x \bigvee y(\bigwedge z(z \in x \rightarrow z \in y) \wedge \bigwedge z(z \in y \rightarrow \xi z \in y))$$

Again we have an axiom schema, from which instances may be obtained by proper substitution on the schematic operation symbol 'ξ'. This schema also serves as an Axiom of Infinity.

A-3: *Union Axiom*

$$\bigwedge w \bigvee y \bigwedge x(x \in y \leftrightarrow \bigvee z(x \in z \wedge z \in w))$$

A-4: *Power Set Axiom*

$$\bigwedge w \bigvee y \bigwedge x(x \in y \leftrightarrow \bigwedge z(z \in x \rightarrow z \in w) \wedge \sum x)$$

A-5: *Axiom of Extensionality*

$$\bigwedge x \bigwedge y(\sum x \wedge \sum y \wedge \bigwedge z(z \in x \leftrightarrow z \in y) \rightarrow x = y)$$

A-6: *Axiom of Regularity*

$$\bigwedge w [\bigvee x(x \in w) \rightarrow \bigvee x(x \in w \wedge \sim \bigvee y(y \in x \wedge y \in w))]$$

A-7: *Only Sets Have Members*

$$\bigwedge x \bigwedge y(x \in y \rightarrow \sum y)$$

A-8: *Axiom of Choice*

$$\bigwedge w [\sum w \wedge \bigwedge x \bigwedge y(x \in w \wedge y \in w \rightarrow$$
$$\rightarrow (x = y \leftrightarrow \bigvee z(z \in x \wedge z \in y))) \rightarrow \bigvee y \bigwedge x(x \in w \rightarrow$$
$$\rightarrow \bigvee z \bigwedge u(u \in x \wedge u \in y \leftrightarrow u = z))]$$

The set of all x such that Fx, or the set of all x satisfying F, will be denoted by '$\{x:Fx\}$'; if such a set does not exist, i.e., if $\sim\bigvee y \bigwedge x(Fx\leftrightarrow \leftrightarrow x\in y)$, '$\{x:Fx\}$' will be taken to denote the empty set. The assertion that such a set does exist will be rendered by '$\{x!Fx\}$'. Let A represent an operation expression of n places; then the set of all those objects $A(x_0,\ldots,x_{n-1})$ such that $x_0,\ldots,x_{n-1}$, satisfy the n-place schematic predicate F is denoted by '$\{A(x_0,\ldots,x_{n-1}):F(x_0,\ldots,x_{n-1})\}$'. The set whose only member is x will be denoted by '$\{x\}$'; that whose only members are x and y by '$\{x, y\}$', etc. The natural number 0 will be identified with the empty set; the successor of a natural number n will be taken to be $n\cup\{n\}$, and the set of all natural numbers will be denoted by omega, 'ω'. Note that the natural number n is just the set of all natural numbers less than n.

Set inclusion will be expressed by '$\subset$', proper inclusion by '$\not\subset$'. The union of two sets x and y will be denoted by '$x\cup y$', their intersection by '$x\cap y$', and their difference by '$x-y$'.

D-0.1 $\quad x\subset y\leftrightarrow(\sum x\wedge \bigwedge z(z\in x\rightarrow z\in y))$

D-0.2 $\quad \bigcup x=\{y:\bigvee z(z\in x\wedge y\in z)\}$

$$x\cup y=\bigcup\{x, y\}$$

$$\bigcap x=\{y:\bigwedge z(z\in x\rightarrow y\in z)\}$$

If x is not a set, then $\{y:\bigwedge z(z\in x\rightarrow y\in z)\}=\{y:y=y\}=0$.

$$x\cap y=\bigcap\{x, y\}$$

D-0.4 $\quad x-y=\{z:z\in x\wedge\sim z\in y\}$

The power set of x is the set of all its subacts and is denoted by Px.

D-0.5 $\quad \mathscr{P}x=\{y:y\subset x\}$

The ordered pair consisting of x followed by y is defined by

D-0.6 $\quad \langle x, y\rangle=\{\{x\}, \{x, y\}\}$

We define sequences as sets of ordered pairs $\langle x, y\rangle$ in which the first object is, a member of an index set I characteristic of the sequence (usually I is ω, or a subset of ω), and in which each member of the index set belongs to exactly one ordered pair:

D-0.7 S *is I-sequence* $\leftrightarrow \sum I \wedge \bigwedge x(x \in I \rightarrow \overset{1}{\bigvee} y\,(\ll\langle x, y\rangle\gg \in S) \wedge$
$\wedge \bigwedge_{x \in S} \bigvee y \bigvee z(y \in I \wedge x = \ll\langle y, z\rangle\gg))$

Sequences indexed by some or all of the natural numbers are of particular interest and deserve a special notation. Note that a sequence whose only term is x, e.g., $\{\ll\langle 0, x\rangle\gg\}$, is to be distinguished from x.

D-0.8 $\langle x\rangle = \{\ll\langle 0, x\rangle\gg\}$
$\langle x_0, \ldots, x_{n-1}\rangle = \{\ll\langle i, x_1\rangle\gg : n \in \omega \wedge i \in n\}$
$\langle x_0, \ldots\rangle = \{\ll\langle i, x_1\rangle\gg : i \in \omega\}$
$x_i = \imath y[\bigwedge z(\langle i, z\rangle \in x \leftrightarrow z = y) \vee (\sim \overset{1}{\bigvee} z(\langle i, z\rangle \in x) \wedge y = 0)]$
$\langle x_i\rangle_{i \in I} = \imath y(y$ *is I-sequence* $\wedge \bigwedge_{i \in I} \ll\langle i, x_i\rangle\gg \in y)$

The concatenation of two sequences f and g, in symbols '$\widehat{fg}$' will turn out to be useful:

D-0.9 $\widehat{fg} = f \cup \{\ll\langle x+i, g_i\rangle\gg : f$ *is x-sequence* $\wedge \bigvee_z \ll\langle i, z\rangle\gg \in g\}$

Relations and functions are defined as sets of sequences; a two place relation, for example, is construed as a set of 2-sequences, that is, as a set of sets which are of the form: $\{\ll\langle 0, x\rangle\gg, \ll\langle 1, y\rangle\gg\}$ rather than as a set of ordered pairs of the form: $\{\{x\}, \{x, y\}\}$. Thus:

D-0.10 R is n-ary relation $\leftrightarrow \bigwedge_{x \in R} x$ *is n-sequence* $\wedge\, n \in \omega$
f is n-place function $\leftrightarrow f$ is $(n+1)$-ary relation $\wedge$
$\wedge \bigwedge_{x, y, z} (\widehat{x\langle y\rangle} \in f \wedge \widehat{x\langle z\rangle} \in f \rightarrow y = z)$
f is fun $\leftrightarrow \bigvee_n (0 < n \wedge f$ is n-place fun)
$\breve{f} = \{\langle x, y\rangle : \langle y, x\rangle \in f\}$
f is 1-1 $\leftrightarrow f$ is fun $\wedge\, \breve{f}$ is fun.

The value of the function f for the argument x (which is a sequence of some sort) is that unique item y such that the concatenate of x and $\langle y\rangle$ belongs to f; the range and domain of a function are defined straightforwardly:

D-0.11 $\imath x = y[\bigwedge_z (x\frown\langle z\rangle \in f \rightarrow z = y) \vee \sim \bigvee_z^1 (x\frown\langle z\rangle \in f \wedge y = 0)]$

$\mathscr{R}f = \{x : \bigvee y \bigvee_{n\in\omega} y \text{ is } n\text{-sequence} \wedge y\frown\langle x\rangle \in f\}$

$\mathscr{D}f = \{x : \bigvee y \bigvee_{n\in\omega} x \text{ is } n\text{-sequence} \wedge x\frown\langle y\rangle \in f\}.$

We write '$x \approx y$' – x is equivalent to y – to express the fact that there is a 1-1 function whose domain is x and whose range is y; intuitively that x and y have the same number of members.

We write '$x \prec y$' – the power of y is at least that of x, or, intuitively, y has at least as many members as x – when there is a subset z of y which is equivalent to x.

D-0.12 $x \approx y \leftrightarrow \bigvee f(f \text{ is fun} \wedge f \text{ is 1-1} \wedge \mathscr{D}f = x \wedge \mathscr{R}f = y)$

$x \preccurlyeq y \leftrightarrow \bigvee z(z \subset y \wedge x \approx z)$

The general Cartesian product will be very useful:

D-0.13 $\prod_{i\in I} x_i = \{s : s \text{ is } I\text{-sequence} \wedge \bigwedge_{i\in I} s_i \in x_i\}$;

$\prod \langle x_0, \ldots, x_{n-1}\rangle = \prod_{i\in n} x_i$

There is a slight (and useful) departure here from ordinary usage, in that the elements of a Cartesian product are not ordered pairs, but sequences – i.e., sets of ordered pairs. Thus

$$\prod_{i\in 2} x_i = \{\langle s_0, s_1\rangle : s_0 \in x_0 \wedge s_1 \in x_1\}$$

The Index set I may be any set at all – e.g., the even natural numbers. Thus if x is an ω-sequence,

$$\prod_{i\in\{2j : j\in\omega\}} x_i = \{\langle s_0, \ldots\rangle : s_0 \in x_0 \wedge s_2 \in x_2 \wedge \ldots\}$$

The set x raised to the power y is the set of all functions with domain y and range included in x:

D-0.14 $x^y = \{f : f \text{ is fun} \wedge \mathscr{D}f = y \wedge \mathscr{R}f \subset x\}$

This is related to the general Cartesian Product as follows:

T-0.1 $z \text{ is } y\text{-sequence} \wedge \bigwedge_{i\in y} z_i = x \rightarrow x^y = \prod_{i\in y} z_i$

A form of Cartesian product which is associative (as the general Cartesian product is not) is of particular interest to us; it applies only to sets of finite sequences:

D-0.15 $w \times y = \imath k[\bigvee_{n,m\in\omega}(\bigwedge_{x\in w} x \textit{ is n-sequence} \wedge \bigwedge_{x\in y} x \textit{ is m-sequence} \wedge$
$\wedge\, k = \{\widehat{xz} : x \in w \wedge z \in y\}) \vee (\sim \bigvee_{n,m\in\omega}(\bigwedge_{x\in w} x \textit{ is n-sequence} \wedge$
$\wedge \bigwedge_{x\in y} x \textit{ is m-sequence}) \wedge k = 0)]$

T-0.2 $w \times (y \times z) = (w \times y) \times z$

T-0.3 $\bigwedge_{n,m\in\omega} x^n \times x^m = x^{n+m}$

The set of real numbers, $\mathbb{R}$, can be defined in our set theory, together with the usual operations on them. The scheme is a conventional one: we take the natural number 0 to be the empty set; the successor of a natural number n to be $n \cup \{n\}$. The non-negative ratios are then taken to be equivalence classes of pairs of natural numbers, and the positive reals non-empty sets of ratios that are bounded above and contain all the ratios less than any ratio they contain. Finally, the (signed) real numbers are obtained as equivalence classes of pairs of positive reals.

Note that the natural numbers (and in particular the empty set) are not real numbers; when we want to refer to the real counterpart of a natural number n, we should affix a subscript unless the context makes clear what we are doing. Thus '2' denotes the natural number and '2_r' denotes the corresponding real number. Often, where the context is clear, the subscript will be suppressed. Thus '0' will denote the empty set, or what is the same thing, the first natural number, or, when the context makes it clear that this is intended, the additive identity of the reals, 0_r.

The set of real numbers, defined as suggested, obey the usual laws, including the least upper bound principle, as stated in the following theorem.

T-0.4 $\{x : Fx\} \neq 0 \wedge \{x : Fx\} \subset \mathbb{R} \wedge \bigvee_{y\in\mathbb{R}} \bigwedge_{x} (Fx \rightarrow x \leqslant y) \rightarrow$
$\rightarrow \bigvee_{y\in\mathbb{R}} [\bigwedge_{x} (Fx \rightarrow x \leqslant y) \wedge \bigwedge_{z\in\mathbb{R}} (\bigwedge_{x} (Fx \rightarrow x \leqslant z) \rightarrow y \leqslant z)]$

For simplicity we also adopt the convention that the quotient of a real number and 0_r is 0_r.

We define upper and lower bounds of a set S of real numbers as follows:

D-0.16 x is u.b. of $S \leftrightarrow x \in \mathbb{R} \wedge S \subset \mathbb{R} \wedge \bigwedge_{y \in S} y \leqslant x \wedge S \neq 0$

x is l.b. of $S \leftrightarrow x \in \mathbb{R} \wedge S \subset \mathbb{R} \wedge \bigwedge_{y \in S} x \leqslant y \wedge S \neq 0$

Least upper and greatest lower bounds of a set S of real numbers may then be defined thus:

D-0.17 l.u.b. of $S = \imath x[x$ is u.b. of $S \wedge \bigwedge_{y} (y$ is u.b. of $S \rightarrow x \leqslant y) \vee$
$\vee \sim \bigvee_{z} (z$ is u.b. of $S \wedge x = 0)]$

D-0.18 g.l.b. of $S = \imath x[x$ is l.b. of $S \wedge \bigwedge_{y} (y$ is l.b. of $S \rightarrow y \leqslant x) \vee$
$\vee \sim \bigvee_{z} (z$ is l.b. of $S \wedge x = 0)]$

Observe that we can now state a slight generalization of T-4:

T-0.5 $\bigvee_{y} (y$ is u.b. of $S) \rightarrow$ l.u.b. of $S \in \mathbb{R}$

The reason that we can state the theorem thus simply is that the empty set has not been identified with any real number and in particular not with the additive identity of the reals.

On occasion we shall need to supplement the set $\mathbb{R}$ of real numbers with two special objects, ∞ and $-\infty$. We may define them simply enough:

D-0.19 $\infty = \{\omega\} \qquad -\infty = \{\{0\}\}$

The supplemented set of real numbers is

D-0.20 $\mathbb{R}_+ = \mathbb{R} \cup \{\{\{0\}\}, \{\omega\}\}$

Our definitions of the operations and relations on $\mathbb{R}_+$ must be supplemented to yield:

T-0.6 $\bigwedge_{x \in \mathbb{R}} [(x + \infty = \infty \wedge x + -\infty = -\infty) \wedge (x < 0_r \rightarrow x \cdot \infty = \infty \cdot x =$
$= -\infty \wedge x \cdot -\infty = -\infty \cdot x = \infty) \wedge (0_r < x \rightarrow x \cdot \infty = \infty \cdot x =$
$= \infty \wedge x \cdot -\infty = -\infty \cdot x = -\infty) \wedge 0_r \cdot \infty = \infty \cdot 0_r = 0_r \cdot -\infty =$

$$= -\infty \cdot 0_r = 0_r \wedge \infty \cdot \infty = \infty + \infty = -\infty \cdot -\infty = \infty \wedge \infty \cdot -\infty =$$
$$= -\infty \cdot \infty = -\infty + -\infty = -\infty \wedge -\infty + \infty = \infty + -\infty = 0_r]$$

4. We may introduce the usual notations for the sum and product of a finite sequence of real numbers.

D-0.21 $\sum_{i<n} x_i = \imath k [(\bigwedge_{i<n} x_i \in \mathbb{R} \wedge \bigvee f(f \text{ is 1-place function} \wedge \mathscr{D}f =$
$= n+1 \wedge f\langle 0\rangle = 0_r \wedge \bigwedge_{m<n} (f\langle m+1\rangle = f\langle m\rangle + x_m) \wedge k =$
$= f\langle n\rangle)) \vee ((\sim \bigwedge_{i<n} x_i \in \mathbb{R} \vee \sim n \in \omega) \wedge k = 0)]$

D-0.22 $\mathbb{P}_{i<n} x_i = \imath k [(\bigwedge_{i<n} x_i \in \mathbb{R} \wedge \bigvee f(f \text{ is 1-place function} \wedge \mathscr{D}f =$
$= n+1 \wedge f\langle 0\rangle = 1_r \wedge \bigwedge_{i<n} (f\langle i+1\rangle = f\langle i\rangle \cdot x_i) \wedge k =$
$= f\langle n\rangle)) \vee ((\sim \bigwedge_{i<n} x_1 \in \mathbb{R} \vee \sim n \in \omega) \wedge k = 0)]$

The n-fold (associative) Cartesian product of $\mathbb{R}^1$ with itself, $\mathbb{R}^n$, will be one of the two kinds of entities that particularly concern us. From D-0.14 we see that the elements of $\mathbb{R}^n$ are n-sequences of real numbers. Thus the elements of $\mathbb{R}^2$, for example, are 2-sequences of the form

$$\langle \prec 0, x \succ, \prec 1, y \succ \rangle$$

rather than ordered pairs $\prec x, y \succ$ as they are sometimes taken to be. We are also interested in the n-fold Cartesian product of the compact space $\mathbb{R}_+$ with itself, $(\mathbb{R}_+)^n$.

T-0.7 $\mathbb{R}^n = \{\langle x_0, x_1, \ldots, x_{n-1}\rangle : \bigwedge_{i<n} x_i \in \mathbb{R}\}$
$\mathbb{R}^n_+ = \{\langle x_0, x_1, \ldots, x_{n-1}\rangle : \bigwedge_{i<n} x_i \in \mathbb{R}_+\}$

The product of two real spaces $\mathbb{R}^m$ and $\mathbb{R}^n$ then turns out very simply to be the set of concatenations of their elements:

$$\mathbb{R}^n \times \mathbb{R}^n = \{\langle x_0, \ldots, x_{m+n-1}\rangle : \langle x_0, \ldots, x_{n-1}\rangle \in \mathbb{R}^n \wedge$$
$$\wedge \langle x_n, \ldots, x_{m+n-1}\rangle \in \mathbb{R}^m\}$$

Addition, subtraction, and multiplication by a constant follow the usual laws for vectors; we use the same notation as for the corresponding operations on the reals when the context makes the nature of the objects clear; otherwise we add a subscript 'v'.

D-0.23 $$x +_v y = \imath k[\bigvee_{n\in\omega} (x, y \in \mathbb{R}^n_+ \wedge k \in \mathbb{R}^n_+ \wedge \bigwedge_{i<n} k_i = x_i + y_i) \vee \vee (\sim \bigvee_{n\in\omega} x, y \in \mathbb{R}^n_+ \wedge k = 0)]$$

Note that $\mathbb{R}^1$, being a set of 1-sequences, is not the same as $\mathbb{R}$; addition in the reals is not a 'special case' of D-0.23.

D-0.24 $$x \cdot_v y = \imath k[(\bigvee_{n\in\omega} (x \in \mathbb{R}^n_+ \wedge y \in \mathbb{R}_+ \wedge k \in \mathbb{R}^n_+ \wedge \bigwedge_{i<n} k_i = y \cdot x_i) \vee \vee \bigvee_{n\in\omega} (y \in \mathbb{R}^n_+ \wedge x \in \mathbb{R}_+ \wedge k \in \mathbb{R}^n_+ \wedge \bigwedge_{i<n} k_i = x \cdot y_i)) \vee \vee (\sim \bigvee_{n\in\omega} ((y \in \mathbb{R}^n_+ \wedge x \in \mathbb{R}_+) \vee (x \in \mathbb{R}^n_+ \wedge y \in \mathbb{R}_+)) \wedge k = 0)]$$

D-0.25 $$x -_v y = x +_v (-1_r) \cdot_v y$$

We suppose the elements of $\mathbb{R}^n_+$ to have the natural partial ordering:

D-0.26 $$x <_v y \leftrightarrow \bigvee_{n\in\omega} (x \in \mathbb{R}^n_+ \wedge y \in \mathbb{R}^n_+ \wedge \bigwedge_{i<n} x_i < y_i)$$

We use the standard Euclidean distance function in $\mathbb{R}^n_+$; note that in $\mathbb{R}^1_+$ it corresponds to absolute value in $\mathbb{R}_+$.

D-0.27 $$|x - y| = \imath k[\bigvee_{n\in\omega} (x, y \in \mathbb{R}^n_+ \wedge k \in \mathbb{R}_+ \wedge k \geqslant 0_r \wedge k^2 = = \sum_{i<n} (x_i - y_i)^2) \vee (\sim \bigvee_{n\in\omega} x, y \in \mathbb{R}^n_+ \wedge k = 0)]$$

5. We define limit points (points of accumulation) simultaneously for sets of reals and for sets of sequences of reals:

D-0.28 $$x \text{ is l.p. of } S \leftrightarrow [S \subset \mathbb{R}_+ \wedge \bigwedge_{y\in\mathbb{R}} (0_r < y \rightarrow \bigvee_{z\in S} 0 < |\langle x\rangle - \langle z\rangle| < < y)] \vee \bigvee_{n\in\omega} [S \subset \mathbb{R}^n_+ \wedge \bigwedge_{y\in\mathbb{R}} (0_r < y \rightarrow \bigvee_{z\in S} 0 < |x - z| < y)]$$

Observe that if x is a limit point of S, then for every positive y there is not only one, but an infinite number of z's within the distance y of x; furthermore, observe that x itself need not belong to S.

The Bolzano-Weierstrass theorem assepts that every bounded infinite set of real numbers has at least one limit point:

T-0.8 $S \subset \mathbb{R} \wedge \omega \preccurlyeq S \wedge \bigvee_{z, w \in \mathbb{R}} (z \text{ is u.b. of } S \wedge w \text{ is l.b. of } S) \rightarrow$
$\rightarrow \bigvee x(x \text{ is l.p. of } S)$

We say that the ω-sequence x converges to a when for any real $y>0$ there is an n such that for all m greater than n the difference between x_m and a is less than y:

D-0.29 $\text{Conv } x \text{ to } a \leftrightarrow x \textit{ is } \omega\textit{-sequence} \wedge \bigwedge_{i \in \omega} x_i \in \mathbb{R} \wedge \bigwedge_{y \in \mathbb{R}} (0<y \rightarrow$
$\rightarrow \bigvee_{n \in \omega} \bigwedge_{m \in \omega} (n<m \rightarrow |\langle x_n \rangle - \langle a \rangle| < y))$

If a sequence coverges to a and to b then of course a and b must be equal; and the set of terms of the sequence must be bounded.

T-0.9 $\text{Conv } x \text{ to } a \rightarrow \bigvee_{z, w \in \mathbb{R}} (z \text{ is u.b. of } \{x_i : i \in \omega\} \wedge w \text{ is l.b. of}$
$\{x_i : i \in \omega\})$

If an ω-sequence is bounded, its limit is just that number, if any, to which it converges; if it is bounded below and not above, we take its limit to be ∞; if it is bounded above and not below we take its limit to be $-\infty$.

D-0.30 $\lim_{i \rightarrow \infty} x = \imath k\, [(x \textit{ is } \omega\textit{-sequence} \wedge \bigwedge_{i \in \omega} x_i \in \mathbb{R} \wedge (\text{Conv } x \text{ to } k \vee$
$\vee (\bigvee z(z \text{ is u.b. of } \{x_i : i \in \omega\}) \wedge \sim \bigvee z(z \text{ is l.b. of } \{x_i : i \in \omega\}) \wedge$
$\wedge k = -\infty) \vee (\bigvee z(z \text{ is l.b. of } \{x_1 : i \in \omega\}) \wedge \sim \bigvee z(z \text{ is u.b. of}$
$\{x_i : i \in \omega\}) \wedge k = \infty))) \vee ((\sim x \textit{ is } \omega\textit{-sequence} \vee \sim \bigvee z(z \text{ is u.b.}$
$\text{of } \{x_i : i \in \omega\} \vee z \text{ is l.b. of } \{x_i : i \in \omega\}) \vee \bigvee (z \text{ is u.b. of}$
$\{x_i : i \in \omega\} \wedge w \text{ is l.b. of } \{x_i : i \in \omega\} \wedge \sim \bigvee y \text{ Conv } x \text{ to } y)) \wedge$
$\wedge k = 0)]$

The value of an infinite sum can now be defined in the conventional way as the limit (if it exists) of the infinite sequence of partial sums:

D-0.31 $\sum_{i \in \omega} x_i = \imath k\, [\bigvee_f f \text{ is 1-place function} \wedge \mathscr{D} f = \omega \wedge f\langle 0 \rangle = 0 \wedge$
$\wedge \bigwedge_{n \in \omega} f\langle n+1 \rangle = f\langle n \rangle + x_n \wedge k = \lim_{i \rightarrow \infty} \{\langle i, f\langle i \rangle \rangle : i \in \omega\}]$

Observe, with respect to this definition, that if anything goes wrong – e.g., if x is not an infinite sequence of real numbers, or if the sequence of partial sums fails to converge or to increase indefinitely or to decrease indefinitely, k turns out to be the empty set 0; on the other hand, if x is an infinite sequence of reals, the function f exists.

We now define convergence generally for functions whose domain is included in $\mathbb{R}^n$, and whose range is included in the reals. We say that fx converges to a as x approaches b, within the set I (b need not belong to I), in symbols, '$\underset{\substack{x\to b\\ x\in I}}{\mathrm{Conv}}\, fx$ to a', when, given any positive real number y, we can find a positive real number z such that as long as the distance between x and b is less than z, for x belonging to I, the distance between fx and a is less than y.

D-0.32 $$\underset{\substack{x\to b\\ x\in I}}{\mathrm{Conv}}\, f \text{ to } a \leftrightarrow a\in\mathbb{R}\wedge \bigvee_{n\in\omega}(I\subset\mathbb{R}^n\wedge b\in\mathbb{R}^n)\wedge$$
$$\wedge \bigwedge_{y\in\mathbb{R}}(0<y\rightarrow \bigvee_{z\in\mathbb{R}}(0<z\wedge \bigwedge_{x\in I}(|x-b|<z\wedge 0<|x-b|\rightarrow$$
$$\rightarrow |\langle fx\rangle - \langle a\rangle| < y)))$$

Again, observe that for a given y there is not merely one z but an infinite number satisfying the last conditional of D-0.32.

Continuity at a point, continuity on the right, and continuity in general receive ordinary definitions:

D-0.33 $$\underset{x\in I}{\mathrm{Cont}}\, f \text{ at } a \leftrightarrow \bigvee_{n\in\omega}(I\subset\mathbb{R}^n\wedge a\in\mathbb{R}^n\wedge \underset{\substack{x\to a\\ x\in I}}{\mathrm{Conv}}\, f \text{ to } fa)$$

D-0.34 $$\mathrm{Cont}\, f \text{ on Rt} \leftrightarrow \bigwedge_{a\in\mathscr{D}f}(\underset{\substack{x\to a\\ x\in\{y:\, y\in\mathscr{D}f\wedge y>a\}}}{\mathrm{Conv}}\, f \text{ to } fa)$$

D-0.35 $$\mathrm{Cont}\, f \leftrightarrow \bigwedge_{a\in\mathscr{D}f}\underset{\substack{x\to a\\ x\in\mathscr{D}f}}{\mathrm{Conv}}\, f \text{ to } fa$$

Finally, we wish to define the limit of a function as its argument approaches a certain point. Clearly if the function converges to a certain value at that point, that will be its limit. Two other cases are also of interest: that in which the function increases without bound as we approach the point in question (we say then that its limit is ∞), and that in which the function decreases without bound as we approach the point in question (we say then that its limit is $-\infty$). And then there is the garbage case, in which something goes wrong; in that case we take the object to be the empty set, as usual.

D-0.36
$$\lim_{\substack{x\to b\\ x\in I}} fx = \imath k[\overset{1}{\bigvee_{w}}(\mathrm{Conv}_{\substack{x\to b\\ x\in I}} f \text{ to } w \vee \bigvee_{n\in\omega}(I\subset\mathbb{R}^n \wedge b\in\mathbb{R}^n_+ \wedge$$
$$\wedge ((\bigwedge_{z\in\mathbb{R}} \bigvee_{y\in\mathbb{R}} (y>0 \wedge (|x-b|<y \to fx>z)) \wedge w=\infty) \vee$$
$$\vee (\bigwedge_{z\in\mathbb{R}} \bigvee_{y\in\mathbb{R}} (y>0 \wedge (|x-b|<y \to fx<z)) \wedge w=-\infty)) \wedge$$
$$\wedge k=w) \vee (\sim \overset{1}{\bigvee_{w}}((\mathrm{Conv}_{\substack{x\to b\\ x\in I}} f \text{ to } w \vee \bigvee_{n\in\omega}(I\subset\mathbb{R}^n \wedge b\subset\mathbb{R}^n_+ \wedge$$
$$\wedge ((\bigwedge_{z\in\mathbb{R}} \bigvee_{y\in\mathbb{R}} (y>0 \wedge (|x-b|<y \to fx>z)) \wedge w=\infty) \vee$$
$$\vee (\bigwedge_{z\in\mathbb{R}} \bigvee_{y\in\mathbb{R}} (y>0 \wedge (|x-b|<y \to fx<z))\, w=-\infty))))\, k=0)]$$

6. In addition to being concerned with *points* in the various spaces $\mathbb{R}^n$, we shall also be concerned with a certain set of subsets in each of these spaces. The set of all subsets of $\mathbb{R}^n$ is too big – that is, we cannot define an interesting measure (e.g., a probability function) on *all* subsets of $\mathbb{R}^n$. But we can define one for the set of all Borel sets in $\mathbb{R}^n$.

We begin by considering open intervals in $\mathbb{R}^n_+$:

D-0.37
$$(a, b) = \imath k(\bigvee_{n\in\omega}(a\in\mathbb{R}^n_+ \wedge b\in\mathbb{R}^n_+ \wedge k=\{x : x\in\mathbb{R}^n_+ \wedge a<x<b\}) \vee$$
$$\vee(\sim \bigvee_{n\in\omega}(a\in\mathbb{R}^n_+ \wedge b\in\mathbb{R}^n_+) \wedge k=0))$$

The closed and half-open intervals, $[a, b]$, $(a, b]$, and $[a, b)$ are defined similarly.

By an additive class of sets in $\mathbb{R}^n$ we mean a class to which $\mathbb{R}^n$ itself belongs, to which every countable union of members of it belongs, and to which the difference between $\mathbb{R}^n$ and any member of it belongs. The set of Borel sets is just the smallest additive class of subsets of $\mathbb{R}^n$ that contains all open intervals; we denote this set by '$\mathbb{B}^n$':

D-0.38
$$\mathbb{B}^n = \bigcap\{x : x\subset\mathscr{P}(\mathbb{R}^n) \wedge \mathbb{R}^n\in x \wedge \bigwedge_{y\in x} \mathbb{R}^n - y\in x \wedge (\bigwedge_{i\in\omega}(y_i\in x) \to$$
$$\to \bigcup\{y_i : i\in\omega\}\in x \wedge \bigwedge_{y,z\in\mathbb{R}} (y, z)\in x\}$$

T-0.10 $0\in\mathbb{B}^n$ (Since $0=\mathbb{R}^n-\mathbb{R}^n$.)

T-0.11 $x, y \in \mathbb{B}^n \rightarrow x - y \in \mathbb{B}^n$ (Since $x - y = \mathbb{R}^n - [(\mathbb{R}^n - x) \cup y]$.)

T-0.12 $a, b \in \mathbb{R}^n \wedge a \leqslant b \rightarrow \{x : x \in \mathbb{R}^n \wedge a \leqslant x \leqslant b\} \in \mathbb{B}^n \wedge$
$\wedge \{x : x \in \mathbb{R}^n \wedge a < x \leqslant b\} \in \mathbb{B}^n \wedge \{x : x \in \mathbb{R}^n \wedge a \leqslant x < b\} \in \mathbb{B}^n$

In particular, the unit set of any point in $\mathbb{R}^n$ is a Borel set:

T-0.13 $x \in \mathbb{R}^n \rightarrow \{x\} \in \mathbb{B}^n$

The following theorem schema (instances are obtained by proper substitution of a formula for the one-place schematic predicate letter F) will be useful in the future:

T-0.14 $\bigwedge_a (a \in \mathbb{R}^n \rightarrow F(\{-\infty\}^n, a]) \wedge \bigwedge_x (x \in \mathscr{P}\mathbb{R}^n \wedge Fx \rightarrow F\, \mathbb{R}^n - x) \wedge$
$\wedge F\, \mathbb{R}^n \wedge \bigwedge_y (y \text{ is } \omega\text{-sequence} \wedge \bigwedge_{i \in \omega} (y_i \in \mathscr{P}\mathbb{R}^n \wedge$
$\wedge F\, y_i \rightarrow F \bigcup \mathscr{R}\, y) \rightarrow \bigwedge_x (x \in \mathbb{B}^n \rightarrow Fx)$

Proof:

Let $z = \{y : y \in \mathscr{P}\mathbb{R}^n \wedge Fy\}$

Assume antecedent.

Show $z \subset \mathscr{P}\mathbb{R}^n$ (obvious)

Show $\mathbb{R}^n \in z$ (clear)

Show $\bigwedge_y (y \in z \rightarrow \mathbb{R}^n - y \in z)$ (naturally)

Show $\bigwedge_{i \in \omega} y_i \in z \rightarrow \bigcup \{y_i : i \in \omega\} \in z$ (of course)

Show $\bigwedge_{y, w \in \mathbb{R}} (y, w) \in z$. This follows from earlier clauses, since

$$(y,w) = \mathbb{R}^n - ((\{-\infty\}^n, y] \cup \left(\mathbb{R}^n - \bigcup_{m \in \omega} \left(\{-\infty\}^n, w - \left\{\frac{1}{m}\right\}^n\right]\right)).$$

Thus z is an additive class of subsets of $\mathbb{R}^n$ to which all open intervals belong, and thus

$\mathbb{B}^n \subset z$

$\therefore \bigwedge_x (x \in \mathbb{B}^n \rightarrow Fx)$

7. Let F be a function whose domain is $\mathbb{R}^2$. We can obtain a correspond-

ing set function whose domain is included in $\mathscr{P}\mathbb{R}^2$ by construing $F\langle x_1, x_2\rangle$ as the value of this set function for the infinite interval $\{\langle y_1, y_2\rangle : y_1 \leqslant x_1 \wedge \wedge\, y_2 \leqslant x_2\}$. We may then compute the value of this set function for the half open interval, $(a, b] = \{\langle y_1, y_2\rangle : a_1 < y_1 \leqslant b_1 \wedge a_2 < y_2 \leqslant b_2\}$, to be:

$$F\langle b_1, b_2\rangle - F\langle a_1, b_2\rangle - F\langle b_1, a_2\rangle + F\langle a_1, a_2\rangle$$

We call this the differential value of F on the interval $(a, b]$, and denote it by '$\Delta F_{a,b}$'. The rationale behind the computation is clear from the accompanying Figure 10.

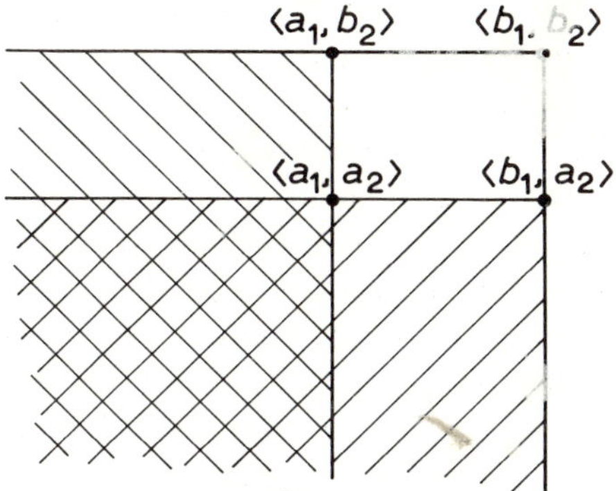

Fig. 10.

The more formal and general definition of the differential value of F on the interval $(a, b]$ is as follows (we use $\binom{n}{i}$ to denote the number of of ways of selecting i objects out of n):

D-0.39 $\Delta F_{a,b} = \imath k[\bigvee_{n\in\omega} (F \text{ is fun } \wedge \mathscr{D}F = \mathbb{R}^n \wedge a, b \in \mathbb{R}^n \wedge$

$\bigvee_{r} (r \textit{ is } n+1\textit{-sequence} \wedge \bigwedge_{i<n+1} [r_i \text{ is } \binom{n}{i}\textit{-sequence} \wedge$

$\wedge \bigwedge_{j<\binom{n}{i}} (r_i\langle j\rangle \textit{ is } n\textit{-sequence} \wedge \bigwedge_{t<n} (r_{ij}\langle t\rangle =$

$= a_t \vee r_{ij}\langle t\rangle = b_t) \wedge r_{ij} \cap a \approx i) \wedge \bigwedge_{s,t<\binom{n}{i}} (s \neq t \rightarrow r_i\langle s\rangle \neq$

$\neq r_i\langle t\rangle)] \wedge k = \sum_{m<n+1} (-1)^m \sum_{j<\binom{n}{m}} Fr_m\langle j\rangle)) \vee$

$$\vee(\sim \bigvee_{n\in\omega} (F \text{ is fun } \wedge \mathscr{D}F = \mathbb{R}^n \wedge a, b \in \mathbb{R}^n) \wedge k = 0)]$$

The usefulness of the definition would be established by the following theorem, which in turn would be proved by largely combinatorial arguments.

T-0.15 $$\bigvee_{n\in\omega} (F \text{ is fun } \wedge \mathscr{D}F = \mathbb{R}^n \wedge a, b \in \mathbb{R}^n) \rightarrow \Delta F_{a,b} \in \mathbb{R}$$

(Note that for any $0 < i < n$, every sequence r satisfying the conditions mentioned yields the same sum, and that at least one such sequence exists.)

We are now in a position to define the class of functions fundamental to our whole approach: distribution functions. In the conventional exposition of statistical concepts probability functions come first; but since in the model theoretic part of our work we shall take distribution functions as yielding the fundamental objects for which interpretations are required, the slightly unconventional procedure of defining distribution functions first and probability functions second, seems desirable.

We say that F is a distribution function in $\mathbb{R}^n$, in symbols, 'F is dist fun in $\mathbb{R}^n$', just in case F is a function whose domain is $\mathbb{R}^n$, whose range is included in the closed interval $[0, 1]$, whose differential value on any half open interval $(a, b]$ in $\mathbb{R}^n$ is non-negative, which is continuous on the right, and which approaches 0_r as its argument approaches $\{-\infty\}^n$, and 1_r as its argument approaches $\{\infty\}^n$.

D-0.40 $$F \text{ is dist fun in } \mathbb{R}^n \leftrightarrow F \text{ is fun } \wedge \mathscr{D}F = \mathbb{R}^n \wedge$$
$$\wedge \mathscr{R}F \subset [0, 1] \wedge \bigwedge_{a, b \in \mathbb{R}^n} \Delta F_{a,b} \geqslant 0 \wedge F \text{ is cont on Rt } \wedge$$
$$\wedge \lim_{\substack{x \to \{-\infty\}^n \\ x \in \mathbb{R}}} Fx = 0_r \wedge \lim_{\substack{x \to \{\infty\}^n \\ x \in \mathbb{R}}} Fx = 1$$

It can be shown that any such function F gives rise to an additive set function whose values are non-negative and whose value for $\mathbb{R}^n$ is 1. (A set function is additive just in case its value for the union of a denumerable sequence of disjoint sets is the sum of its values for each of the sets.) Any two such functions (for given F) with their domains restricted to Borel sets will coincide. Conversely, given any additive, non-negative, set function

P, for which $P\langle\mathbb{R}^n\rangle=1$, we may find a distribution function F in $\mathbb{R}^n$ such that $P_F=P$. This distribution function is clearly unique. We shall not go through the rather long process of developing the set function P_F from a distribution function F, but simply offer the following informal definition, followed by some explanatory theorems.

D-0.41 $P_F = \imath k((\bigvee_{n\in\omega}$ F is dist fun in $\mathbb{R}^n \wedge k$ is the (unique) set function with domain restricted to $\mathbb{B}^n$ constructed from $F) \vee (\sim \bigvee_{n\in\omega}$ F is dist fun in $\mathbb{R}^n \wedge k=0))$

T-0.16 $(F$ is dist fun in $\mathbb{R}^n \rightarrow P_F \neq 0) \wedge (P_F \neq P_G \rightarrow F \neq G)$

T-0.17 F is dist fun in $\mathbb{R}^n \rightarrow (\bigwedge_{x\in\mathbb{B}^n} \bigvee_{y\in\mathbb{R}} (P_F\langle x\rangle = y \wedge 0 \leqslant y \leqslant 1) \wedge$
$\wedge \bigwedge_{a,b\in\mathbb{R}^n} P_F\langle(a,b]\rangle = \Delta F_{a,b})$

T-0.18 F is dist fun in $\mathbb{R}^n \wedge z$ *is ω-sequence* $\wedge \bigwedge_{i\in\omega} z_i \in \mathbb{B}^n \wedge$
$\wedge\, y = \bigcup\{z_i : i\in\omega\} \wedge \bigwedge_{i,j\in\mathbb{R}^n} (i\neq j \rightarrow z_i \cap z_j = 0) \rightarrow$
$\rightarrow P_F\langle y\rangle = \sum_{i\in\omega} P_F\langle z_i\rangle$

T-0.19 F is dist fun in $\mathbb{R}^n \wedge y\in\mathbb{B}^n \rightarrow P_F\langle y\rangle = 1 - P_F\langle\mathbb{R}^n - y\rangle$
F is dist fun in $\mathbb{R}^n \wedge x, y\in\mathbb{B}^n \wedge x\subset y \rightarrow P_F\langle x\rangle \leqslant P_F\langle y\rangle$

A set x in $\mathbb{R}^n$ is said to be *P_F-measurable* if there is a real number y such that $P_F(x)=y$; the set of P_F-measurable sets always includes the Borel sets $\mathbb{B}^n$, and in general is broader yet.

If F is a distribution function in $\mathbb{R}^n$, ${}_{\langle i\rangle}F$ will be the marginal distribution function of the ith component. It is the projection of F on the ith axis.

D-0.42 ${}_{\langle i\rangle}F = \imath k[\bigvee_n (F$ is dist fun in $\mathbb{R}^n \wedge i<n \wedge \mathscr{D}k = \mathbb{R}^1 \wedge$
$\wedge \bigwedge_{y\in\mathbb{R}^1} ky = \lim_{\substack{x\rightarrow 0 \\ x\in\{z:z\in\mathbb{R}\wedge 0<z\}}} F\imath w(w$ *is n-sequence* $\wedge$
$\wedge \bigwedge_{j<n} [(j=i \wedge w_j = y) \vee (j\neq i \wedge w_j = x^{-1})])) \vee$
$\vee (\sim \bigvee_n (F$ is dist fun in $\mathbb{R}^n \wedge i<n) \wedge k=0)]$

We shall also have occasion to refer to marginal distributions in a wider sense: the projection of a distribution in $\mathbb{R}^n$ onto a space of j dimensions.

D-0.43 $_sF = \iota k[\bigvee_{n\in\omega} (F \text{ is dist fun in } \mathbb{R}^n \wedge \bigvee_{m\in\omega} (s \textit{ is m-sequence} \wedge$
$\wedge\, m<n \wedge \bigwedge_{i,j<m} s_i<n \wedge (i<j \rightarrow s_i<s_j)) \wedge \mathscr{D}k=\mathbb{R}^m \wedge \bigwedge_{y\in\mathbb{R}^m} ky=$
$= \lim_{\substack{x\to 0\\ x\in\{z:z\in\mathbb{R}\wedge z>0\}}} F\iota w(w \textit{ is n-sequence} \wedge \bigwedge_{j<n} [(j\in\mathscr{R}s \wedge w_j=$
$=y_j) \vee (\sim j\in\mathscr{R}s \wedge w_j=x^{-1})]])) \vee (\sim \bigvee_{n\in\omega} (F \text{ is dist fun in } \mathbb{R}^n \wedge$
$\wedge \bigvee_{m\in\omega} s \textit{ is m-sequence} \wedge m<n \wedge \bigwedge_{i,j<m} (s_i<n \wedge$
$\wedge (i<j \rightarrow s_i<s_j)) \wedge k=0)]$

Not surprisingly, these marginal distribution functions are distribution functions:

T-0.20 F is dist fun in $\mathbb{R}^n \wedge i<n \rightarrow {}_{\langle i\rangle}F$ is dist fun in $\mathbb{R}^1$
F is dist fun in $\mathbb{R}^n \wedge s$ *is m-sequence* $\wedge\, m<n \wedge \bigwedge_{i,j\in\omega} (s_i<n \wedge$
$\wedge (i<j \rightarrow s_i<s_j)) \rightarrow {}_sF$ is dist fun in $\mathbb{R}^n$

T-0.21 F is dist fun in $\mathbb{R}^1 \rightarrow \{y: \sim F \text{ is cont at } y\} \preccurlyeq \omega$

T-0.22 F is dist fun in $\mathbb{R}^n \rightarrow \{y: \bigvee_{x\in\mathbb{R}} (x>0 \wedge \lim_{\substack{z\to y\\ z\in\mathbb{R}^n}} |Fy-Fz|>x)\} \prec \omega$

An *excluded point* for a distribution F is a point of discontinuity in a marginal distribution of F:

D-0.44 x_i is excluded point of $F \leftrightarrow \sim {}_{\langle i\rangle}F$ is cont at x_i

T-0.23 $\{x_i : i\in\omega \wedge x_i \text{ is excluded point of } F\} \preccurlyeq \omega$

T-0.24 $\sim \bigvee_{i\in n} y_i$ is excluded point of $F \wedge F$ is dist fun in $\mathbb{R}^n \wedge y\in\mathbb{R}^n \rightarrow$
$\rightarrow F$ is cont at y

Given a distribution in n dimensions and one in m dimensions, we may construct a distribution in $m+n$ dimensions – indeed, an infinite number of them:

T-0.25 F_1 is dist fun in $\mathbb{R}^n \wedge F_2$ is dist fun in $\mathbb{R}^m \rightarrow$
$\rightarrow \omega \preccurlyeq \{F: F$ is dist fun in $\mathbb{R}^{n+m} \wedge {}_{\langle 0,\ldots,n-1\rangle}F = F_1 \wedge$
$\wedge {}_{\langle n,\ldots n+m-1\rangle}F = F_2\} \wedge \imath F\,[\bigwedge_{x\in\mathbb{R}^n} \bigwedge_{y\in\mathbb{R}^m} F_1 x \cdot F_2 y =$
$= F\widehat{xy} \wedge F$ is fun $\wedge\, \mathscr{D}F = \mathbb{R}^{n+m}]$ is dist fun in $\mathbb{R}^{n+m}$

Although there are an infinite number of distributions in $\mathbb{R}^{m+n}$ whose marginal distributions are F_1 and F_2, the one that is of special interest is the one whose value for any member of $\mathbb{R}^{m+n}$ is the product of the corresponding values of F_1 and F_2. In this situation we say that the distributions characterized by the distribution functions F_1 and F_2 are *independent*. The concept of independence is more familiarly expressed in terms of the set function on P_F than in terms of the corresponding distribution function F. The connection is given by the following theorem:

T-0.26 F is dist fun in $\mathbb{R}^{n+m} \wedge F_1$ is dist fun in $\mathbb{R}^n \wedge F_2$ is dist fun in $\mathbb{R}^m \rightarrow [\bigwedge_{S\in\mathbb{B}^{m+n}} \bigwedge_{S_1\in\mathbb{B}^n} \bigwedge_{S_2\in\mathbb{B}^m} (S = (S_1 \times \mathbb{R}^m) \cap (\mathbb{R}^n \times S_2) \rightarrow$
$\rightarrow P_F(S) = P_{F_1}(S_1) \cdot P_{F_2}(S_2)) \leftrightarrow \bigwedge_{x\in\mathbb{R}^{m+n}} \bigwedge_{y\in\mathbb{R}^n} \bigwedge_{z\in\mathbb{R}^m} (x = \widehat{yz} \rightarrow$
$\rightarrow Fx = F_1 y \cdot F_2 z)]$

8. Let S be a Borel set; R is a *division* of S, in symbols, R div S, if R is a finite sequence of disjoint Borel sets whose union is S:

D-0.45 R div $S \leftrightarrow \bigvee_{n,m\in\omega} (S \in \mathbb{B}^n \wedge R$ *is m-sequence* $\wedge \bigwedge_{i<m} R_i \in \mathbb{B}^n \wedge$
$\wedge \bigwedge_{i,j<m} (i \neq j \rightarrow R_i \cap R_j = 0) \wedge \bigcup \{R_i : i < m\} = S)$

Corresponding to any function g and distribution F and any division R of S, we may define upper and lower Darboux sums, UDS (g, F, S, R) and LDS (g, F, S, R):

D-0.46 $\mathrm{UDS}(g, F, S, R) = \imath k[(g$ is fun $\wedge S \subset \mathscr{D}g \wedge \mathscr{R}g \subset \mathbb{R} \wedge$
$\wedge R$ div $S \wedge \bigvee_{r,s\in\mathbb{R}} \bigwedge_{w\in S} r < gw < s \wedge k =$
$\sum_{i<\imath m(R \text{ is } m\text{-sequence})} (\text{l.u.b}\,\{gw : w \in R_i\})\, P_F(R_i)$
$\vee (\sim (g$ is fun $\wedge S \subset \mathscr{D}g \wedge$
$\mathscr{R}g \subset \mathbb{R} \wedge R$ div $S \wedge \bigvee_{r,s\in\mathbb{R}} \bigwedge_{w\in S} r < gw < s) \wedge k = 0)]$

D-0.47 $\mathrm{LDS}(g, F, S, R) = \imath k[(g \text{ is fun } \wedge S \subset \mathscr{D}g \wedge \mathscr{R}g \subset \mathbb{R} \wedge \wedge R \text{ div } S \wedge \bigvee_{r,\, s \in \mathbb{R}} \bigwedge_{w \in S} r < gw < s \wedge k = \sum_{i < \imath m(R \text{ is } m\text{-sequence})} (g.l.b. \{gw : w \in R_i\}) \cdot P_F(R_i) \vee (\sim(g \text{ is fun } \wedge S \subset \mathscr{D}g \wedge \wedge \mathscr{R}g \subset \mathbb{R} \wedge R \text{ div } S \wedge \bigvee_{r,\, s \in \mathbb{R}} \bigwedge_{w \in S} r < gw < s) \wedge k = 0)]$

T-0.27 $g \text{ is function } \wedge S \subset \mathscr{D}g \wedge \mathscr{R}g \subset \mathbb{R} \wedge \langle R_0, R_1, \ldots, R_{m-1}\rangle \text{ div } S \wedge \wedge \bigwedge_{w \in S} r < gw < s \rightarrow \mathrm{UDS}(g, F, S, R) = \sum_{i<m} (\text{l.u.b.} \{gw : w \in R_i\} \cdot \cdot P_F(R_i)) \wedge \mathrm{LDS}(g, F, S, R) = \sum_{i<m} \text{g.l.b.} \{gw : w \in R_i\} \cdot \cdot P_F(R_i)) \wedge \mathrm{UDS}(g, F, S, R) < s \wedge \mathrm{LDS}(g, F, S, R) > r \wedge \wedge \mathrm{LDS}(g, F, S, R) \leqslant \mathrm{UDS}(g, F, S, R)$

One division is superimposed on another if every set in the second division is equal to a finite union of sets in the first division. If T is superimposed on R, then the upper Darboux sum corresponding to T can be no greater than that corresponding to R, and the lower Darboux sum corresponding to T can be no less than the lower Darboux sum corresponding to R.

T-0.28 $\mathrm{UDS}(g, F, S, R) \in \mathbb{R} \wedge R = \langle R_0, \ldots, R_{m-1}\rangle \wedge \bigwedge_{i \in m} T_i \text{ div } R_i \wedge \wedge T = T_0 ^\frown T_1 ^\frown \ldots ^\frown T_{m-1} \rightarrow T \text{ div } S \wedge \mathrm{UDS}(g, F, S, R) \geqslant \geqslant \mathrm{UDS}(g, F, S, T) \wedge \mathrm{LDS}(g, F, S, R) \leqslant \mathrm{LDS}(g, F, S, T)$

The upper and lower integrals of the bounded function g over the set S with respect to the distribution function F, in symbols '$\overset{*}{\int_S} g \, \mathrm{d}F$' and '$\int_{*S} g \, \mathrm{d}F$' respectively, are simply the greatest lower bound of the set of all upper Darboux sums, and the least upper bound of the set of all lower Darboux sums.

D-0.48 $\overset{*}{\int_S} g \, \mathrm{d}F = \text{g.l.b.} \{x : \bigvee_R x = \mathrm{UDS}(g, F, S, R)\}$

D-0.49 $\displaystyle {}_*\!\!\int_S g\,\mathrm{d}F = \text{l.u.b.}\{x: \bigvee_R x = \mathrm{LDS}(g, F, S, R)\}$

Since Darboux sums exist (i.e., are real numbers) only when g is bounded on S, S is a Borel set, and so on, the upper and lower integrals automatically turn out to be the empty set if something goes wrong:

T-0.29 $$\sim \bigvee_{r,s\in\mathbb{R}} \bigwedge_{w\in S} r < gw < s \vee \sim g \text{ is fun} \vee \sim S \subset \mathscr{D}g \vee$$
$$\sim \mathscr{R}g \subset \mathbb{R} \vee \sim \bigvee_{n\in\omega} (F \text{ is dist fun in } \mathbb{R}^n \wedge S \in \mathbb{B}^n) \rightarrow$$
$$\rightarrow {}^*\!\!\int_S g\,\mathrm{d}F = {}_*\!\!\int_S g\,\mathrm{d}F = 0$$

But if nothing goes wrong both integrals denote real numbers and the upper integral is at least as great as the lower integral.

T-0.30 $$\bigvee_{r,s\in\mathbb{R}} \bigwedge_{w\in S} r < gw < s \wedge g \text{ is fun} \wedge S \subset \mathscr{D}g \wedge \mathscr{R}g \subset \mathbb{R} \wedge$$
$$\bigvee_{n\in\omega} (F \text{ is dist fun in } \mathbb{R}^n \wedge S \in \mathbb{B}^n) \rightarrow$$
$$\rightarrow {}^*\!\!\int_S g\,\mathrm{d}F \in \mathbb{R} \wedge {}_*\!\!\int_S g\,\mathrm{d}F \in \mathbb{R} \wedge {}^*\!\!\int_S g\,\mathrm{d}F \geqslant {}_*\!\!\int_S g\,\mathrm{d}F$$

If these upper and lower integrals are equal real numbers, then we take their common value as the integral of the bounded function g over the Borel set S with respect to the distribution function F.

D-0.50 $$*\!\!\int_S g\,\mathrm{d}F = \imath k \, [(\bigvee_{t\in\mathbb{R}} {}^*\!\!\int_S g\,\mathrm{d}F = t \wedge {}_*\!\!\int_S g\,\mathrm{d}F = t \wedge k = t) \vee$$
$$\vee (\sim \bigvee_{t\in\mathbb{R}} ({}^*\!\!\int_S g\,\mathrm{d}F = t \wedge {}_*\!\!\int_S g\,\mathrm{d}F = t) \wedge k = 0)]$$

We finally obtain the most general forms of Lebesgue-Stieltjes integral we need by considering also unbounded functions:

D-0.51 $$\int_S g\,\mathrm{d}F = \imath k[(* \int_S g\,\mathrm{d}F = k \wedge k \in \mathbb{R}) \vee \bigvee_n (S \in \mathbb{B}^n \wedge$$
$$\wedge \bigwedge_{a,\,b \in \mathbb{R}^n} \bigvee_h h \text{ is fun} \wedge \mathscr{D}h = S \wedge \bigwedge_{w \in S} [(w < a \wedge hw =$$
$$= a_0) \vee (w > b \wedge hw = b_0) \vee (a \leqslant w \leqslant b \wedge hw = gw)] \wedge$$
$$\wedge * \int_S h\,\mathrm{d}F \in \mathbb{R} \wedge k = \lim_{\substack{a \to \{-\infty\}^n \\ a \in \mathbb{R}^n}} \lim_{\substack{b \to \{\infty\}^n \\ b \in \mathbb{R}^n}} * \int_S h\,\mathrm{d}F) \vee$$
$$\vee \sim \bigvee_n (S \in \mathbb{B}^n \wedge \bigwedge_{a,\,b \in \mathbb{R}^n} \bigvee_h (h \text{ is fun} \wedge \mathscr{D}h = S \wedge$$
$$\wedge \bigwedge_{w \in S} [(w < a \wedge hw = a_0) \vee (w > b \wedge hw = b_0) \vee (a \leqslant w \leqslant b \wedge$$
$$\wedge hw = gw)] \wedge * \int_S h\,\mathrm{d}F \in \mathbb{R}) \wedge k = 0)]$$

A real valued function g is Borel measurable in a Borel set S if the set of all points x in S such that $gx \leqslant r$ is a Borel set for every $r \in \mathbb{R}$.

D-0.52 $$g \text{ is } B\text{-measurable in } S \leftrightarrow \bigvee_{n \in \omega} (g \text{ is fun} \wedge S \subset \mathscr{D}g \wedge$$
$$\wedge S \in \mathbb{B}^n \wedge \mathscr{R}g \subset \mathbb{R} \wedge \bigwedge_{r \in \mathbb{R}} \{x : x \in S \wedge gx < r\} \in \mathbb{B}^n$$

The set of Borel measurable functions includes all those we are at all likely to be interested in here.

T-0.31 $\text{Cont } f \wedge S \in \mathbb{B}^n \wedge S \subset \mathscr{D}f \to f$ is B-measurable in S

T-0.32 $$g \textit{ is } \omega\textit{-sequence} \wedge \bigwedge_{i \in \omega} (\text{cont } g_i \wedge \mathscr{D}g_i = \mathbb{R}^n \wedge \mathscr{R}g_i \in \mathbb{R}) \wedge$$
$$\bigwedge_{x \in \mathbb{R}^n} fx = \lim_{i \to \infty} g_i x \to f \text{ is } B\text{-measurable}$$

An interesting and important fact is that if g is bounded in S and is B-measurable in S, then it is integrable with respect to F over S.

T-0.33 $$\bigvee_{r,\,s \in \mathbb{R}} \bigwedge_{w \in S} r < gw < s \wedge g \text{ is } B\text{-measurable in } S \wedge S \in \mathbb{B}^n \wedge$$
$$F \text{ is dist fun in } \mathbb{R}^n \to \int_S g\,\mathrm{d}F \in \mathbb{R}$$

The following theorems represent familiar and useful properties of integrals.

T-0.34 $c \in \mathbb{R} \rightarrow \int_S c \cdot g \, \mathrm{d}F = c \cdot \int_S g \, \mathrm{d}F$

T-0.35 $S_1, S_2 \in \mathbb{B}^n \wedge S_1 \cap S_2 = 0 \rightarrow \int_{S_1 \cup S_2} g \, \mathrm{d}F = \int_{S_1} g \, \mathrm{d}F + \int_{S_2} g \, \mathrm{d}F$

T-0.36 $\left| \int_S g \, \mathrm{d}F \right| \leqslant \int_S |g| \, \mathrm{d}F$

T-0.37 $\int_S (g_1 + g_2) \, \mathrm{d}F = \int_S g_1 \, \mathrm{d}F + \int_S g_2 \, \mathrm{d}F$

T-0.38 $S \in \mathbb{B}^n \rightarrow \int_S \mathrm{d}F = P_F(S)$

9. There are three basic kinds of distributions: discrete, absolutely continuous, and singular. A discrete distribution is one in which the whole measure is carried by a countable number of points in $\mathbb{R}^n$

D-0.53 F is discrete in $\mathbb{R}^n \leftrightarrow F$ is dist fun in $\mathbb{R}^n \wedge \bigvee_m \bigvee_s (m \leqslant \omega \wedge$

$\wedge\, s$ *is m-sequence* $\wedge\, \mathscr{R}s \subset \mathbb{R}^n \wedge \sum_{i<m} P_F(\{s_i\}) = 1)$

An absolutely continuous distribution is one for which there exists a function f whose value is always non-negative, whose integral over the whole space is 1, and which has the property that $Fx = \int_{\{-\infty\}^n}^{x} fx \, \mathrm{d}x$ for every x in $\mathbb{R}^n$, where $\int_{\{-\infty\}^n}^{x} fx \, \mathrm{d}x$ is the ordinary Riemann integral.

D-0.54 F is absolutely continuous in $\mathbb{R}^n \leftrightarrow F$ is dist fun in $\mathbb{R}^n \wedge$

$$\wedge \bigvee_{f} \bigwedge_{x \in \mathbb{R}^n} (fx \geqslant 0_r \wedge Fx = \int_{\{-\infty\}^n}^{x} fx \, dx)$$

A distribution is singular if the whole measure is concentrated on a subset of the space which has the cardinality of the continuum, but is of dimensionality smaller than that of the whole space. Put otherwise, we may say that F is singular if it is continuous, but has a non-vanishing derivative only on a subset of $\mathbb{R}^n$ of Lebesgues measure 0. Casually denoting Lebesgues measure by 'Vol' (for 'volume'), we have

D-0.55 F is singular in $\mathbb{R}^n \leftrightarrow \bigwedge_{y \in \mathbb{R}^n}$ (F is cont at $y \wedge \text{Vol}\{y : y \in \mathbb{R}^n \wedge F'y \neq 0 \wedge F'y \neq 0_r\} = 0_r$)

For example, let x have a rectangular distribution on $[0, 1]$. Then the joint distribution of x and y in $\mathbb{R}^2$, where $y = x$, is:

$F\langle u, v\rangle =$ minimum of u and v if $\langle u, v\rangle \in [\langle 0, 0\rangle, \langle 1, 1\rangle]$
$F\langle u, v\rangle = 0$ if $u < 0$ or $v < 0$
$F\langle u, v\rangle = 1$ if $1 <$ minimum of u and v.

Singular distributions are of relatively little practical importance, but they are of theoretical importance because every distribution in $\mathbb{R}^n$ can be expressed as the weighted sum of a discrete distribution, an absolutely continuous distribution and a singular distribution.

T-0.39 F is dist fun in $\mathbb{R}^n \rightarrow \bigvee_{a, b, c} \bigvee_{F_1, F_2, F_0}$ (F_0 is discrete in $\mathbb{R}^n \wedge F_1$ is absolutely continuous in $\mathbb{R}^n \wedge F_2$ is singular in $\mathbb{R}^n \wedge a, b, c \in$ $\in [0, 1] \wedge a + b + c = 1 \wedge \bigwedge_{x \in \mathbb{R}^n} Fx = aF_1x + bF_2x + cF_0x$)

The one example I know of that involves a distribution containing all three components which is not purely abstract and artificial is due to Professor Jack Hall. In a certain dining hall each person receives one pat of butter and one piece of bread. Let x be the fraction of the pat of butter consumed and y the fraction of the piece of bread. The joint distribution of x and y may plausibly the regarded as having all three components: a certain fraction c of the people will consume the same fraction of their butter as of their bread; a certain fraction a will eat no butter, but consume all their bread, or will eat no bread but put their butter on their potatoes, or eat everything, or never eat bread and butter; and a certain fraction b will consume continuously varying fractions of their bread and butter. The first group gives rise to the singular component of the distribution, the second to the discrete component and the third (perhaps a bit imaginatively), to the absolutely continuous component.

Let us denote the partial derivative of F with respect to each of its coordinates (if it exists) by 'F'' – otherwise we take 'F'' to denote the empty set. Then we have

T-0.40 F is absolutely continuous in $\mathbb{R}^n \rightarrow \bigwedge_{x \in \mathbb{R}^n} (F'x > 0 \wedge Fx =$

$$= \int_{\{-\infty\}^n}^{x} F'x \, \mathrm{d}x) \wedge \int_{\{-\infty\}^n}^{\{\infty\}^n} F'x\mathrm{d}x = 1 \wedge \int_S g\mathrm{d}F = \int_S gxF'xdx$$

Let s be an increasing m-sequence of integers less than n; we denote the mth partial derivative of F with respect to the s_0th, the s_1th, ... the s_mth coordinate by '$\partial^m F/\partial s$'. If that derivative does not exist, we take $\partial^m F/\partial s$ to be the empty set. Thus

$$F' = \frac{\partial^m F}{\partial \prod_{i<n} \{i\}}, \quad \text{if } F \text{ is } n\text{-dimensional.}$$

Let s and t be increasing sequences of integers less than n, such that the sets of their terms are exclusive and exhaustive of the integers less than n. An n-sequence of reals $\langle z_0, \ldots, z_{n-1} \rangle$ may be broken up into two subsequences x and y – one with coordinates corresponding to the terms of s, and one with coordinates corresponding to the terms of t. It is convenient to have a brief expression for the result of putting x and y back together again. Thus if $x = \langle a_0, a_1 \rangle$ and $y = \langle b_0, b_1, b_2 \rangle$, $s = \langle 0, 3 \rangle$ and $t = \langle 1, 2, 4 \rangle$, then the result of putting x and y back together, $0(x, y, s, t)$ is to be $\langle a_0, b_0, b_1, a_1, b_2 \rangle$.

D-0.56 $O(x, y, s, t) = \imath f(\bigvee_{n_1, n_2}$ (x *is* n_1*-sequence* $\wedge\, y$ *is* n_2*-seqience* $\wedge$

$$s \text{ is } n_1\text{-sequence} \wedge t \text{ is } n_2\text{-sequence} \wedge f \text{ is } n_1+n_2\text{-sequence} \wedge$$
$$\wedge \bigwedge_{i<n_1+n_2} (\bigvee_j (f\ i=xj \wedge i=sj) \vee \bigvee_j (f\ i=yj \wedge i=tj))) \vee$$
$$\vee (\sim \bigvee_{n_1, n_2, g} (x \text{ is } n_1\text{-sequence} \wedge y \text{ is } n_2\text{-sequence} \wedge s \text{ is } n_1\text{-se-}$$
$$\text{quence} \wedge t \text{ is } n_2\text{-sequence} \wedge g \text{ is } n_1+n_2\text{-sequence} \wedge$$
$$\wedge \bigwedge_{i<n_1+n_2} (\bigvee_j (gi=xj \wedge i=sj) \vee \bigvee_j (gi=yj \wedge i=tj))) \wedge f=0))$$

It will also be helpful to use the following abbreviation: we will write '$A(F, n, n_1, n_2, s, t)$' to mean that F is an n-dimensional distribution function, s and t are increasing sequences of n_1 and n_2 integers, the terms of the two sequences being exclusive and jointly comprising the set of integers less than n.

D-0.57 $A(F, n, n_1, n_2, s, t) \leftrightarrow F$ is dist fun in $\mathbb{R}^n \wedge s$ *is* n_1*-sequence* $\wedge$
$$\wedge t \text{ is } n_2\text{-sequence} \wedge \bigwedge_{i,j<n_1} (i<j \rightarrow s_i < s_j) \wedge$$
$$\wedge \bigwedge_{i,j<n_2} (i<j \rightarrow t_i < t_j) \wedge \mathscr{R}s \cup \mathscr{R}t = n$$

In the case of absolutely continuous and discrete distribution functions, and their mixtures, we may define *density functions* that are very useful. In the discrete case a density function f assigns positive values to a countable set of points in $\mathbb{R}^n$ – namely, for every point x, it assigns the value $P_F\{x\}$. In the absolutely continuous case f is the derivative of the distribution function – namely, a function whose existence is asserted in D-0.54. In the mixed case f is the product of the derivative of F with respect to the variables for which the marginal (joint) distribution is absolutely continuous, and, with respect to the variables discretely distributed, the value of the corresponding set function P_F for those variables. Thus if ${}_{\langle 0 \rangle}F$ is discrete and ${}_{\langle 1 \rangle}F$ is continuous, the density function f_F corresponding to F will have for the sequence $\langle x_0, y_0 \rangle$ the value

$$P_{\langle 0 \rangle F}\{x_0\} \left[\frac{\partial^1 F}{\partial \langle 1 \rangle}\right] \langle x_0, y_0 \rangle$$

D-0.58 $f_F = \imath k[\bigvee_{n, n_1, n_2, s, t} (A(F, n, n_1, n_2, s, t) \wedge {}_sF$ is discrete in $\mathbb{R}^{n_1} \wedge$
$$\wedge {}_tF \text{ is absolutely continuous in } \mathbb{R}^{n_2} \wedge \bigwedge_{x,y} (x \in \mathbb{R}^{n_1} \wedge y \in \mathbb{R}^{n_2} \rightarrow$$

$$\rightarrow k0(x, y, s, t) = P_{_sF}\{x\} \cdot \frac{\partial^{n_2} F}{\partial t} 0(x, y, s, t)) \wedge k \text{ is fun} \wedge \mathscr{D}k =$$

$$= \{0(x', y', s, t) : x' \in \mathbb{R}^{n_1} \wedge y' \in \mathbb{R}^{n_2}\}) \vee$$
$$\vee (\sim \bigvee_{n_1, n_2, n, s, t} (A(F, n, n_1, n_2, s, t) \wedge {_sF} \text{ is discrete in } \mathbb{R}^{n_1} \wedge$$

$$\wedge {_tF} \text{ is absolutely continuous in } \mathbb{R}^{n_2}) \wedge k = 0).$$

The following theorems illustrate the usefulness of the density function.

T-0.41 $\int_S g \, \mathrm{d}F \in \mathbb{R} \wedge \bigvee_n F \text{ is discrete in } \mathbb{R}^n \rightarrow \int_S g \, \mathrm{d}F = \sum_{\{y : y \in S\}} gyf_F y$

T-0.42 $\int_S g \, \mathrm{d}F \in \mathbb{R} \wedge \bigvee_n F \text{ is absolutely continuous in } \mathbb{R}^n \rightarrow$

$$\rightarrow \int_S g \, \mathrm{d}F = \int_S gxf_F x \, \mathrm{d}x$$

T-0.43 $A(F, n, n_1, n_2, s, t) \wedge {_sF} \text{ is discrete in } \mathbb{R}^{n_1} \wedge {_tF} \text{ is absolutely continuous in } \mathbb{R}^{n_2} \rightarrow \int_S g \, \mathrm{d}F =$

$$\sum_{\{x : \bigvee_y (0(x, y, s, t) \in S\}} \left(\int_{\{y : 0(x, y, s, t) \in S\}} gyf_F 0(x, y, s, t) \, \mathrm{d}y \right)$$

T-0.44 $F = aF_0 + bF_1 + cF_2 \wedge F \text{ is dist fun in } \mathbb{R}^n \wedge F \text{ is discrete in } \mathbb{R}^n \wedge F_1 \text{ is absolutely continuous in } \mathbb{R}^n \text{ and } F_2 \text{ is singular in } \mathbb{R}^n \wedge \int_S g \, \mathrm{d}F \in \mathbb{R} \rightarrow \int_S g \, \mathrm{d}F = a \sum_{\{y : y \in S\}} gyf_{F_0} y +$

$$+ b \int_S gyf_{F_1} y \, \mathrm{d}y + c \int_S g \, \mathrm{d}F_2$$

10. Conditional distributions are of fundamental importance in statistical inference. Let F be a distribution function in $\mathbb{R}^n$. Let s and t be two sequences of integers occurring in order of magnitude such that they are disjoint and such that the union of their ranges is n. Let n_1 and n_2 be the

number of terms in s and t respectively, and let $B_1 \in \mathbb{B}^{n_1}$. Then, according to the Radon-Nicodym theorem, there exists a Borel-measurable function Φ whose domain is $\mathbb{R}^{n_2}$ such that for every $B_2 \in \mathbb{B}^{n_2}$,

$$P_F(B_1 \times \mathbb{R}^{n_2} \cap \mathbb{R}^{n_1} \times B_2) = \int_{B_1 \times B_2} \mathrm{d}F = \int_{B_2} \Phi \, \mathrm{d}\,{}_tF$$

The value of the function Φy is called 'a version of the conditional probability of $x \in B_1$ given $z = y$' when x and z are the random variables of the marginal distributions ${}_sF$ and ${}_tF$ respectively. If Φ_1 and Φ_2 are two versions of the conditional probability of $x \in B_1$ given $z = y$ and $D = \{w : w \in \mathbb{R}^n \wedge \Phi_1 w \neq \Phi_2 w\}$, $\int_D \mathrm{d}\,{}_tF = 0$ – i.e., Φ_1 and Φ_2 differ only on a set of points of marginal ${}_tF$ measure 0.

Formally, we have the following theorem:

T-0.45 $A(F, n, n_1, n_2, s, t) \wedge B_1 \in \mathbb{B}^{n_1} \rightarrow \bigvee_{\Phi} (\Phi$ is B-measurable $\wedge$

$$\wedge\, \mathscr{D}\Phi = \mathbb{R}^{n_2} \wedge \bigwedge_{B_2 \in \mathbb{R}^{n_2}} \int_{B_1 \times B_2} \mathrm{d}F = \int_{B_2} \Phi \, \mathrm{d}\,{}_tF)$$

A number of special cases are of particular interest. Let the marginal distribution ${}_tF$ be discrete. Then Φ may be taken as $\int_{B \times \{y\}} \mathrm{d}F / f_{{}_tF}\, y$.

T-0.46 $A(F, n, n_1, n_2, s, t) \wedge {}_tF$ is discrete in $\mathbb{R}^n \wedge B_1 \in \mathbb{B}^{n_1} \wedge$

$$\wedge\, B_2 \in \mathbb{B}^{n_2} \rightarrow \int_{B_1 \times B_2} \mathrm{d}F = \sum_{y \in B_2} \left(\int_{B \times \{y\}} \mathrm{d}F / f_{{}_tF}\, y \right) f_{{}_tF}\, y$$

Proof:

$$\int_{B_1 \times B_2} \mathrm{d}F = \sum_{y \in B_2} \int_{B_1 \times \{y\}} \mathrm{d}F \qquad \text{T-0.43}$$

If F is discrete, then both ${}_tF$ and ${}_sF$ are discrete, and we have:

T-0.47 $A(F, n, n_1, n_2, s, t) \wedge F$ is discrete in $\mathbb{R}^n \wedge B_1 \in \mathbb{B}^{n_1} \wedge$

$$\wedge B_2 \in \mathbb{B}^{n_2} \rightarrow \int_{B_1 \times B_2} \mathrm{d}F = \sum_{y \in B_1} \sum_{x \in B_2} \left[\frac{f_F O(x, y, s, t)}{f_{{}_tF} y}\right] f_{{}_tF} y$$

Proof:

(1) $$\int_{B_1 \times B_2} \mathrm{d}F = \sum_{y \in B_2} \sum_{z \in B_1 \times \{y\}} f_F z$$

(2) $$\sum_{z \in B_1 \times \{y\}} f_F z = \sum_{x \in B_1} f_F O(x, y, s, t)$$

Now let the marginal distribution ${}_sF$ be absolutely continuous. Then for Φ we have $\int_{B_1} f_F O(x, y, s, t)/f_{{}_tF} y$:

T-0.48 $A(F, n, n_1, n_2, s, t) \wedge {}_tF$ is discrete in $\mathbb{R}^{n_2} \wedge {}_sF$ is absolutely continuous in $\mathbb{R}^{n_1} \wedge \mathrm{B}_1 \in \mathbb{B}^{n_1} \wedge \mathrm{B}_2 \in \mathbb{B}^{n_2} \rightarrow$

$$\rightarrow \int_{B_1 \times B_2} \mathrm{d}F = \sum_{y \in B_2} \left(\int_{B_1} \frac{f_F O(x, y, s, t)}{f_{{}_tF} y} \mathrm{d}x\right) f_{{}_tF}\, y$$

In the general case where the marginal distribution ${}_tF$ is absolutely continuous, Φ may be taken as $\int_{B_1 \times \{y\}} \mathrm{d}F/f_{{}_tF} y$.

T-0.49 $A(F, n, n_1, n_2, s, t) \wedge {}_tF$ is absolutely continuous in $\mathbb{R}^{n_2} \wedge$

$$\wedge B_1 \in \mathbb{B}^{n_1} \wedge B_2 \in \mathbb{B}^{n_2} \rightarrow \int_{B_1 \times B_2} \mathrm{d}F =$$

$$\int_{B_2} \left(\int_{B_1 \times \{y\}} \mathrm{d}F/f_{{}_tF} y\right) f_{{}_tF} y\, \mathrm{d}y$$

If F is absolutely continuous, then both ${}_sF$ and ${}_tF$ are absolutely continuous, and we have, parallel to T-0.47.

T-0.50 $A(F, n, n_1, n_2, s, t) \wedge F$ is absolutely continuous in $\mathbb{R}^n \wedge$

$$\wedge B_1 \in \mathbb{B}^{n_1} \wedge B_2 \in \mathbb{B}^{n_2} \rightarrow \int_{B_1 \times B_2} \mathrm{d}F =$$

$$\int_{B_2} \left(\int_{B_1} \left(\frac{f_F O(x, y, s, t)}{f_{{}_tF} y} \right) \mathrm{d}x \right) f_{{}_tF} y \, \mathrm{d}y$$

If ${}_sF$ is discrete, we have, parallel to T-0.48,

T-0.51 $A(F, n, n_1, n_2, s, t) \wedge {}_sF$ is discrete in $\mathbb{R}^{n_1} \wedge {}_tF$ is absolutely continuous in $\mathbb{R}^{n_2} \wedge B_1 \in \mathbb{B}^{n_1} \wedge B_2 \in \mathbb{B}^{n_2} \rightarrow \int_{B_1 \times B_2} \mathrm{d}F =$

$$= \int_{B_2} \left(\sum_{x \in B_1} \frac{f_F O(x, y, s, t)}{f_{{}_tF} y} \right) f_{{}_tF} y \, \mathrm{d}y$$

In the extra special cases represented by T-0.47, T-0.48, T-0.50, and T-0.51, we can define *conditional* density functions and conditional distribution functions. Interpreting s and t as before, we shall write ${}_sF_y$ for the conditional distribution of the s variables on the t variables taking the value y, and $f_{{}_sF_y}$ as the corresponding density function.

D-0.59 ${}_sF_y = \imath k[\bigvee_{t, n, n_1, n_2} (A(F, n, n_1, n_2, s, t) \wedge y \in \mathbb{R}^{n_2} \wedge ({}_sF$ is discrete in $\mathbb{R}^{n_1} \vee {}_sF$ is absolutely continuous in $\mathbb{R}^{n_1}) \wedge ({}_tF$ is discrete in $\mathbb{R}^{n_2} \vee {}_tF$ is absolutely continuous in $\mathbb{R}^{n_2}) \wedge (({}_sF$ is discrete in $\mathbb{R}^{n_1} \wedge k = \sum_{w < x} \frac{f_F O(w, y, s, t)}{f_{{}_tF} y}) \vee ({}_sF$ is absolutely continuous in $\mathbb{R}^{n_1} \wedge k = \int_{\{-\infty\}^{n_1}}^{x} [f_F O(w, y, s, t)/f_{{}_tF} y] \, \mathrm{d}w))) \vee (\sim \bigvee_{t, n, n_1, n_2}$ $(A(F, n, n_1, n_2, s, t) \wedge y \in \mathbb{R}^{n_2} \wedge ({}_sF$ is discrete in $\mathbb{R}^{n_1} \vee {}_sF$ is absolutely continuous in $\mathbb{R}^{n_1}) \wedge ({}_tF$ is discrete in $\mathbb{R}^{n_2} \vee$ $\vee ({}_tF$ is absolutely continuous in $\mathbb{R}^{n_2})) \wedge k = 0)]$

The conditional density functions turn out to be just what we would expect:

T-0.52 $A(F, n, n_1, n_2, s, t) \wedge y \in \mathbb{R}^{n_1} \wedge x \in \mathbb{R}^{n_2} \wedge ({}_sF$ is discrete in $\mathbb{R}^{n_1} \vee$ $\vee\, {}_sF$ is absolutely continuous in $\mathbb{R}^{n_1}) \wedge ({}_tF$ is discrete in $\mathbb{R}^{n_2}$ $\vee\, {}_tF$ is absolutely continuous in $\mathbb{R}^{n_2}) \rightarrow f_{{}_sF_y}\, x = \dfrac{f_F O(x, y, s, t)}{f_{{}_tF}\, y}$

We also have a Bayes' theorem for conditional densities:

T-0.53 $A(F, n, n_1, n_2, s, t) \wedge y \in \mathbb{R}^{n_2} \wedge x \in \mathbb{R}^{n_1} \wedge ({}_sF$ is discrete in $\mathbb{R}^{n_1} \vee$ $\vee\, {}_sF$ is absolutely continuous in $\mathbb{R}^{n_1}) \wedge ({}_tF$ is discrete in $\mathbb{R}^{n_2} \vee {}_tF$ is absolutely continuous in $\mathbb{R}^{n_2}) \rightarrow$

$$\rightarrow f_{{}_tF_x}\, y = \frac{f_{{}_tF}\, y \cdot f_{{}_sF_y}\, x}{f_{{}_sF}\, x}$$

We may define independence between the variables whose subscripts belong to the range of s and those whose subscripts belong to the range of t quite generally. We write '$sIFt$' to mean that, according to the distribution F, those quantities corresponding to the variables with subscripts in s are independent of those whose subscripts are in t.

D-0.60 $sIFt \leftrightarrow \bigvee_{n, n_1, n_2} (A(F, n, n_1, n_2, s, t) \wedge$

$$\wedge \bigwedge_{x \in \mathbb{R}^{n_1}} \bigwedge_{y \in \mathbb{R}^{n_1}} F\, O(x, y, s, t) = {}_sF\, x \cdot {}_tF\, y)$$

The following two theorems give connections between independence and density functions and between independence and conditional density functions:

T-0.54 $sIFt \wedge A(F, n, n_1, n_2, s, t) \wedge f_{{}_sF} \neq 0 \rightarrow$

$$\rightarrow \bigwedge_{y \in \mathbb{R}^{n_2}} \bigwedge_{x \in \mathbb{R}^{n_1}} f_F O(x, y, s, t) = f_{{}_sF} x \cdot f_{{}_tF} y$$

T-0.55 $sIFt \wedge A(F, n, n_1, n_2, s, t) \wedge f_{{}_sF} \neq 0 \wedge f_{{}_tF} \neq 0 \rightarrow$

$$\rightarrow \bigwedge_{y \in \mathbb{R}^{n_2}} ({}_sF = {}_sF_y \wedge f_{{}_sF} = f_{{}_sF_y}).$$

INDEX

SYNTHESE LIBRARY

Monographs on Epistemology, Logic, Methodology,
Philosophy of Science, Sociology of Science and of Knowledge, and on the
Mathematical Methods of Social and Behavioral Sciences

1. J. M. BOCHEŃSKI, *A Precis of Mathematical Logic.* 1959, X + 100 pp.
2. P. L. GUIRAUD, *Problèmes et méthodes de la statistique linguistique.* 1960, VI + 146 pp.
3. HANS FREUDENTHAL (ed.), *The Concept and the Role of the Model in Mathematics and Natural and Social Sciences. Proceedings of a Colloquium held at Utrecht, The Netherlands, January 1960.* 1961, VI + 194 pp.
4. EVERT W. BETH, *Formal Methods. An Introduction to Symbolic Logic and the Study of Effective Operations in Arithmetic and Logic.* 1962, XIV + 170 pp.
5. B. H. KAZEMIER and D. VUYSJE (eds.), *Logic and Language. Studies dedicated to Professor Rudolf Carnap on the Occasion of his Seventieth Birthday.* 1962, VI + 256 pp.
6. MARX W. WARTOFSKY (ed.), *Proceedings of the Boston Colloquium for the Philosophy of Science, 1961–1962,* Boston Studies in the Philosophy of Science (ed. by Robert S. Cohen and Marx W. Wartofsky), Volume I. 1963, VIII + 212 pp.
7. A. A. ZINOV'EV, *Philosophical Problems of Many-Valued Logic.* 1963, XIV + 155 pp.
8. GEORGES GURVITCH, *The Spectrum of Social Time.* 1964, XXVI + 152 pp.
9. PAUL LORENZEN, *Formal Logic.* 1965, VIII + 123 pp.
10. ROBERT S. COHEN and MARX W. WARTOFSKY (eds.), *In Honor of Philipp Frank,* Boston Studies in the Philosophy of Science (ed. by Robert S. Cohen and Marx W. Wartofsky), Volume II. 1965, XXXIV + 475 pp.
11. EVERT W. BETH, *Mathematical Thought. An Introduction to the Philosophy of Mathematics.* 1965, XII + 208 pp.
12. EVERT W. BETH and JEAN PIAGET, *Mathematical Epistemology and Psychology.* 1966, XXII + 326 pp.
13. GUIDO KÜNG, *Ontology and the Logistic Analysis of Language. An Enquiry into the Contemporary Views on Universals.* 1967, XI + 210 pp.
14. ROBERT S. COHEN and MARX W. WARTOFSKY (eds.), *Proceedings of the Boston Colloquium for the Philosophy of Science 1964–1966, in Memory of Norwood Russell Hanson,* Boston Studies in the Philosophy of Science (ed. by Robert S. Cohen and Marx W. Wartofsky), Volume III. 1967, XLIX + 489 pp.
15. C. D. BROAD, *Induction, Probability, and Causation. Selected Papers.* 1968, XI + 296 pp.
16. GÜNTHER PATZIG, *Aristotle's Theory of the Syllogism. A Logical-Philosophical Study of Book A of the Prior Analytics.* 1968, XVII + 215 pp.
17. NICHOLAS RESCHER, *Topics in Philosophical Logic.* 1968, XIV + 347 pp.

18. Robert S. Cohen and Marx W. Wartofsky (eds.), *Proceedings of the Boston Colloquium for the Philosophy of Science 1966–1968*, Boston Studies in the Philosophy of Science (ed. by Robert S. Cohen and Marx W. Wartofsky), Volume IV. 1969, VIII + 537 pp.
19. Robert S. Cohen and Marx W. Wartofsky (eds.), *Proceedings of the Boston Colloquium for the Philosophy of Science 1966–1968*, Boston Studies in the Philosophy of Science (ed. by Robert S. Cohen and Marx W. Wartofsky), Volume V. 1969, VIII + 482 pp.
20. J. W. Davis, D. J. Hockney, and W. K. Wilson (eds.), *Philosophical Logic*. 1969, VIII + 277 pp.
21. D. Davidson and J. Hintikka (eds.), *Words and Objections: Essays on the Work of W. V. Quine*, 1969, VIII + 366 pp.
22. Patrick Suppes, *Studies in the Methodology and Foundations of Science. Selected Papers from 1911 to 1969*. 1969, XII + 473 pp.
23. Jaakko Hintikka, *Models for Modalities. Selected Essays*. 1969, IX + 220 pp.
24. Nicholas Rescher *et al.* (eds.), *Essay in Honor of Carl G. Hempel. A Tribute on the Occasion of his Sixty-Fifth Birthday*. 1969, VII + 272 pp.
25. P. V. Tavanec (ed.), *Problems of the Logic of Scientific Knowledge*. 1969, XII + 429 pp.
26. Marshall Swain (ed.), *Induction, Acceptance, and Rational Belief*. 1970, VII + 232 pp.
27. Robert S. Cohen and Raymond J. Seeger (eds.), *Ernst Mach: Physicist and Philosopher*, Boston Studies in the Philosophy of Science (ed. by Robert S. Cohen and Marx W. Wartofsky), Volume VI. 1970, VIII + 295 pp.
28. Jaakko Hintikka and Patrick Suppes, *Information and Inference*. 1970, X + 336 pp.
29. Karel Lambert, *Philosophical Problems in Logic. Some Recent Developments*. 1970, VII + 176 pp.
30. Rolf A. Eberle, *Nominalistic Systems*. 1970, IX + 217 pp.
31. Paul Weingartner and Gerhard Zecha (eds.), *Induction, Physics, and Ethics, Proceedings and Discussions of the 1968 Salzburg Colloquium in the Philosophy of Science*. 1970, X + 382 pp.
32. Evert W. Beth, *Aspects of Modern Logic*. 1970, XI + 176 pp.
33. Risto Hilpinen (ed.), *Deontic Logic: Introductory and Systematic Readings*. 1971, VII + 182 pp.
34. Jean-Louis Krivine, *Introduction to Axiomatic Set Theory*. 1971, VII + 98 pp.
35. Joseph D. Sneed, *The Logical Structure of Mathematical Physics*. 1971, XV + 311 pp.
36. Carl R. Kordig, *The Justification of Scientific Change*. 1971, XIV + 119 pp.
37. Milič Čapek, *Bergson and Modern Physics*, Boston Studies in the Philosophy of Science (ed. by Robert S. Cohen and Marx W. Wartofsky), Volume VII. 1971, XV + 414 pp.
38. Norwood Russell Hanson, *What I do not Believe, and other Essays* (ed. by Stephen Toulmin and Harry Woolf). 1971, XII + 390 pp.
39. Roger C. Buck and Robert S. Cohen (eds.), *PSA 1970. In Memory of Rudolf Carnap*, Boston Studies in the Philosophy of Science (ed. by Robert S. Cohen and Marx W. Wartofsky), Volume VIII. 1971, LXVI + 615 pp. Also available as a paperback.
40. Donald Davidson and Gilbert Harman (eds.), *Semantics of Natural Language*. 1972, X + 769 pp. Also available as a paperback.
41. Yehosua Bar-Hillel (ed.), *Pragmatics of Natural Languages*. 1971, VII + 231 pp.
42. Sören Stenlund, *Combinators, λ-Terms and Proof Theory*. 1972, 184 pp.
43. Martin Strauss, *Modern Physics and Its Philosophy. Selected Papers in the Logic, History, and Philosophy of Science*. 1972, X + 297 pp.

44. MARIO BUNGE, *Method, Model and Matter*. 1973, VII + 196 pp.
45. MARIO BUNGE, *Philosophy of Physics*. 1973, IX + 248 pp.
46. A. A. ZINOV'EV, *Foundations of the Logical Theory of Scientific Knowledge (Complex Logic)*, Boston Studies in the Philosophy of Science (ed. by Robert S. Cohen and Marx W. Wartofsky), Volume IX. Revised and enlarged English edition with an appendix, by G. A. Smirnov, E. A. Sidorenka, A. M. Fedina, and L. A. Bobrova. 1973, XXII + 301 pp. Also available as a paperback.
47. LADISLAV TONDL, *Scientific Procedures*, Boston Studies in the Philosophy of Science (ed. by Robert S. Cohen and Marx W. Wartofsky), Volume X. 1973, XII + 268 pp. Also available as a paperback.
48. NORWOOD RUSSELL HANSON, *Constellations and Conjectures* (ed. by Willard C. Humphreys, Jr.). 1973, X + 282 pp.
49. K. J. J. HINTIKKA, J. M. E. MORAVCSIK, and P. SUPPES (eds.), *Approaches to Natural Language. Proceedings of the 1970 Stanford Workshop on Grammar and Semantics.* 1973, VIII + 526 pp. Also available as a paperback.
50. MARIO BUNGE (ed.), *Exact Philosophy – Problems, Tools, and Goals*. 1973, X + 214 pp.
51. RADU J. BOGDAN and ILKKA NIINILUOTO (eds.), *Logic, Language, and Probability*. A selection of papers contributed to Sections IV, VI, and XI of the Fourth International Congress for Logic, Methodology, and Philosophy of Science, Bucharest, September 1971. 1973, X + 323 pp.
52. GLENN PEARCE and PATRICK MAYNARD (eds.), *Conceptual Change*. 1973, XII + 282 pp.
53. ILKKA NIINILUOTO and RAIMO TUOMELA, *Theoretical Concepts and Hypothetico-Inductive Inference*. 1973, VII + 264 pp.
54. ROLAND FRAÏSSÉ, *Course of Mathematical Logic* – Volume I: *Relation and Logical Formula*. 1973, XVI + 186 pp. Also available as a paperback.
55. ADOLF GRÜNBAUM, *Philosophical Problems of Space and Time*. Second, enlarged edition, Boston Studies in the Philosophy of Science (ed. by Robert S. Cohen and Marx W. Wartofsky), Volume XII. 1973, XXIII + 884 pp. Also available as a paperback.
56. PATRICK SUPPES (ed.), *Space, Time, and Geometry*. 1973, XI + 424 pp.
57. HANS KELSEN, *Essays in Legal and Moral Philosophy*, selected and introduced by Ota Weinberger, 1973, XXVIII + 300 pp.
58. R. J. SEEGER and ROBERT S. COHEN (eds.), *Philosophical Foundations of Science, Proceedings of an AAAS Program, 1969*. Boston Studies in the Philosophy of Science (ed. by Robert S. Cohen and Marx W. Wartofsky), Volume XI. 1974, IX + 545 pp. Also available as paperback.
59. ROBERT S. COHEN and MARX W. WARTOFSKY (eds.), *Logical and Epistemological Studies in Contemporary Physics*, Boston Studies in the Philosophy of Science (ed. by Robert S. Cohen and Marx W. Wartofsky), Volume XIII. 1973, VIII + 462 pp. Also available as a paperback.
60. ROBERT S. COHEN and MARX W. WARTOFSKY (eds.), *Methodological and Historical Essays in the Natural and Social Sciences. Proceedings of the Boston Colloquium for the Philosophy of Science, 1969–1972*, Boston Studies in the Philosophy of Science (ed. by Robert S. Cohen and Marx W. Wartofsky), Volume XIV. 1974, VIII + 405 pp. Also available as a paperback.
63. SÖREN STENLUND (ed.), *Logical Theory and Semantic Analysis. Essays Dedicated to Stig Kanger on His Fiftieth Birthday*. 1974, V + 217 pp.
64. KENNETH SCHAFFNER and ROBERT S. COHEN (eds.), *Proceedings of the 1972 Biennial Meeting, Philosophy of Science Association*, Boston Studies in the Philosophy of Science (ed. by Robert S. Cohen and Marx W. Wartofsky), Volume XX. 1974, VIII + 445 pp. Also available as a paperback.

65. HENRY E. KYBURG, JR., *The Logical Foundations of Statistical Inference*. 1974, IX + 421 pp.
66. MARJORIE GRENE, *The Understanding of Nature: Essays in the Philosophy of Biology*, Boston Studies in the Philosophy of Science (ed. by Robert S. Cohen and Marx W. Wartofsky), Volume XXIII. 1974, XII + 360 pp. Also available as a paperback.

In Preparation

61. ROBERT S. COHEN and MARX W. WARTOFSKY (eds.), *For Dirk Struik. Scientific, Historical, and Political Essays in Honor of Dirk J. Struik*, Boston Studies in the Philosophy of Science (ed. by Robert S. Cohen and Marx W. Wartofsky), Volume XV. Also available as a paperback.
62. KAZIMIERZ AJDUKIEWICZ, *Pragmatic Logic*, transl. from the Polish by Olgierd Wojtasiewicz.
67. JAN M. BROEKMAN, *Structuralism: Moscow, Prague, Paris.*
68. NORMAN GESCHWIND, *Selected Papers on Language and the Brain*, Boston Studies in the Philosophy of Science (ed. by Robert S. Cohen and Marx W. Wartofsky) Volume XVI. Also available as a paperback.
69. ROLAND FRAÏSSÉ, *Course of Mathematical Logic* – Volume II: *Model Theory*.

SYNTHESE HISTORICAL LIBRARY

Texts and Studies
in the History of Logic and Philosophy

Editors:

1. M. T. BEONIO-BROCCHIERI FUMAGALLI, *The Logic of Abelard*. Translated from the Italian. 1969, IX + 101 pp.
2. GOTTFRIED WILHELM LEIBNITZ, *Philosophical Papers and Letters*. A selection translated and edited, with an introduction by Leroy E. Loemker. 1969, XII + 736 pp.
3. ERNST MALLY, *Logische Schriften* (ed. by Karl Wolf and Paul Weingartner). 1971, X + 340 pp.
4. LEWIS WHITE BECK (ed.), *Proceedings of the Third International Kant Congress*. 1972, XI + 718 pp.
5. BERNARD BOLZANO, *Theory of Science* (ed. by Jan Berg). 1973, XV + 398 pp.
6. J. M. E. MORAVCSIK (ed.), *Patterns in Plato's Thought. Papers arising out of the 1971 West Coast Greek Philosophy Conference*, 1973, VIII + 212 pp.
7. NABIL SHEHABY, *The Propositional Logic of Avicenna: A Translation from al-Shifā': al-Qiyās*, with Introduction, Commentary and Glossary. 1973, XIII + 296 pp.
8. DESMOND PAUL HENRY, *Commentary on De Grammatico: The Historical-Logical Dimensions of a Dialogue of St. Anselm's*. 1974, IX + 345 pp.
9. JOHN CORCORAN, *Ancient Logic and Its Modern Interpretations*. 1974, X + 208 pp.

In Preparation

10. E. M. BARTH, *The Logic of the Articles in Traditional Philosophy*.
11. JAAKKO HINTIKKA, *Knowledge and the Known*.
12. E. J. ASHWORTH, *Language and Logic in the Post-Medieval Period*.